New Science of Boron Allotropes, Compounds, and Nanomaterials

New Science of Boron Allotropes, Compounds, and Nanomaterials

Editors

Takahiro Kondo
Iwao Matsuda
Josep M. Oliva-Enrich

MDPI • Basel • Beijing • Wuhan • Barcelona • Belgrade • Manchester • Tokyo • Cluj • Tianjin

Editors
Takahiro Kondo
Institute of Pure and
Applied Sciences
University of Tsukuba
Tsukuba
Japan

Iwao Matsuda
Institute for Solid
State Physics
The University of Tokyo
Chiba
Japan

Josep M. Oliva-Enrich
"Blas Cabrera" Institute
of Physical-Chemistry
CSIC
Madrid
Spain

Editorial Office
MDPI
St. Alban-Anlage 66
4052 Basel, Switzerland

This is a reprint of articles from the Special Issue published online in the open access journal *Molecules* (ISSN 1420-3049) (available at: www.mdpi.com/journal/molecules/special_issues/boron_compounds_nanomaterials).

For citation purposes, cite each article independently as indicated on the article page online and as indicated below:

LastName, A.A.; LastName, B.B.; LastName, C.C. Article Title. *Journal Name* **Year**, *Volume Number*, Page Range.

ISBN 978-3-0365-8327-3 (Hbk)
ISBN 978-3-0365-8326-6 (PDF)

Cover image courtesy of Takahiro Kondo and Iwao Matsuda

Contents

About the Editors vii

Preface to "New Science of Boron Allotropes, Compounds, and Nanomaterials" ix

Yasunobu Ando, Takeru Nakashima, Heming Yin, Ikuma Tateishi, Xiaoni Zhang and Yuki Tsujikawa et al.
Prediction of a Cyclic Hydrogenated Boron Molecule as a Promising Building Block for Borophane
Reprinted from: *Molecules* **2023**, *28*, 1225, doi:10.3390/molecules28031225 1

Levan Chkhartishvili
Relative Stability of Boron Planar Clusters in Diatomic Molecular Model
Reprinted from: *Molecules* **2022**, *27*, 1469, doi:10.3390/molecules27051469 15

Yuki Tsujikawa, Makoto Shoji, Masashi Hamada, Tomoya Takeda, Izumi Mochizuki and Toshio Hyodo et al.
Structure of χ_3-Borophene Studied by Total-Reflection High-Energy Positron Diffraction (TRHEPD)
Reprinted from: *Molecules* **2022**, *27*, 4219, doi:10.3390/molecules27134219 35

Ping Zhang, Xirui Tian, Shaoxiang Sheng, Chen Ma, Linjie Chen and Baojie Feng et al.
Vibrational Property of α-Borophene Determined by Tip-Enhanced Raman Spectroscopy
Reprinted from: *Molecules* **2022**, *27*, 834, doi:10.3390/molecules27030834 43

T. Hirabayashi, S. Yasuhara, S. Shoji, A. Yamaguchi, H. Abe and S. Ueda et al.
Fabrication of Hydrogen Boride Thin Film by Ion Exchange in MgB_2
Reprinted from: *Molecules* **2021**, *26*, 6212, doi:10.3390/molecules26206212 51

Natsumi Noguchi, Shin-ichi Ito, Miwa Hikichi, Yohei Cho, Kazuho Goto and Atsushi Kubo et al.
Highly Dispersed Ni Nanoclusters Spontaneously Formed on Hydrogen Boride Sheets
Reprinted from: *Molecules* **2022**, *27*, 8261, doi:10.3390/molecules27238261 63

Xiaoni Zhang, Miwa Hikichi, Takushi Iimori, Yuki Tsujikawa, Mei Yuan and Masafumi Horio et al.
Accelerated Synthesis of Borophane (HB) Sheets through HCl-Assisted Ion-Exchange Reaction with $YCrB_4$
Reprinted from: *Molecules* **2023**, *28*, 2985, doi:10.3390/molecules28072985 75

Ikuma Tateishi, Xiaoni Zhang and Iwao Matsuda
Electronic Structures of Polymorphic Layers of Borophane
Reprinted from: *Molecules* **2022**, *27*, 1808, doi:10.3390/molecules27061808 91

Anca-Iulia Stoica, Clara Viñas and Francesc Teixidor
History of Cobaltabis(dicarbollide) in Potentiometry, No Need for Ionophores to Get an Excellent Selectivity
Reprinted from: *Molecules* **2022**, *27*, 8312, doi:10.3390/molecules27238312 105

Lukáš Fojt, Bohumír Grüner, Jan Nekvinda, Ece Zeynep Tűzűn, Luděk Havran and Miroslav Fojta
Electrochemistry of Cobalta Bis(dicarbollide) Ions Substituted at Carbon Atoms with Hydrophilic Alkylhydroxy and Carboxy Groups
Reprinted from: *Molecules* **2022**, *27*, 1761, doi:10.3390/molecules27061761 119

Magdalena Jakubczyk, Slawomir Michalkiewicz, Agata Skorupa and Kinga Krajcarz
Electrochemical Characterization and Voltammetric Determination of Methylisothiazolinone on a Boron-Doped Diamond Electrode
Reprinted from: *Molecules* **2022**, *27*, 9013, doi:10.3390/molecules27249013 . . . . **133**

Norinobu Watanabe, Keisuke Miyazaki, Masayuki Toyoda, Kotaro Takeyasu, Naohito Tsujii and Haruki Kusaka et al.
Rhombohedral Boron Monosulfide as a p-Type Semiconductor
Reprinted from: *Molecules* **2023**, *28*, 1896, doi:10.3390/molecules28041896 . . . . **149**

Artem L. Khoroshilov, Kirill M. Krasikov, Andrey N. Azarevich, Alexey V. Bogach, Vladimir V. Glushkov and Vladimir N. Krasnorussky et al.
Hall Effect Anisotropy in the Paramagnetic Phase of $Ho_{0.8}Lu_{0.2}B_{12}$ Induced by Dynamic Charge Stripes
Reprinted from: *Molecules* **2023**, *28*, 676, doi:10.3390/molecules28020676 . . . . **159**

About the Editors

Takahiro Kondo

Dr. Takahiro Kondo is Professor at the Institute of Pure and Applied Sciences and Director of the R&D Center for Zero CO_2 Emission with Functional Materials at the University of Tsukuba from 2022. He also holds a specially appointed Professor position at the Advanced Institute for Materials Research (AIMR) at Tohoku University. He received his Doctorate of Philosophy in Engineering from the University of Tsukuba in 2003. He spent 4 years at the Institute of Physical and Chemical Research (RIKEN) as a Postdoctoral researcher. He then moved to the University of Tsukuba in 2007. His research focuses on creating new materials, technologies, sciences, and concepts to solve energy, environmental, and resource problems.

Iwao Matsuda

Dr. Iwao Matsuda is a professor at the Institute for Solid State Physics at the University of Tokyo. He has published over 250 papers in international journals, and he has been awarded prizes eight times in the fields of Physics and Surface Science. He has developed two books, namely "*2D Boron: Boraphene, Borophene, and Boronene*" (Springer, 2021) and "*Monatomic Two-Dimensional Layers: Modern Experimental Approaches for Structure, Properties, and Industrial Use*" (Elsevier, 2019). His research interests are fundamental physics and chemistry in materials science.

Josep M. Oliva-Enrich

Dr. Josep M. Oliva-Enrich is a Titular Scientist at the "Blas Cabrera" Physical Chemistry Institute of the Spanish National Research Council (CSIC). He has published more than 120 papers in international journals, contributed to six book chapters and two books, and presented more than 100 contributions at international conferences. He was a Marie-Curie Fellow at the University of Bristol (UK), and he was awarded the IBM/Löwdin prize from the Quantum Theory Project of the University of Florida, USA. His research interests include the electronic structure of matter, theoretical chemistry, computational chemistry, boron chemistry, and valence-bond theory.

Preface to "New Science of Boron Allotropes, Compounds, and Nanomaterials"

Elemental boron has attracted researchers because of its large number of allotropes, thus providing a variety of scientific aspects within physics and chemistry. Boron compounds and boron nanomaterials have also shown various polymorphs that have been investigated as functional materials for our society. The uniqueness of the boron atom is due to its high frequency of electron multi-center bonding with neighboring atoms, which is still a significant topic in theoretical and computational research. Recently, there have been reports on the realization of two-dimensional (2D) or planar boron nanomaterials, the so-called borophenes. As expected, observations of the various 2D polymorphisms have triggered a considerable number of research lines worldwide. This reprint presents a collection of such papers at the forefront. The contributed papers are transferred from the Special Issue of "New Science of Boron Allotropes, Compounds, and Nanomaterials" in the journal *Molecules*. The issue was launched in 2021 and is aimed at providing a forum for the dissemination of the latest information on boron allotropes, compounds, and nanomaterials.

This reprint contains nine articles, two communications, and two reviews, covering experimental and theoretical research on borophene and various nanostructures of boron compounds. Two theoretical articles report intriguing planar boron molecules, which would be potential building blocks in borophene. Two experimental articles present the atomic structure and vibrational properties of fabricated borophene on the substrate, respectively. There are three articles on hydrogenated borophene (HB), borophane, reporting the thin-film fabrication, the support application of the highly dispersed nickel nanoclusters, and the synthesis acceleration of the novel 5,7-ring type sheets. Focusing on borophane (HB), there is also a review that comprehensively describes the electronic structures of polymorphic layers of borophane. Some topics, especially electrochemistry, on boron materials are featured, for example, cobalt bis(dicarbollide), rhombohedral boron monosulfide, $Ho_{0.8}Lu_{0.2}B_{12}$, and boron-doped diamond.

We hope that this reprint will provide useful insight to readers who have been working on boron-related materials and also on materials science.

Takahiro Kondo, Iwao Matsuda, and Josep M. Oliva-Enrich
Editors

molecules MDPI

Article

Prediction of a Cyclic Hydrogenated Boron Molecule as a Promising Building Block for Borophane

Yasunobu Ando [1,*], Takeru Nakashima [1], Heming Yin [2], Ikuma Tateishi [3], Xiaoni Zhang [2], Yuki Tsujikawa [2], Masafumi Horio [2], Nguyen Thanh Cuong [4], Susumu Okada [4], Takahiro Kondo [4] and Iwao Matsuda [2,*]

[1] CD-FMat, National Institute of Advanced Industrial Science and Technology (AIST), Tsukuba 305-8560, Ibaraki, Japan
[2] Institute for Solid State Physics, The University of Tokyo, Kashiwa 277-8581, Chiba, Japan
[3] RIKEN Center for Emergent Matter Science, Wako 351-0198, Saitama, Japan
[4] Faculty of Pure and Applied Sciences, University of Tsukuba, Tsukuba 305-8573, Ibaraki, Japan
[*] Correspondence: yasunobu.ando@aist.go.jp (Y.A.); imatsuda@issp.u-tokyo.ac.jp (I.M.)

Abstract: We have extensively searched for a cyclic hydrogenated boron molecule that has a three-center two-electron bond at the center. Using first-principles calculations, we discovered a stable molecule of 2:4:6:8:-2H-1,5:1,5-μH-B_8H_{10} and propose its existence. This molecule can be regarded as a building block for sheets of topological hydrogen boride (borophane), which was recently theoretically proposed and experimentally discovered. The electronic structure of the cyclic hydrogenated boron molecule is discussed in comparison with that of cyclic hydrogenated carbon molecules.

Keywords: boron; hydrogenated boron; calculation

1. Introduction

The theoretical design of new molecules is a central issue in chemistry. Such methods have led to the syntheses of new functional materials such as catalysts or device components that have become technologically significant [1]. Molecules with unique bonding schemes have attracted academic interest because they often have unexpected chemical properties that violate well-established models and trigger innovations.

Boron is an electron-deficient element, and it has been found to form unique bonding configurations with multi-center atoms. Various boron-based molecules have been designed and synthesized [2]. Some of these molecules have shown anti-Van't Hoff/Le Bel features [3–6]. The predicted existence of a 2D or planar boron material, borophene, has triggered vigorous explorations of its molecules or atomic sheets [7]. Researchers have reported growths of various borophene layers on metal substrates [8–10]. Theoretical and experimental searches for 2D boron have been extended to seeking low-dimensional materials composed of boron and other elements [7]. An example is the recent synthesis of sheets of hydrogen boride (HB) or borophane [11–13], which show intriguing functions such as hydrogen storage [14]. Boron and hydrogen atoms create a three-center two-electron (3c-2e) bond with unique chemical characteristics that allow us to design new 2D HB structures [15–17].

In this study, we searched for a possible planar structure of hydrogenated boron molecules (0D HB) through theoretical calculations. On the basis of the well-known boron hydride molecule diborone, shown in Figure 1a, we designed various molecular models by adding cyclic structures with different numbers of member atoms to different types of well-known boron-hydride (borane) molecules [18]. Among the twelve candidates, we found that only the molecule 2:4:6:8:-2H-1,5:1,5-μH-B_8H_{10} was stable both thermodynamically and kinetically. The structural formula and atomic structure of the molecule are drawn in Figure 1b,c, respectively. The molecule has a 3c-2e bond at the center and two five-membered rings on the sides. The nomenclature of the molecule is

Citation: Ando, Y.; Nakashima, T.; Yin, H.; Tateishi, I.; Zhang, X.; Tsujikawa, Y.; Horio, M.; Cuong, N.T.; Okada, S.; Kondo, T.; et al. Prediction of a Cyclic Hydrogenated Boron Molecule as a Promising Building Block for Borophane. *Molecules* **2023**, *28*, 1225. https://doi.org/10.3390/molecules28031225

Academic Editor: Tian Wang

Received: 20 December 2022
Revised: 20 January 2023
Accepted: 21 January 2023
Published: 26 January 2023

based on references [19,20]; its atomic structure can be regarded as a building block of a topological borophane layer, discovered recently [12,13,15,17]. This research shows intriguing relationships between the planar molecules and atomic sheets of the hydrogenated boron compound. This cyclic hydrogenated boron molecule provides new perspectives in boron chemistry.

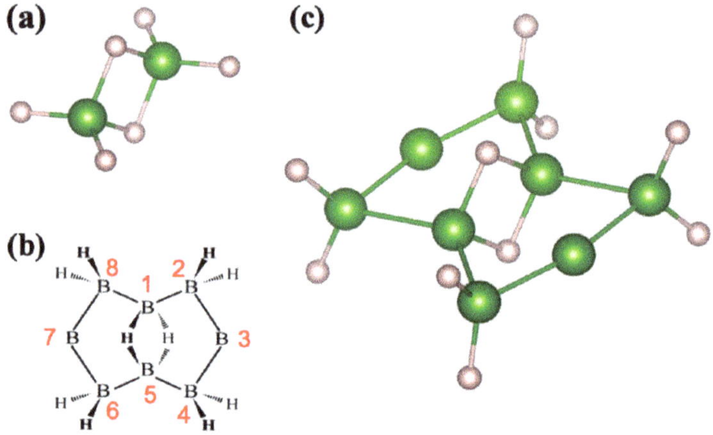

Figure 1. Molecules of hydrogenated boron compounds. (**a**) Atomic structure of diborane, B_2H_6. (**b**) A structural formula and (**c**) the optimized molecular structure of 2:4:6:8:-2H-1,5:1,5-μH-B_8H_{10}. Numbers of the boron atoms for nomenclature are labeled in (**b**).

2. Results and Discussion

2.1. Candidates from the Cyclic Hydrogenated Boron Compounds

In searching for a new hydrogenated boron molecule, we examined the 12 candidates displayed in Figure 2 and listed in Table 1. We followed the same nomenclature as for the B_8H_{10} molecule [19,20]. The atomic structures of the molecular candidates are schematically presented in Appendix A (Figure A1). The stable molecule was determined by filtering values of cohesive energy and signs of vibration frequency. All the candidates were found to gain cohesive energy during molecular formation. However, most of them were excluded because they had negative phonon energies or imaginary vibrational modes, indicating breakdowns in their molecular structure. It is apparent in Table 1 that only the molecule 2:4:6:8:-2H-1,5:1,5-μH-B_8H_{10} (Figure 1b,c) is stable and could possibly exist in nature. The formation energy of this molecule is E_{form} = 194.5 kcal/mol, from four molecules of diborane—B_2H_6 (Table 2). Comparing this with E_{form} = 156.3 kcal/mol for eight molecules of borane (BH_3) shows that the formation energy of the proposed molecule is 38.3 kcal/mol, which is slightly higher than that of the borane molecule.

A molecule of 2:4:6:8:-2H-1,5:1,5-μH-B_8H_{10} has two types of boron atoms: those that make chemical bonds with hydrogen and those that do not. Two boron atoms in the 3c-2e bond at the center share hydrogen atoms. Besides the two boron atoms that connect with their neighbors, the other four boron atoms are individually terminated by two hydrogen atoms. To help understand the electronic states of the molecule, Figure 3 shows the schematics of the highest occupied and the lowest unoccupied molecular orbitals, HOMO and LUMO, respectively. Both orbitals extend inside the cyclic plane and have σ bonds.

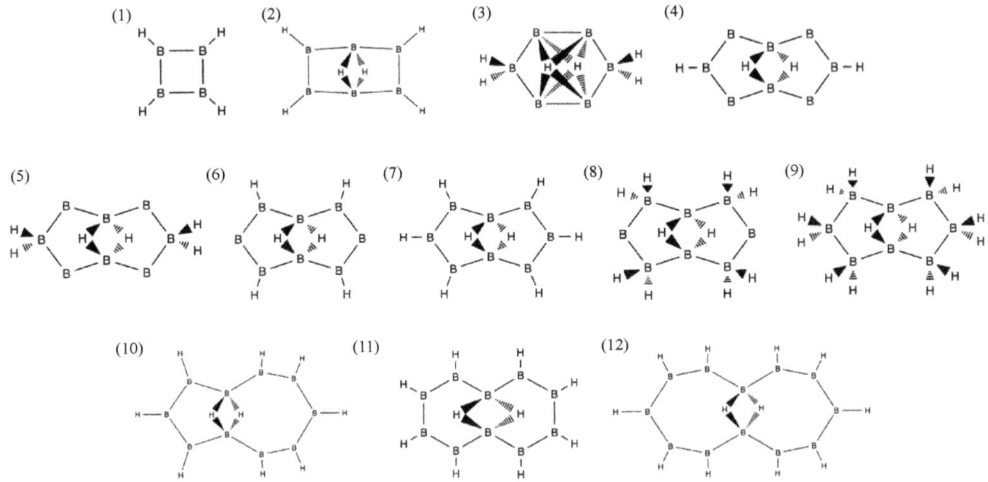

Figure 2. Structural formula of 12 candidates for a new cyclic hydrogenated boron molecule. A top-left number of each formula corresponds to the index in Table 1.

Table 1. Filtration of the molecule candidates, TRUE or FALSE. Stability is determined by the absence (stability: TRUE) or presence (Stability: FALSE) of an imaginary vibrational mode. Atomic structures of the cyclic hydrogenated boron molecules are shown in Figure 2 and Appendix A.

Index	Molecule Candidate	Total Energy (eV)	HOMO Level (eV)	LUMO Level (eV)	Stability
1	1:2:3:4:-H-B_4H_4	−2768.32	−6.64	−2.50	FALSE
2	2:3:5:6:-H-1,4:1,4-μH-B_6H_6	−4148.76	−6.56	−3.73	FALSE
3	2:5:-2H-1,3,4,6:1,3,4,6-μH-B_6H_6	−4149.17	−7.65	−4.22	FALSE
4	3:7:-H-1,5:1,5-μH-B_8H_4	−5466.93	−5.93	−4.30	FALSE
5	3:7:-2H-1,5:1,5-μH-B_8H_6	−5499.07	−5.20	−4.54	FALSE
6	2:4:6:8:-H-1,5:1,5-μH-B_8H_6	−5501.27	−5.93	−4.16	FALSE
7	2:3:4:6:7:8:-H-1,5:1,5-μH-B_8H_8	−5533.90	−6.64	−4.24	FALSE
8	2:4:6:8:-2H-1,5:1,5-μH-B_8H_{10}	−5568.84	−8.14	−4.05	TRUE
9	2:3:4:6:7:8:-2H-1,5:1,5-μH-B_8H_{14}	−5635.07	−7.10	−3.70	FALSE
10	2:3:4:6:7:8:9:10:-H-1,5:1,5-μH-$B_{10}H_{10}$	−6915.86	−6.18	−4.24	FALSE
11	2:3:4:5:7:8:9:10:-H-1,6:1,6-μH-$B_{10}H_{10}$	−6917.17	−6.18	−4.46	FALSE
12	2:3:4:5:6:8:9:10:11:12:-H-1,7:1,7-μH-$B_{12}H_{12}$	−8299.69	−6.31	−4.57	FALSE

Table 2. B_2H_6, 8 BH_3, and B_8H_{10} + 7H_2. All the systems have 8 boron and 24 hydrogen atoms.

System	Total Energy E (Hartree)	E − 4E(B_2H_6) (Hartree)	E − 4E(B_2H_6) (kcal/mol)
4 B_2H_6	−213.218	0.000	0.000
8 BH_3	−212.969	0.249	156.250
B_8H_{10} + 7H_2	−212.908	0.310	194.528

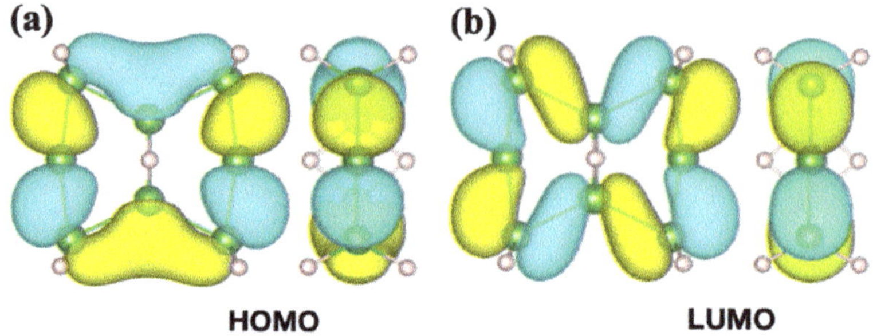

Figure 3. Top and side views of (**a**) the highest occupied molecular orbitals (HOMO) and (**b**) the lowest unoccupied molecular orbitals (LUMO) of 2:4:6:8:-2H-1,5:1,5-μH-B_8H_{10}. The HOMO and LUMO energy levels were −8.14 eV and −4.05 eV with reference to the vacuum level, respectively. The colors of the molecular orbitals correspond to the sign of their wave functions.

2.2. Relation to the Borophane Layer

The molecule 2:4:6:8:-2H-1,5:1,5-μH-B_8H_{10} has a five-membered boron ring on each of the two sides of the 3c-2e bond, as shown in Figure 1b,c. The topological borophane layer (α-HB sheet) reported recently also has such a pair of five-membered rings in its sheet structure, as shown in Figure 4a,b [12,13]. Among the different arrangements, the pair unit is the tiled β-HB sheet (Figure 4c,d). In this section, we discuss the physical and chemical relationships between the proposed molecule and borophane layers.

The α-HB sheet in Figure 4a,b was predicted by topological band theory and confirmed to be thermodynamically and kinetically stable by first-principles calculations [15,17]. It is noteworthy that its kinetic stability was supported by the positive vibrational energy of the phonon modes. This material was successfully synthesized through an ion-exchange reaction [12,13]. The α-HB sheet is a semimetal, with electronic states of the Dirac nodal loop (DNL) at the Fermi level. X-ray spectroscopic measurements have revealed a gapless electronic structure [12,13]. In Figure 4, the wave functions of (a) the lower and (b) the higher DNL bands at the point schematically overlap in the atomic structure model [15]. Focusing on the spatial distributions of the electronic states, one finds that the borophane layer can be described in terms of tiling with the molecular orbitals of the pair of five-membered rings. Wave functions of the energetically higher and lower DNL bands have out-of-plane (π) and in-plane (σ) characteristics, respectively. Instead of the molecular orbitals of 2:4:6:8:-2H-1,5:1,5-μH-B_8H_{10}, the shapes match the HOMO and LUMO of 2:3:4:6:7:8:-H-1,5:1,5-μH-B_8H_8, which is shown in Figure 4e,f.

A case of the β-HB sheet is given in Figure 4c,d; as the electronic structure and stability of the borophane layer have not been derived yet, we present our calculation results in Appendix B. The β sheet is also semimetallic, with a DNL at the Fermi level (Figure A2 in Appendix B). The phonon dispersion diagram for β-type borophane is shown in Figure A3. The phonon dispersion diagrams for β_1-borophane show the imaginary phonon, while the one for β_2-borophane shows that all the phonon energy is positive. The absence of an imaginary phonon modes implies kinematic stability, but it may not be sufficient to conclude their actual existence, as quadratic phonon bands appear near zero phonon energy. As shown in Appendix B, the electronic states of the β-HB sheet presented in Figure 4c,d are built with blocks of molecular orbitals of the pair of five-membered rings, as in the case of the α-sheet. The wave functions of the energetically higher and lower DNL bands have out-of-plane (π) and in-plane (σ) characteristics, respectively. The molecular orbital shapes are also similar to those of 2:3:4:6:7:8:-H-1,5:1,5-μH-B_8H_8 in Figure 4e,f.

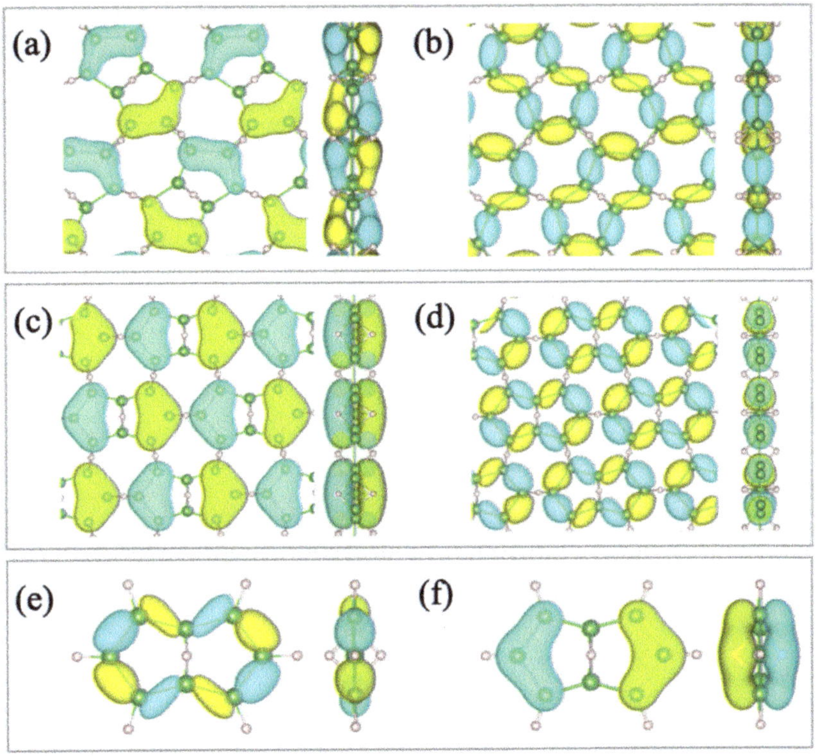

Figure 4. Atomic structures of the borophane layer (α_1, β_1), and the molecule (2:3:4:6:7:8:-H-1,5:1,5-μH-B_8H_8). Wave functions and molecular orbitals at point are schematically overlapped in the atomic models. (**a**) Lower and (**b**) higher bands of a Dirac nodal loop of the α_1-borophane [15,17]. (**c**) Lower and (**d**) higher bands of a Dirac nodal loop of the β_1-borophane (see Appendix A). (**e**) HOMO and (**f**) LUMO of the 2:3:4:6:7:8:-H-1,5:1,5-μH-B_8H_8 molecule. The color of each molecular orbital corresponds to the sign of its wave functions.

Comparing Figures 3 and 4 indicates that the electronic states of the borophane layer agree better with the HOMO and LUMO of 2:3:4:6:7:8:-H-1,5:1,5-μH-B_8H_8 than those of 2:4:6:8:-2H-1,5:1,5-μH-B_8H_{10}. The electronic properties can naturally be understood in terms of the bonding scheme between boron and hydrogen atoms. Besides the 3c-2e bond at the center, the rest of the B–H bonds in the B_8H_8 molecule are associated with one hydrogen atom, while those in B_8H_{10} have two or no hydrogen atoms. The chemical environment of the B_8H_8 molecule generates the same π/σ LUMO/HOMO states as those of the borophane layer. It is worth mentioning that—in Table 1—2:3:4:6:7:8:-H-1,5:1,5-μH-B_8H_8 is unstable, although the HB sheets themselves are stable. We infer that the π electronic state is unfavorable for a single molecule of hydrogenated boron and that the molecule 2:4:6:8:-2H-1,5:1,5-μH-B_8H_{10} consists of σ bonding states.

By further comparing wave functions between the B_8H_8 molecule and the DNL sheets of HB, one finds that the energy levels of the molecular orbitals are reversed between the π and σ types. It is intriguing to find this contrast although the molecule corresponds to a building block of the layer. To systematically investigate their electronic properties, this study calculated the energy difference $E(\pi) - E(\sigma)$ for the molecules and the sheets. As shown in Figure 5a, the two types of low-dimensional structures—0D (molecule) and 2D (sheets)—have an opposite sign of energy difference $E(\pi) - E(\sigma)$. This difference is apparently due to interactions between the building blocks in the 2D network. To examine the issue quantitatively, the tight-binding model was calculated using the hopping parameters between the composing boron atoms. An α-HB sheet structure was adopted in the simulation. The critical parameters were the transfer integrals—t_b and t_i—at the two types of inequivalent 3c-2e bonds, where t_b and t_i correspond to the bonds at the unit center and those at the linkage between the units, respectively (Figure 5b). In the calculation, the transfer integrals at the single bonds between the boron atoms were uniformly set as t_s. Under the condition $t_b/t_s = 1$, the energy level of the π-orbital decreases by $\Delta E = -0.114\ t_s$ when t_i/t_s is changed only from 0 to 0.1. The electronic change in the molecular orbital through the t_i interaction is shown in Figure 5c. The isolated molecular orbitals ($t_i/t_{s.} = 0$) interact with each other and form a delocalized wave function ($t_i/t_s = 0.1$) similar to the one in Figure 4a. The model calculation demonstrates that the energy level of the π molecular orbitals can become lower than that of a σ orbital through interactions between the building units. The development of the delocalized π states is likely responsible for forming the 2D network in the HB sheet.

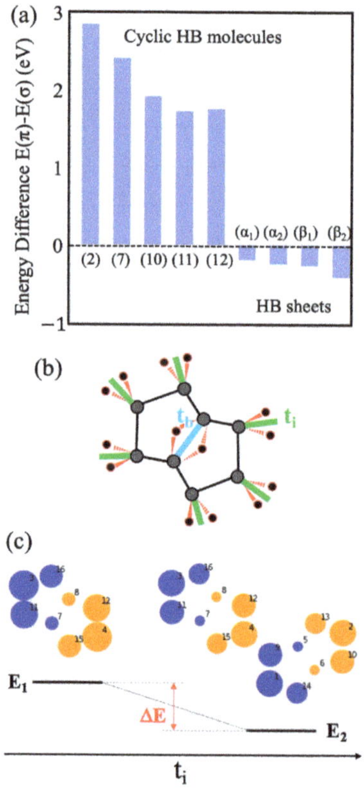

Figure 5. (**a**) Energy differences E(π) − E(σ) between boron materials, corresponding to the HOMO-LUMO gap (gap between lower-band and higher band of DNL at point) for the HB molecules (the HB sheets). Molecules: (2) 2:3:5:6:-H-1,4:1,4-μH-B_6H_6, (7) 2:3:4:6:7:8:-H-1,5:1,5-μH-B_8H_8, (10) 2:3:4:6:7:8:9:10:-H-1,5:1,5-μH-$B_{10}H_{10}$, (11) 2:3:4:5:7:8:9:10:-H-1,6:1,6-μH-$B_{10}H_{10}$, (12) 2:3:4:5:6:8:9:10:11:12:-H-1,7:1,7-μH-$B_{12}H_{12}$. The index numbers correspond to those in Figure 2 and Table 1. Sheets (at point): (α_1) α_1-borophane, (α_2) α_2-borophane, (β_1) β_1-borophane, (β_2) β_2-borophane. (**b**) Descriptions of the tight-binding parameters t_b and t_i. (**c**) Energy diagram calculated from the tight-binding model with wave function forms before and after the interaction. Only one of the neighboring molecular orbitals is shown for the case $t_i/t_s = 0.1$ for comparison with the results in Figure 4a.

The calculation results showed a significant role of the 3c-2e bond in the change in molecular orbitals between the cyclic molecule and topological borophane. Besides the electronic structure, it is interesting to consider a possible reaction path for building an HB sheet from HB molecules. Figure 6a recalls the molecular structure of 2:4:6:8:-2H-1,5:1,5-μH-B_8H_{10}. There are three types of boron atoms: B_C, B_H, and B_A. The B_C atoms take part in the center 3c-2e B–H–B bonds and, thus, are not involved in linkages between molecules. The B_H atoms individually bond with two hydrogen atoms, while the B_A atoms are absent from the B–H bonding. Thus, we infer that a combination of B_H and B_A atoms naturally contributes to the two 3c-2e B–H–B bonds that connect the molecules that form the 2D network of the HB sheet.

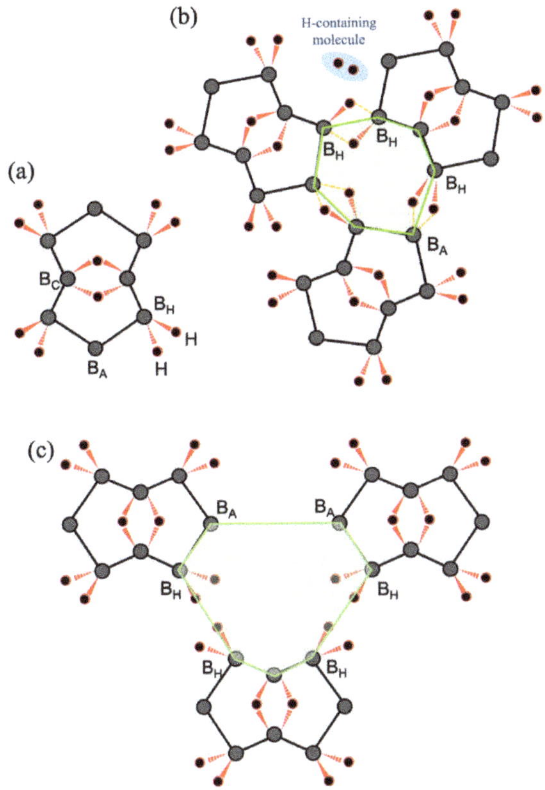

Figure 6. Construction of the borophane layer with the 2:4:6:8:-2H-1,5:1,5-μH-B_8H_{10} molecules. (**a**) Molecular structures with labels on the different boron sites. (**b**) Building the α-borophane. (**c**) Building the β-borophane. Intermediate seven-membered rings are shaded in green in (**b**,**c**).

To build an α-sheet composed of five-membered and seven-membered rings, one can assemble B_8H_{10} molecules to form an intermediate seven-membered ring structure, as shown in Figure 6b. The heptagonal network is associated with two pairs of B_H and B_A atoms that naturally lead to the formation of 3c-2e bonds. There is also a pair of B_H atoms that can induce intermolecular B–H–B bonding by releasing two hydrogen atoms or H_2. However, for a β-sheet, the intermediate seven-membered ring is formed by combinations of B_H atoms with combinations of B_A atoms, as illustrated in Figure 6c. Pairs of B_A atoms lack a hydrogen atom and are unlikely to generate a 3c-2e B–H–B bond themselves; thus, the B_8H_{10} molecules cannot be used as blocks for building a β-structure. The argument for B–H–B bond formation favors α-sheet rather than β-sheet formation.

It is intriguing that β-borophane (Figure 4c,d) is not stable according to the calculations (Appendix B). However, the cyclic hydrogenated boron molecule 2:4:6:8:-2H-1,5:1,5-μH-B_8H_{10} in Figure 1b,c and the topological α-borophane in Figure 1a,b were found to be stable. It is coincidental but intriguing to find that the reaction path of the molecular assembly favors a combination of the stable hydrogenated boron species.

Layers of topological borophane have been synthesized by liquid exfoliation associated with an ion-exchange reaction [11–13]. This is a top-down approach and typically results in powders of microscopic flakes. However, various 2D materials such as graphene and transition-metal dichalcogenide have been grown by chemical vapor deposition (CVD). This bottom-up approach has led to the synthesis of wide-area and single-crystalline films [21]. The cyclic B_8H_{10} molecule proposed in this research has an atomic structure that corresponds to the building blocks of topological α-borophane. Thus, this molecule is the ideal precursor or CVD reacting gas for preparing single-crystalline topological borophane with a macroscopically wide area.

2.3. Orbitals of the Cyclic Boron Molecule in the Hückel Model

To examine the electronic states of a cyclic hydrogenated boron molecule, the classical Hückel model was applied and the results were compared with those for carbon molecules. The Hückel model has been used to simply calculate the energies of the π molecular orbitals in a conjugated carbon system by regarding all molecular orbitals as linear combinations of atomic orbitals [22–27]. Figure 7a shows the example of the cyclic conjugated carbon compound naphthalene ($C_{10}H_8$). This molecule consists of two loops of p_z-orbitals, and each molecular orbital belongs to the genus-2 topological classification.

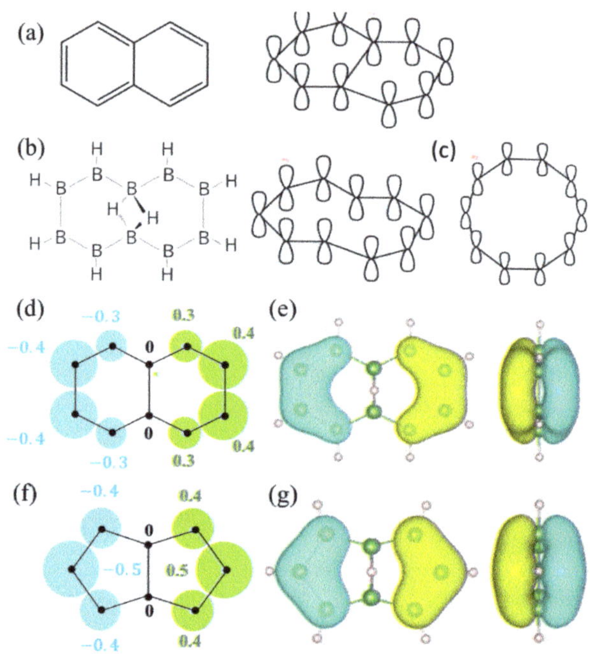

Figure 7. Application of the Hückel model to cyclic molecules. (**a**) Naphthalene molecule and the p_z configuration. (**b**) 1,6-1,6-decadiborane and the p_z configuration. (**c**) p_z configuration of a cyclic decagon molecule. (**d**,**e**) Molecular orbitals of 1,6-1,6-decadiborane, calculated from (**d**) the Hückel model and (**e**) first-principles calculations. (**f**,**g**) Molecular orbitals of 1,5-1,5-octadiborane, calculated from the (**g**) Hückel model and (**f**) first-principles calculations. The color of each molecular orbital corresponds to the sign of its wave functions.

For comparison, we prepared a cyclic hydrogenated boron molecule with 10 boron atoms, corresponding to the 10 carbon atoms in naphthalene. The molecule is 1,6-1,6-decadiborane, $B_{10}H_{10}$, as illustrated in Figure 7b. As the two central boron atoms connect to each other through 3c-2e bonds, the p_z-orbitals of the constituent boron atoms formed one molecular orbital loop, as shown in Figure 7b. This loop was topologically equivalent to a circle (genus-1), as illustrated in Figure 6c. This made it simple to calculate the Hückel model of the cyclic hydrogenated boron molecule. In this research, the Hückel model was calculated with a constant onsite Coulomb integral and constant resonance integral between neighboring atoms. The overlap integral was set to zero. A wave function calculated from the Hückel model is shown in Figure 7d for comparison with the LUMO in Figure 7e obtained from first-principles calculations. The molecular orbitals agreed well. Figure 7f,g also compares molecular orbitals between the Hückel model and the first-principles calculations for 2:4:6:8:-2H-1,5:1,5-μH-B_8H_{10}. Again, the molecular orbitals of the two models were similar. This similarity could be a theoretical test ground for deepening our understandings of quantum chemistry-based topological structures and electronic states.

3. Calculation Methods

Calculations of optimized atomic structures, molecular orbitals, and vibrational properties were all made via density functional theory (DFT), implemented using the Python package "Psi4 1.5.0" [28]. Becke three-parameter exchange and Lee–Yang–Parr correlations (B3LYP) were selected as an exchange-correlation functional [29,30]. The Gaussian basis set 6-311g(d,p) was applied to describe the electronic structures. Molecular geometry was optimized under Gaussian-level criteria.

For 2D HB or borophane sheets, calculations of wave functions and band diagrams were carried out via "Quantum ESPRESSO 7.1", which is a first-principles code based on DFT, the plane-wave method, and the pseudopotential method [31]. The present calculations used the GGA–PBE exchange-correlation functional [32] with ultra-soft pseudopotentials in the library of standard solid-state pseudopotentials called "SSSP" [33]. The structure stabilities were evaluated from the phonon dispersion obtained from "Phonopy 2.16.3" [34], which can be used as post-process code for Quantum ESPRESSO.

The atomic structures and wave functions were visualized via Visualization of Electronic and STructural Analysis (VESTA) [35].

4. Conclusions

In summary, we predict the existence of a new cyclic hydrogenated boron molecule, 2:4:6:8:-2H-1,5:1,5-μH-B_8H_{10}, on the basis of first-principles calculations. This molecule can be regarded as a building block of topological hydrogen boride (borophane) sheets. This research shows intriguing relationships between the planar molecule and the atomic sheet forms of the hydrogenated boron compound. The cyclic B_8H_{10} molecule provides new perspectives in boron chemistry. This molecule could also be the ideal precursor or CVD reaction gas for preparing wide-area topological borophane, which could lead to technological innovations.

Author Contributions: Conceptualization, Y.A. and I.M.; calculation, Y.A., T.N., H.Y., I.T. and N.T.C.; validation, X.Z., Y.T., M.H., S.O., T.K. and I.M.; writing—original draft preparation, Y.A. and I.M.; writing—review and editing, Y.A. and I.M.; visualization, Y.A., T.N., H.Y. and I.M.; supervision, S.O., T.K. and I.M.; project administration, I.M.; funding acquisition, I.M. All authors have read and agreed to the published version of the manuscript.

Funding: This research was funded by JST, CREST Grant Number JPMJCR21O4, Japan, and by JSPS KAKENHI Grant Numbers JP19H04398 and JP18H03874.

Data Availability Statement: The data is contained in the article.

Acknowledgments: We acknowledge M. Sato, T. Sumi, and F. Komori for their discussions with us.

Conflicts of Interest: The authors declare no conflict of interest.

Appendix A

Figure A1 shows candidates for cyclic hydrogen boron molecules that were theoretically examined in this research.

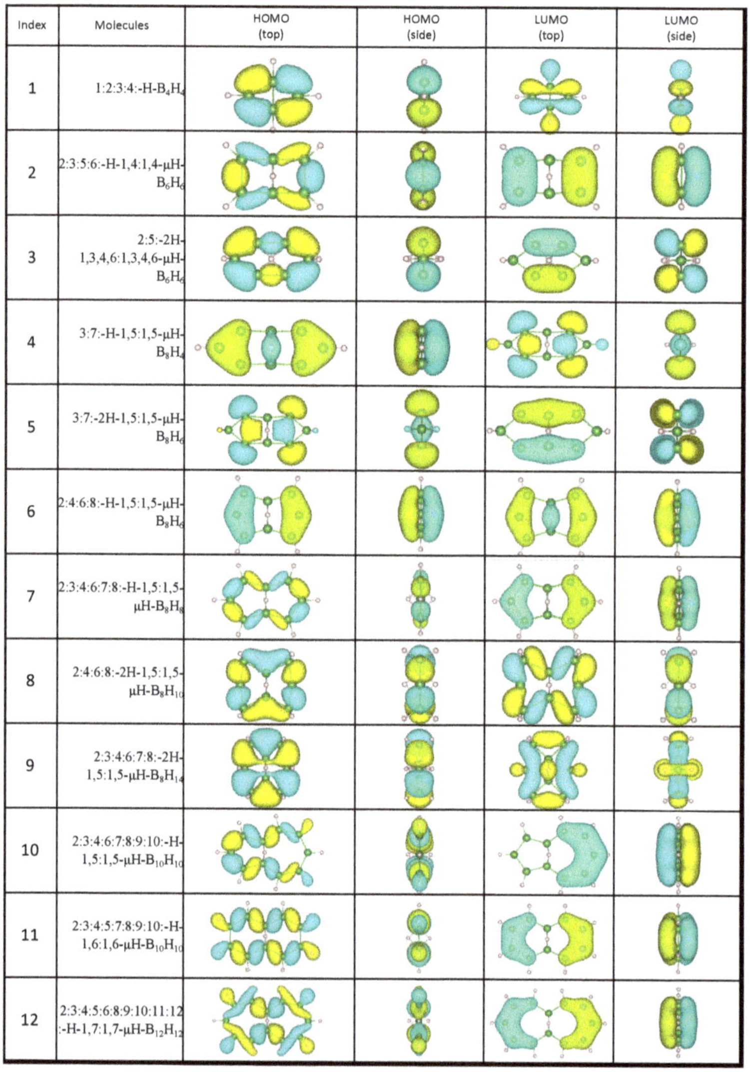

Figure A1. Candidates for the cyclic hydrogenated boron molecules. The structures are overlapped with molecular orbitals of HOMOs and LUMOs. The color of each molecular orbitals corresponds to the sign of its wave functions.

Appendix B

Figure A2 summarizes atomic models and electronic structures of the β_1- and β_2-borophane layers. The phonon dispersion curves of the individual HB sheets are given in Figure A3.

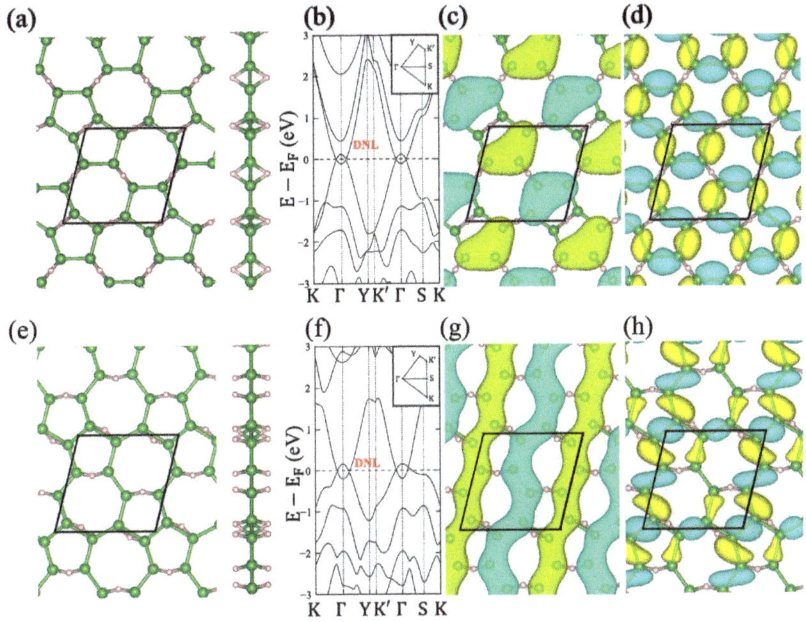

Figure A2. The β borophane. The β_1-type: (**a**) atomic structure, (**b**) band diagram, (**c**) π wave functions (lower bands of a DNL), and (**d**) σ wave functions (higher bands of a DNL). The β_2-type: (**e**) atomic structure, (**f**) band diagram, (**g**) π wave functions (lower bands of a DNL), and (**h**) σ wave functions (higher bands of a DNL). The color of each molecular orbitals corresponds to the sign of its wave functions. The unit cell is represented as a rhombus in each subfigure corresponding to the atomic structures and orbitals. The letters shown in (**b**,**f**) represent the high symmetry points in the first Brillouin zone shown as insets of (**b**,**f**).

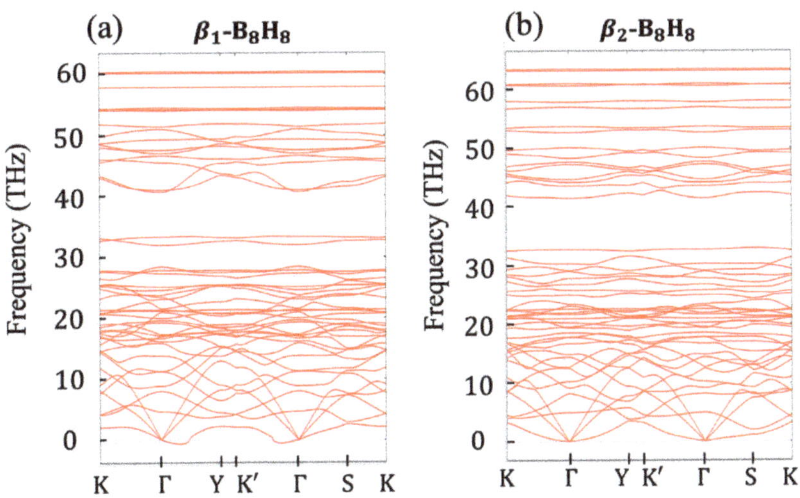

Figure A3. Phonon dispersion diagrams of the (**a**) β_1- and (**b**) β_2-borophane. The letters shown in (**a**,**b**) represent the high symmetry points in the first Brillouin zone shown as insets of Figure A2b,f.

The β_1-borophane layer is tiled with the pseudo-2:3:4:6:7:8:-H-1,5:1,5-μH-B_8H_8 structures as shown in Figure A2a which has the 3c-2e bond at a shared edge of pentagons. The band structure of β_1-borophane is shown in Figure A2b, which has a DNL. The wave functions that form a DNL are periodic HOMO (σ bonding; Figure A2c) and LUMO (π bonding; Figure A2d) arrangements of the 2:3:4:6:7:8:-H-1,5:1,5-μH-B_8H_8 molecule (Figure 4e,f). On an isolated molecule, the π bond is at a higher energy level than the σ bond. When the 2:3:4:6:7:8:-H-1,5:1,5-μH-B_8H_8 molecule is crystallized in 2D by forming intermolecular 3c-2e B–H–B bonds, both molecular orbitals overlap with those on the neighboring molecules, and hopping integrals are introduced. Because of the out-of-plane structure of the 3c-2e bond, the sign of the hopping integral is different for the σ and π bonds, as discussed in reference [16]. The hopping integral for the π (σ) bond is negative (positive), so the band dispersion of the 2D crystal shows a cosine band with a lower (higher) energy at the point. These hopping integrals have energy scales comparable with those of the molecular orbitals, and thus the band is inverted at the point (Figure A2b). Because the inverted bands have different eigenvalues with respect to the mirror operation about the z = 0 plane, hybridization between them is prohibited and a gapless band crossing is produced, which is the DNL.

The atomics structure of β_2-borophane layer is shown in Figure A2e which has a different 3c-2e bonds arrangement from that of β_1-borophane layer. The band structure of β_2-borophane is shown in Figure A2f, which also has a DNL. The wave functions that form the lower (higher) bands of a DNL is π (σ) bonding states as shown in Figure 4g,h. The π bonding states of β_2-borophane has nodes at 3c-2e bonds like that of β_1-borophane, though the character of the wave function differs from the isolated 2:3:4:6:7:8:-H-1,5:1,5-μH-B_8H_8 molecule due to the different 3c-2e bonding arrangement.

The phonon dispersion curves of both β_1- and β_2-borophane layers are given in Figure A3 to evaluate their stability. The phonon dispersion of the β_1-borophane shows the imaginary phonon, which implies that the β_1-borophane layers are unstable. On the other hand, the β_2-borophane shows that all the band dispersion is positive. Though the absence of imaginary phonon modes implies kinematic stability, but it may not be sufficient to conclude their actual existence because quadratic phonon bands appear near zero phonon energy.

References

1. Callister, W.D., Jr.; Rethwisch, D.G. *Materials Science and Engineering: An Introduction*, 10th ed.; Wiley: Hoboken, NJ, USA, 2018.
2. Hagemann, H. Boron Hydrogen Compounds: Hydrogen Storage and Battery Applications. *Molecules* **2021**, *26*, 7425. [CrossRef] [PubMed]
3. van't Hoff, J.H. A suggestion looking to the extension into space of the structural formulas at present used in chemistry, and a note upon the relation between the optical activity and the chemical constitution of organic compounds. *Arch. Neerl. Sci. Exactes Nat.* **1874**, *9*, 445–454.
4. Le Bel, J.A. Sur les Relations qui existent entre les Formules atomiques des Corps organiques et le Pouvoir rotatoire de leurs Dissolutions. *Bell. Soc. Chim. Fr.* **1874**, *22*, 337–347.
5. Fau, S.; Frenking, G. Anti van't Hoff/Le Bel geometries of boron compounds. A theoretical study of classical and nonclassical isomers of B_2CH_4, B_2NH_3 and B_2OH_2. *J. Mol. Struct. THEOCHEM* **1995**, *338*, 117–130. [CrossRef]
6. Song, B.; Zhou, Y.; Yang, H.-M.; Liao, J.-H.; Yang, L.-M.; Yang, X.-B.; Ganz, E. Two-dimensional anti-Van't Hoff/Le Bel array AlB_6 with high stability, unique motif, triple dirac cones, and superconductivity. *J. Am. Chem. Soc.* **2019**, *141*, 3630–3640. [CrossRef]
7. Matsuda, I.; Wu, K. *2D Boron: Boraphene, Borophene, Boronene*, 1st ed.; Springer: Cham, Switzerland, 2021.
8. Feng, B.; Zhang, J.; Zhong, Q.; Li, W.; Li, S.; Li, H.; Cheng, P.; Meng, S.; Chen, L.; Wu, K. Experimental Realization of Two-Dimensional Boron Sheets. *Nat. Chem.* **2016**, *8*, 563–568. [CrossRef]
9. Feng, B.; Sugino, O.; Liu, R.-Y.; Zhang, J.; Yukawa, R.; Kawamura, M.; Iimori, T.; Kim, H.; Hasegawa, Y.; Li, H.; et al. Dirac Fermions in Borophene. *Phys. Rev. Lett.* **2017**, *118*, 096401. [CrossRef] [PubMed]
10. Feng, B.; Zhang, J.; Ito, S.; Arita, M.; Cheng, C.; Chen, L.; Wu, K.; Komori, F.; Sugino, O.; Miyamoto, K.; et al. Discovery of 2D anisotropic Dirac cones. *Adv. Mater.* **2018**, *30*, 1704025. [CrossRef]
11. Nishino, H.; Fujita, T.; Cuong, N.T.; Tominaka, S.; Miyauchi, M.; Iimura, S.; Hirata, A.; Umezawa, N.; Okada, S.; Nishibori, E.; et al. Formation and Characterization of Hydrogen Boride Sheets Derived from MgB_2 by Cation Exchange. *J. Am. Chem. Soc.* **2017**, *139*, 13761–13769. [CrossRef]

12. Niibe, M.; Cameau, M.; Cuong, N.T.; Sunday, O.I.; Zhang, X.; Tsujikawa, Y.; Okada, S.; Yubuta, K.; Kondo, T.; Matsuda, I. Electronic structure of a borophene layer in rare-earth aluminum/chromium boride and its hydrogenated derivative borophane. *Phys. Rev. Mater.* **2021**, *5*, 084007. [CrossRef]
13. Zhang, X.; Tsujikawa, Y.; Tateishi, I.; Niibe, M.; Wada, T.; Horio, M.; Hikichi, M.; Ando, Y.; Yubuta, K.; Kondo, T.; et al. Electronic Topological Transition of 2D Boron by the Ion Exchange Reaction. *J. Phys. Chem. C* **2022**, *126*, 12802–12808. [CrossRef]
14. Kawamura, R.; Cuong, N.T.; Fujita, T.; Ishibiki, R.; Hirabayashi, T.; Yamaguchi, A.; Matsuda, I.; Okada, S.; Kondo, T.; Miyauchi, M. Photoinduced hydrogen release from hydrogen boride sheets. *Nat. Commun.* **2019**, *10*, 4880. [CrossRef] [PubMed]
15. Cuong, N.T.; Tateishi, I.; Cameau, M.; Niibe, M.; Umezawa, N.; Slater, B.; Yubuta, K.; Kondo, T.; Ogata, M.; Okada, S.; et al. Topological Dirac Nodal Loops in Nonsymmorphic Hydrogenated Monolayer Boron. *Phys. Rev. B* **2020**, *101*, 195412. [CrossRef]
16. Tateishi, I.; Cuong, N.T.; Moura, C.A.S.; Cameau, M.; Ishibiki, R.; Fujino, A.; Okada, S.; Yamamoto, A.; Araki, M.; Ito, S.; et al. Semimetallicity of Free-Standing Hydrogenated Monolayer Boron from MgB_2. *Phys. Rev. Mater.* **2019**, *3*, 024004. [CrossRef]
17. Tateishi, I.; Zhang, X.; Matsuda, I. Electronic Structures of Polymorphic Layers of Borophane. *Molecules* **2022**, *27*, 1808. [CrossRef]
18. *Boron Hydride Chemistry*; Muetterties, E.L. (Ed.) Academic Press: New York, NY, USA, 2012.
19. Connelly, N.G.; Damhus, T.; Hartshorn, R.M.; Hutton, A.T. *Nomenclature of Inorganic Chemistry: IUPAC Recommendations 2005*; Royal Society of Chemistry: Cambridge, UK, 2005.
20. Beckett, M.A.; Brellochs, B.; Chizhevsky, I.T.; Damhus, T.; Hellwich, K.-H.; Kennedy, J.D.; Laitinen, R.; Powella, W.H.; Rabinovich, D.; Viñas, C.; et al. Nomenclature for boranes and related species (IUPAC Recommendations 2019). *Pure Appl. Chem.* **2020**, *92*, 355–381. [CrossRef]
21. Choi, S.H.; Yun, S.J.; Won, Y.S.; Oh, C.S.; Kim, S.M.; Kim, K.K.; Lee, Y.H. Large-scale synthesis of graphene and other 2D materials towards industrialization. *Nat. Commun.* **2022**, *13*, 1484. [CrossRef]
22. Solà, M. Aromaticity rules. *Nat. Chem.* **2022**, *14*, 585–590. [CrossRef]
23. Mutai, K. *Quantum Chemistry, Ryoshi Kagaku*; SHOKODO Co. Ltd.: Tokyo, Japan, 1991. (In Japanese)
24. Hückel, E. Quantum-theoretical contributions to the benzene problem. I. The electron configuration of benzene and related compounds. *Z. Phys.* **1931**, *70*, 204–286. [CrossRef]
25. Krygowski, T.M.; Cyrański, M.K.; Czarnocki, Z.; Häfelinger, G.; Katritzky, A.R. Aromaticity: A Theoretical Concept of Immense Practical Importance. *Tetrahedron* **2000**, *56*, 1783–1796. [CrossRef]
26. Hess, B.A., Jr.; Schaad, L.J. Hueckel molecular orbital .pi. resonance energies. Benzenoid hydrocarbons. *J. Am. Chem. Soc.* **1971**, *93*, 2413–2416. [CrossRef]
27. Schaad, L.J.; Hess, B.A., Jr. Huckel theory and aromatically. *J. Chem. Educ.* **1974**, *51*, 640. [CrossRef]
28. Smith, D.G.A.; Burns, L.A.; Simmonett, A.C.; Parrish, R.M.; Schieber, M.C.; Galvelis, R.; Kraus, P.; Kruse, H.; Remigio, R.D.; Alenaizan, A.; et al. PSI4 1.4: Open-source software for high-throughput quantum chemistry. *J. Chem. Phys.* **2020**, *152*, 184108. [CrossRef] [PubMed]
29. Becke, A.D. Density-functional thermochemistry. III. The role of exact exchange. *J. Chem. Phys.* **1993**, *98*, 5648. [CrossRef]
30. Lee, C.; Yang, W.; Parr, R.G. Development of the Colle-Salvetti correlation-energy formula into a functional of the electron density. *Phys. Rev. B* **1988**, *37*, 785. [CrossRef]
31. Giannozzi, P.; Andreussi, O.; Brumme, T.; Bunau, O.; Nardelli, M.B.; Calandra, M.; Car, R.; Cavazzoni, C.; Ceresoli, D.; Cococcioni, M.; et al. Advanced capabilities for materials modelling with Quantum ESPRESSO. *J. Phys. Condens. Matter* **2017**, *29*, 465901. [CrossRef] [PubMed]
32. Perdew, J.P.; Burke, K.; Ernzerhof, M. Generalized Gradient Approximation Made Simple. *Phys. Rev. Lett.* **1996**, *77*, 3865. [CrossRef]
33. Prandini, G.; Marrazzo, A.; Castelli, I.E.; Mounet, N.; Marzari, N. Precision and efficiency in solid-state pseudopotential calculations. *npj Comput. Mater.* **2018**, *4*, 72. [CrossRef]
34. Togo, A.; Tanaka, I. First principles phonon calculations in materials science. *Scr. Mater.* **2015**, *108*, 1–5. [CrossRef]
35. Momma, K.; Izumi, F. VESTA 3 for three-dimensional visualization of crystal, volumetric and morphology data. *J. Appl. Cryst.* **2011**, *44*, 1272–1276. [CrossRef]

molecules MDPI

Article

Relative Stability of Boron Planar Clusters in Diatomic Molecular Model

Levan Chkhartishvili [1,2]

[1] Department of Engineering Physics, Georgian Technical University, 77 Merab Kostava Avenue, Tbilisi 0160, Georgia; levanchkhartishvili@gtu.ge
[2] Boron and Powder Composite Materials Laboratory, Ferdinand Tavadze Metallurgy and Materials Science Institute, 8b Elizbar Mindeli Street, Tbilisi 0186, Georgia

Abstract: In the recently introduced phenomenological diatomic molecular model imagining the clusters as certain constructions of pair interatomic chemical bonds, there are estimated specific (per atom) binding energies of small all-boron planar clusters B_n, $n = 1$–15, in neutral single-anionic and single-cationic charge states. The theoretically obtained hierarchy of their relative stability/formation probability correlates not only with results of previous calculations, but also with available experimental mass-spectra of boron planar clusters generated in process of evaporation/ablation of boron-rich materials. Some overestimation in binding energies that are characteristic of the diatomic approach could be related to differences in approximations made during previous calculations, as well as measurement errors of these energies. According to the diatomic molecular model, equilibrium binding energies per B atom and B–B bond lengths are expected within ranges 0.37–6.26 eV and 1.58–1.65 Å, respectively.

Keywords: planar cluster; charge state; bond length; specific binding energy; relative stability; formation probability; boron

Citation: Chkhartishvili, L. Relative Stability of Boron Planar Clusters in Diatomic Molecular Model. *Molecules* **2022**, 27, 1469. https://doi.org/10.3390/molecules27051469

Academic Editors: Takahiro Kondo, Iwao Matsuda and Josep M. Oliva-Enrich

Received: 27 December 2021
Accepted: 12 February 2022
Published: 22 February 2022

1. Introduction

Nanoboron and boron-rich nanomaterials are of current academic and practical interests because of their widely variable interatomic bonding mechanism and related unique complex of physical–chemical properties useful in technological applications—see some of the reviews in the last decade [1–9].

Among them, the all-boron clusters B_n, $n = 2, 3, 4, \ldots$, as individual species in the gas phase, play an important role, as they can serve for building blocks in the boron-rich solids chemistry [10]. For example, a quasi-planar boron cluster B_{35} with a double-hexagonal hole at the center has been reported [11] as a flexible structural motif for borophene allotropies, as it can be used to construct atom-thin boron sheets with various hexagonal hole densities.

In this regard, it should be noted that, depending on the number of atoms and also formation kinetics, boron clusters can take several different shapes. Joint experimental studies and computational simulations revealed [12] that boron clusters, which favor (quasi)planar, i.e., 2D, structures up to 18 atoms, prefer 3D structures beginning at 20 atoms. The B_{20} neutral cluster was found to have a double-ring tubular ground structure. As for the B_{20}^- anion, its tubular structure was shown to be almost isoenergetic to 2D structures. Thus, the usually observed 2D-to-3D structural transition suggests that B_{20} may be considered as the embryo of thinnest single-walled boron nanotubes. According to the QC (Quantum Chemical) and DFT (Density Functional Theory) investigations [13], there are two structural transitions that are expected in boron clusters: the second transition from double-ring system into triple-ring one occurs between B_{52} and B_{54}.

At low (namely, from 7 to 20) nuclearities, i.e., for (quasi)planar boron clusters, the separate quantum rules of in- and out-of-plane bonding were obtained [14], using the free-particle-on-disk and rectangle models combined with DFT electronic structure calculations.

In this Introduction, the quite-rich data available on boron clusters synthesis methods, binding parameters and their potential applications are only briefly discussed.

1.1. Synthesizing

Boron quasi-planar clusters can be formed in process of thermal vaporization [15–21], bombardment with high-energy particles [22], even grinding of boron-rich solid materials [23–25] and mainly by their laser ablation [26–29].

An effective method of thermal generating of pure boron cluster-ions for their further use as a plasma-process feed gas was proposed by Becker [18]. Chamber's electrode material is a compound of boron with Me metal(s) thermally decomposable within a suitable temperature range to provide boron in the vapor, but other species are substantially not in the vapor states. Magnetic confinement of the simultaneously released electrons causes numerous collisions, resulting in boron vapor ionization to the plasma state. This plasma is then extracted and accelerated at a suitable energy toward the workpiece. Created in this way, boron clusters can be self-assembled into nanostructures [19].

As early as in References [23,24], electron microscopic study of the elementary boron powder structure revealed that ultrafine particles (of $\leq$200 Å size) of freely grown boron are characterized by a stable 2D shape with almost hexagonal profile and aspect ratio of ~20:1. Computer simulations were performed [25] to model structural relaxation in 2D-clusters mimicking these boron small particles.

Theoretically, using DFT, a growth path for small boron clusters B_n was discussed [30] with a size range and isomers structure. The thermochemical parameters that were determined by using coupled-cluster theory calculations suggested [31] the evolution of geometry and resonance energy of B_n clusters through the number of B–B bonds. Based on the known geometrical characteristics of boron clusters, their general growth mechanism was proposed in Reference [32]. Moreover, a systematic structural investigation of B_n clusters established a picture of their growth behavior [33].

1.2. Structure and Binding

According to References [15,34], the diboron molecule B_2 experimental dissociation energy is within range of 2.82 $\pm$ 0.24 or 2.69 $\pm$ 0.43 eV, respectively. The absorption transition at 3200–3300 Å observed in B_2 indicated [35] that its ground electronic state should be the lower state of Σ-type. Moreover, as B_2 was not observed via ESR (Electron Spin Resonance), the ground electronic state was identified with ${}^3\Sigma_g{}^-$. An accurate CI (Configuration Interaction) calculation confirmed [36] that mentioned transition is from the first excited state of ${}^3\Sigma_u{}^-$ type. Molecular binding energy for B_2 at the HF (Hartree–Fock) level of theory [37] is 2.861 eV, and its scaled ground-state harmonic frequency equals to 0.120 eV. From the DFT calculations [30], these parameters are 2.718 and 0.134 eV, respectively. By using the MO (Molecular Orbitals) method, the dissociation energy of B_2 ground state was calculated to be 2.71 eV [38]. Handbook [39] recommends the diboron molecule ground electron state dissociation energy of 3.02 eV. In this state, its bond length and vibration quantum are 1.590 Å and 0.130 eV, and in the first excited state, they are 1.625 Å and 0.116 eV, respectively. The term of corresponding transition equals to 3.79 eV.

Potential energy curves for the states of B_2 were constructed by the complete-active-space SCF (Self Consistent Field) method at the multi-reference CI level [36]. According to other multi-reference CI study [40], its ground-state curve parameters—bond length, dissociation energy and relative vibration quantum are of 1.600–1.607 Å, 2.70–2.78 and 0.128–0.129 eV, respectively. For the B_2 molecule ground-state interatomic potential energy, P curve $P - d$, where d is the inter-nuclear distance, constructed [41,42] within quasi-classical approach (Figure 1), the curve's parameters are as follows: dissociation energy of 2.80 eV, equilibrium bond length of 1.78 Å and vibration quantum of 0.13 eV. Based on explicit expressions for intersite distances in boron nanotubes of regular geometry in terms of B−B bonds length and using this quasi-classical B−B pair potential, there were estimated some ground-state parameters of boron nanotubes [43–45].

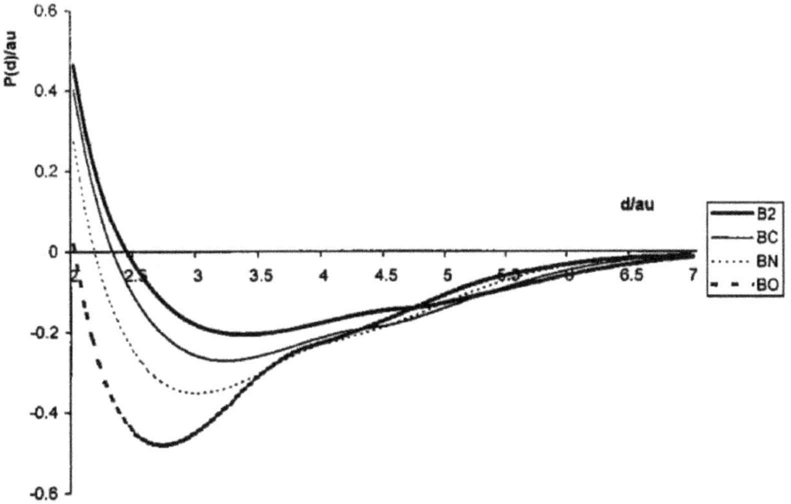

Figure 1. Quasi-classically calculated interatomic potentials for boron-containing diatomic molecules B_2, BC, BN and BO versus inter-nuclear distances [42]. Copyright (2000), with permission from Elsevier.

The potential curves, transition energies, bond lengths and vibration frequencies of ground and some of low-lying excited electronic states of B_2^+ and B_2 were obtained by using a CI approach [46]. The B_2^+ cation ground state, which was found to be $^2\Sigma_g{}^+$, shows a rather shallow potential curve with a bond length of 2.125 Å and vibration quantum of 0.052 eV, when compared with that of the $^3\Sigma_g{}^-$ state of B_2 neutral: 1.592 Å and 0.131 eV. The first excited state of B_2^+, $^2\Pi_u$, lies at 0.30 eV. As a result of loss of bonding electron, the ground-state dissociation energy for B_2^+ with calculated value of 1.94 eV is smaller than that of B_2.

In Reference [47], the local nature of different types of boron–boron bonds from the topological analysis of ELF (Electron Localization Function) perspective was investigated in 23 boron-containing molecules.

Initially, Boustani demonstrated that stable structures of neutral bare boron clusters, B_n, with n = 12, 16, 22, 32, 42 and 46, can easily be constructed with the help of the so-called Aufbau Principle suggested on the basis of HF SCF direct CI [48], as well as QC, DFT and LMTO (Linear Muffin Tin Orbital) [49,50] studies.

Based on different theoretical approaches, such as LDA (Local Density Approximation) and LSD (Local Spin Density) versions of DFT, HF, complete active space and scattered-wave SCF; correlated CI, QC, MO, PES (Potential Energy Surface) coupled-cluster, Born–Oppenheimer and full-potential LMTO versions of MD (Molecular Dynamics); and other methods, there are reported the key structural and binding parameters of boron clusters calculated or scaled from measured ones: B_3 [37,51]; B_3 and B_4 in neutral and anionic forms [52]; B_4 [53]; B_2, B_3 and B_4 [54]; B_5, B_5^+ and B_5^- isomers [55]; B_5^- [56]; B_6 and B_6^- [57]; B_7 and B_7^- [58]; B_2^+–B_8^+ [59]; B_n with n = 2–8 in both the neutral and cationic states [60]; B_n clusters with $4 \leq n \leq 8$ [61]; 8- and 9-atom boron clusters [62]; small boron clusters with up to 10 atoms [63]; $B_{2\text{–}12}$ and $B_{2\text{–}12}^+$ [64]; small neutral B_n clusters with n = 2–12 [30]; icosahedral cluster B_{12} [65]; B_{12} [66,67]; B_2^+–B_{13}^+ [68] (see also Reference [69]); multi-charged clusters B_n with n = 2–13 [70]; set of small-sized neutral B_n and anionic B_n^- boron clusters with n = 5–13 [31]; B_7, B_{10} and B_{13} [71]; B_{12} and B_{13} for neutrals and cations [72]; B_{12}^+ and B_{13}^+ [73]; B_{12} and B_{13}^+ [74]; isomers of boron 13-clusters [75]; cationic, neutral and anionic charge states of B_{13} [76]; planar or quasi-planar structures of B_{13}, B_{13}^+ and B_{13}^- [77]; isomers of planar boron cluster B_{13} [78]; small cationic clusters B_n^+ with n = 2–14 [79]; B_n with n = 2–14 [80,81]; B_n clusters for $n \leq 14$ [82]; B_n with n = 5–14 [83]; boron clusters in the 10- to 15-atom size range [84]; B_{16}^- and B_{16}^{2-} [85]; B_{19}^- [86]; neutral and anionic B_{20} isomers [87]; B_{23}^+ [88]; isomers of 24-atom boron cluster [89]; B_{25}^- [90]; B_{27}^- [91]; B_n with n = 26–29 in both neutral and anionic states [32]; B_{32} [92]; B_{36} and B_{36}^- [93,94]; B_{41}^- and B_{42}^- [95]; neutral B_n clusters with n = 31–50 [33]; and boron cluster-families: spheres, double-rings and quasi-planars containing up to n = 122 atoms [96].

1.3. Applications

The unique behaviors of clusters of elemental boron have been identified: they react readily with metal surfaces; bond covalently to metal atoms; and cover surfaces with a boron-enriched hard, smooth and corrosion-resistant layer, which can be called nano-boron coating. Some parameters, especially feeding gas concentration, substrate temperature and input power, were optimized [97] to prepare high-pure boron (94%B) coating films by plasma-assisted CVD (Chemical Vapor Deposition). However, due to its thermal resilience, elemental boron is a difficult material to work with. To overcome this problem, Becker perfected a technique for generating clusters of elemental boron in plasma [98]. Unique tribological applications and scalability of boron-rich materials are likely to emerge from the combination of high mechanical strength, chemical stability, exceptional hardness and toughness, wear resistance, strong binding to substrates, low density and other promising physical–chemical properties [99].

The durability of boron coatings in sliding friction has been mentioned [100]. In the search ways to enhance the surface hardness of aluminum, the equilibrium structure, stability, elastic properties and formation dynamics of a boron-enriched surface were studied by using DFT [101]. Nano-indentation simulations suggested that the presence of boron nanostructures in the subsurface region significantly enhances the mechanical hardness of aluminum surfaces.

Boron is the only practical solid material with both volumetric and gravimetric energy densities substantially greater than those of hydrocarbons. That is why boron powder is attractive as a fuel or a supplement in propellants and explosives and potential source of secondary energy generation as well. One study [102] aimed to obtain the energy from elemental boron burning as solid fuel, which is synthesized from boron minerals. By experimental investigation [103] of the combustion characteristics of boron nanoparticles in the post-flame region, a two-stage combustion phenomenon was observed. The extended combustion model for single boron particles of sizes relevant for ramjet chambers was introduced [104] and validated [105]; it comprised a consistent formulation of the heat and mass transfer processes in the boron particles' environment. A review on boron powders given in Reference [106] serves as the basis for research on the tendency of nanosized boron particles to group in an oscillating flow and its effect on the combustion process, flame characteristics and pollutants' emission. Nanoboron can be considered as a superior rocket fuel because nano-particles have almost fluid-like properties. To optimize the reactive surface for combustion, nanoparticle size could be shrunk to clusters consisting of several atoms each. One review paper [107] encompassed the status and challenges in the synthesis process of boron nanoparticles, their dispersion and stability of in liquid hydrocarbon fuels, ignition and combustion characteristics of boron loaded liquid fuel, particle combustion and characterization of post-combustion products. The combustion characteristics of nanofluid fuels containing additions of boron and iron particles together were investigated in Reference [108]. Furthermore, mechanical milling was used to prepare a boron-based composite powder containing 5wt.% nanoiron to behave as a catalyst of boron oxidation [109]. The energy density of reactive metal fuel containing Ti, Al and B nanopowders was optimized [110] by varying the Ti:Al:B ratio.

As for information about oxidation of boron clusters, the cross-sections for ionic products formed in reactions of B_{1-13}^+ with oxygen were measured under single collision conditions [111] and three main reaction mechanisms found to be important: oxidative fragmentation, collision induced dissociation and boron atom abstraction. Cross-sections for oxidation reactions of CO_2 with boron cluster ions B_{1-14}^+ were reported as a function of collision energy [112]. At least in some cases, oxidation causes structural rearrangement of the boron clusters. To rationally design and explore a future energy source based on the highly exothermic oxidation of boron, DFT was used to characterize small boron clusters with 0–3 oxygen atoms and, in total, up to 10 atoms [63].

Thin coatings made from the ^{10}B isotopically enriched nanoboron providing the highest possible concentration of neutron capture centers can greatly simplify the problem of protection against thermal neutron irradiation [113–115], as well as neutron-fluence nanosensors [116,117]. Boron 2D metallic crystal is a prospective electromagnetic shielding nanomaterial as well [118], so nanoboron can combine neutron and electromagnetic shielding properties.

Reference [119] presented the concept that an elongated planar boron cluster can serve as a "tank tread" at the sub-nanometer scale, a novel propulsion system for potential nanomachines. Ferromagnetism in all-boron planar clusters, e.g., in B_{34}, has been revealed theoretically [120]. They can be assembled to construct all-boron ferromagnetic monolayers, in which ferromagnetism–paramagnetism and semiconductor–metal transitions are expected to occur around 500 K, indicating their potential applications in nanoelectronic and spintronic devices at room temperature.

The recent review [121] on borophene potential applications discusses in detail its other utilizations, such as alkali metal ion and Li–S batteries, hydrogen storage, supercapacitors, sensor and catalytic in hydrogen evolution, oxygen reduction and evolution and CO_2 electroreduction reaction.

The hierarchy of clusters relative stability/formation probability mainly (together with peculiarities characteristic of formation kinetics) is determined by the cluster-specific binding energy—its binding energy per atom. This work is focused on theoretical estimation of this key energy parameter for boron small planar clusters in frames of recently modified phenomenological diatomic molecular model.

2. Methodology

Specific binding energy or binding energy per chemical formula unit of the substance clustered form serves for important factor determining relative stabilities and, consequently, affects the relative concentrations of clusters with different numbers of formula units synthesized during a formation process (of course, the mentioned concentrations are influenced by the process kinetics as well). Here, we intend to calculate specific binding energies for boron small clusters in (quasi)planar structures, starting from the old diatomic model [122] of bounded multi-atomic structures.

The diatomic model is based on the saturation property of interatomic bonding. In its initial approximation, when binding energy is the sum of energies of pair interactions between only neighboring in the structure atoms, the microscopic theory of expansion allows for the quite correct estimation of the thermal expansion coefficient for crystals [123]. Despite its simplicity, the diatomic model has been successfully used to calculate some other anharmonic effects in solids as well [124].

As for the clusters binding energy, to the best our knowledge, there are no reports on its calculations within diatomic model, unless our previous estimates of small all-boron (quasi)planar clusters relative stabilities [125–128] and also their dipole moments [129,130]. These results have been summarized in the mini-reviews [131,132]. Furthermore, the problem was specially analyzed [133] for the three most abundant clusters, namely B_{11}, B_{12} and B_{13}, in different charge states, while taking into account ionization processes.

As is mentioned above, boron (quasi)planar clusters are of special interest, as they can form a borophene–monatomic boron layer with unique properties. The B–B bond length value obtained for boron finite planar clusters in quasi-classical approximation was used for an input parameter in quasi-classical calculations of electron energy band surfaces and DoS (Density-of-State) for a flat boron sheet with a perfect (i.e., without any type of holes) triangular network [134]. It is expected to have metallic properties. The Fermi curve of the boron flat sheet is found consist of 6 parts of 3 closed curves well-approximated by ellipses and then representing the quadric energy dispersion of the conduction electrons. The effective mass of electrons at the Fermi level is found to be too small compared with the free electron mass and highly anisotropic. The low effective mass of conduction electrons indicates their high mobility and, hence, high conductivity of the boron sheet.

Recently, we introduced [135] the most general formulation of the diatomic model allowing analytical calculation of the clusters binding energy. It is based on the following assumptions:

- Cluster binding energy is the sum of energies of pair interactions between nearest neighboring in its structure atoms;
- Assuming that relative deviations of bond lengths in the multi-atomic cluster structure from their values in corresponding diatomic molecules are small, pair interaction energies between neighboring atoms are approximated by their quadratic functions;
- Valence-electron-density-redistribution-related corrections to the bond energies in the cluster can be expressed through effective static charges localized on pairs of nearest neighboring atoms;
- Interatomic vibrations related corrections can be approximated by ground-state vibrational energies of corresponding diatomic molecules.

In the simplest but of practical interest special case, when all the bond lengths can be assumed to be almost equal each to other, a, cluster symmetry does not lead to any constrain (a relation to be satisfied by bond lengths), and most of other characteristics of valence bonding are equal each to other as well: a_0 is the bond length corresponding to diatomic molecule, E_0 is the diatomic molecule binding/dissociation energy, ω is the cyclic frequency of relative interatomic vibrations and M is the reduced mass of the pair of atoms with masses μ_1 and μ_2 constituting the bond:

$$\frac{1}{\mathrm{M}} = \frac{1}{\mu_1} + \frac{1}{\mu_2} \tag{1}$$

In case of identical atoms, we have the following:

$$\mu_1 = \mu_2 \equiv \mu \tag{2}$$

and

$$\mathrm{M} = \frac{\mu}{2} \tag{3}$$

However, effective atomic charge numbers Z_{i1} and Z_{i2} characterizing electrostatic correction to the valence bonding energy remain different. The point is that the static charges localized on the pair of nearest-neighboring atoms differ for pairs placed at the center and periphery of the cluster, which is a finite structure of atoms. Here, the i index is as follows:

$$i = 1, 2, 3, \ldots, k \tag{4}$$

where the numbers are different types of chemical bonds presented in the cluster, and k denotes their total number. If N_i is the number of bonds of i-type, then we have the following:

$$N = \sum_{i=1}^{i=k} N_i \tag{5}$$

which is the number of bonds in whole the cluster.

When we introduce the following parameter,

$$\mathrm{Z} = \sum_{i=1}^{i=k} N_i Z_{i1} Z_{i2} \tag{6}$$

equilibrium bonds length, a, and corresponding cluster binding energy, E (in Gauss units), are as follows:

$$a = a_0 + \frac{2e^2 Z}{\mu\omega^2 a_0^2 N} \tag{7}$$

and

$$E = \left(E_0 - \frac{\hbar\omega}{2}\right)N - \frac{e^2 Z}{a_0} + \frac{e^4 Z^2}{\mu\omega^2 a_0^4 N} \tag{8}$$

respectively.

From the data available [39] for diboron molecule, $a_0 \approx 1.590$ Å, $E_0 \approx 3.02$ eV and $\omega \approx 1051.3$ cm^{-1} and boron atomic mass weighted for stable isotopes (^{10}B and ^{11}B) natural abundance $\mu \approx 10.811$ amu, we get the following formula for numerical calculations:

$$E[\text{eV}] \approx 2.995\, N - 9.053\, Z + \frac{0.7273\, Z^2}{N} \tag{9}$$

The specific binding energy for cluster of n atoms is calculated as E/n.

To know parameter Z, one needs the estimates of effective atomic charges in the cluster. Below, they are found based on the assumption that the effective number of outer valence shell electrons localized on a given atomic site is proportional to its coordination number. Every neutral boron atom contains only 1 electron in the outer valence shell (2p-state). Then their total number in the all-boron cluster B_n is given as follows:

$$\nu = \begin{cases} n & B_n^0 \\ n-1 & B_n^+ \\ n+1 & B_n^- \end{cases} \tag{10}$$

Moreover, if C_j denotes the coordination number of j-site, then we have the following:

$$Z_j \approx 1 - \frac{\nu C_j}{\sum_{l=1}^{l=n} C_l} \tag{11}$$

The numerical values of these effective charge numbers and a summary of them for parameter Z for the ground-state structures of boron planar clusters with $n = 2$–15 atoms in three charge states are shown below in Table 1.

Table 1. Specific binding energy of boron small planar clusters calculated in diatomic model.

Number of Atoms n	Structure	Number of Bonds N	Charge State	Charge Numbers in Dependence on Coordination Number					Parameter Z	Specific Binding Energy E/n, eV	Bonds Length, a, Å
				1	2	3	4	6			
2		1	$B_2{}^0$	0					0	1.48	1.59
			$B_2{}^+$	$+\frac{1}{2}$					+0.2500	0.37	1.65
			$B_2{}^-$	$-\frac{1}{2}$					+0.2500	0.37	1.65
3	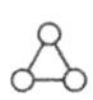	3	$B_3{}^0$		0				0	2.96	1.59
			$B_3{}^+$		$+\frac{1}{3}$				+0.3333	1.96	1.62
			$B_3{}^-$		$-\frac{1}{3}$				+0.3333	1.96	1.62
4	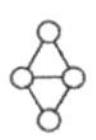	5	$B_4{}^0$		$+\frac{1}{5}$	$-\frac{1}{5}$			−0.1200	3.97	1.58
			$B_4{}^+$		$+\frac{2}{5}$	$+\frac{1}{10}$			+0.1700	3.31	1.60
			$B_4{}^-$		0	$-\frac{1}{2}$			+0.2500	3.13	1.60
5	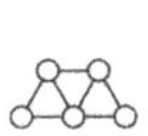	7	$B_5{}^0$		$+\frac{2}{7}$	$-\frac{1}{14}$	$-\frac{3}{7}$		−0.2194	4.54	1.58
			$B_5{}^+$		$+\frac{3}{7}$	$+\frac{1}{7}$	$-\frac{1}{7}$		−0.0204	4.17	1.59
			$B_5{}^-$		$+\frac{1}{7}$	$-\frac{2}{7}$	$-\frac{5}{7}$		+0.2041	3.77	1.60

Table 1. *Cont.*

Number of Atoms n	Structure	Number of Bonds N	Charge State	Charge Numbers in Dependence on Coordination Number					Parameter Z	Specific Binding Energy E/n, eV	Bonds Length, a, Å
				1	2	3	4	6			
6	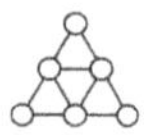	9	B_6^0		$+\frac{1}{3}$		$-\frac{1}{3}$		−0.3333	4.94	1.58
			B_6^+		$+\frac{4}{9}$		$-\frac{1}{9}$		−0.2593	4.82	1.58
			B_6^-		$+\frac{2}{9}$		$-\frac{5}{9}$		+0.1852	4.15	1.60
7	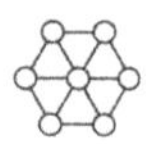	12	B_7^0			$+\frac{1}{8}$		$-\frac{3}{4}$	−0.4688	5.67	1.58
			B_7^+			$+\frac{1}{4}$		$-\frac{1}{2}$	−0.3750	5.55	1.58
			B_7^-			0		−1	0	5.07	1.59
8	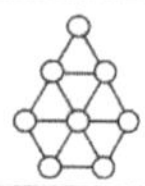	14	B_8^0		$+\frac{3}{7}$	$+\frac{1}{7}$	$-\frac{1}{7}$	$-\frac{5}{7}$	−0.2857	5.50	1.59
			B_8^+		$+\frac{1}{2}$	$+\frac{1}{4}$	0	$-\frac{1}{2}$	−0.3125	5.53	1.58
			B_8^-		$+\frac{5}{14}$	$+\frac{1}{28}$	$-\frac{2}{7}$	$-\frac{13}{14}$	+0.2589	4.88	1.60
9	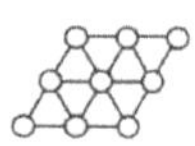	16	B_9^0		$+\frac{7}{16}$	$+\frac{5}{32}$	$-\frac{1}{8}$	$-\frac{11}{16}$	+0.2500	5.00	1.59
			B_9^+		$+\frac{1}{2}$	$+\frac{1}{4}$	0	$-\frac{1}{2}$	0	5.25	1.59
			B_9^-		$+\frac{3}{8}$	$+\frac{1}{16}$	$-\frac{1}{4}$	$-\frac{7}{8}$	+1.0000	4.25	1.61
10	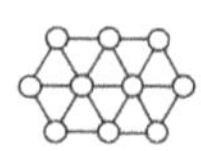	19	B_{10}^0			$+\frac{4}{19}$	$-\frac{1}{19}$	$-\frac{11}{19}$	−0.3850	5.96	1.59
			B_{10}^+			$+\frac{11}{38}$	$+\frac{1}{19}$	$-\frac{8}{19}$	−0.2465	5.84	1.59
			B_{10}^-			$+\frac{5}{38}$	$-\frac{3}{19}$	$-\frac{14}{19}$	+0.4127	5.24	1.60
11	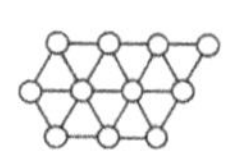	21	B_{11}^0		$+\frac{10}{21}$	$+\frac{9}{42}$	$-\frac{1}{21}$	$-\frac{4}{7}$	+0.0102	5.63	1.59
			B_{11}^+		$+\frac{11}{21}$	$+\frac{2}{7}$	$+\frac{1}{21}$	$-\frac{3}{7}$	+0.2381	5.45	1.59
			B_{11}^-		$+\frac{3}{7}$	$+\frac{1}{7}$	$-\frac{1}{7}$	$-\frac{5}{7}$	−0.7143	6.23	1.58
12	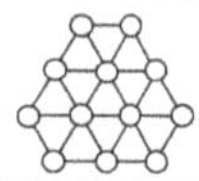	24	B_{12}^0			$+\frac{1}{4}$	0	$-\frac{1}{2}$	+0.1875	5.77	1.59
			B_{12}^+			$+\frac{5}{16}$	$+\frac{1}{12}$	$-\frac{3}{8}$	−0.3125	6.15	1.59
			B_{12}^-			$+\frac{3}{16}$	$-\frac{1}{12}$	$-\frac{5}{8}$	+0.2461	5.72	1.59
13	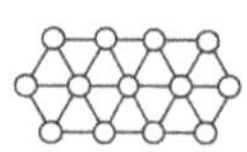	26	B_{13}^0			$+\frac{1}{4}$	0	$-\frac{1}{2}$	−0.5000	6.26	1.59
			B_{13}^+			$+\frac{4}{13}$	$+\frac{1}{13}$	$-\frac{5}{13}$	−0.1657	6.03	1.59
			B_{13}^-			$+\frac{5}{26}$	$-\frac{1}{13}$	$-\frac{8}{13}$	−0.2071	6.05	1.59
14	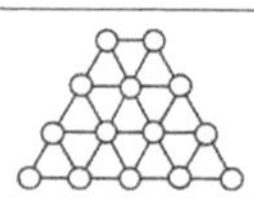	28	B_{14}^0		$+\frac{1}{2}$	$+\frac{1}{4}$	0	$-\frac{1}{2}$	+0.5625	5.55	1.60
			B_{14}^+		$+\frac{15}{28}$	$+\frac{17}{56}$	$+\frac{1}{14}$	$-\frac{11}{28}$	+0.0590	5.87	1.59
			B_{14}^-		$+\frac{13}{28}$	$+\frac{11}{56}$	$-\frac{1}{14}$	$-\frac{17}{28}$	+1.6151	4.87	1.61
15	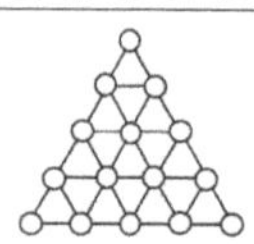	30	B_{15}^0		$+\frac{1}{2}$		0	$-\frac{1}{2}$	+0.7500	5.46	1.60
			B_{15}^+		$+\frac{8}{15}$		$+\frac{1}{15}$	$-\frac{2}{5}$	+0.4133	5.66	1.59
			B_{15}^-		$+\frac{7}{15}$		$-\frac{1}{15}$	$-\frac{3}{5}$	+1.4133	5.06	1.60

3. Results and Discussion

Here, we report the specific binding energy that we calculated for boron small planar clusters only in their ground-state structures. Ground-state structural isomers were chosen based on two criteria leading to maximal binding energy: (1) maximum number of bonds and (2) highest symmetry. We calculated clusters containing up to 15 boron atoms, because, for bigger species, 2D (quasi)planar structures are challenged by 3D ring-like structures (actually fragments of nanotubes), which are characteristic for boron clusters at $n \geq 20$.

Specific binding energies calculated in diatomic model for neutral B_n^0, cationic B_n^+ and anionic B_n^- clusters with the number of atom $n = 2$–15 are listed in Table 1 and presented in Figure 2.

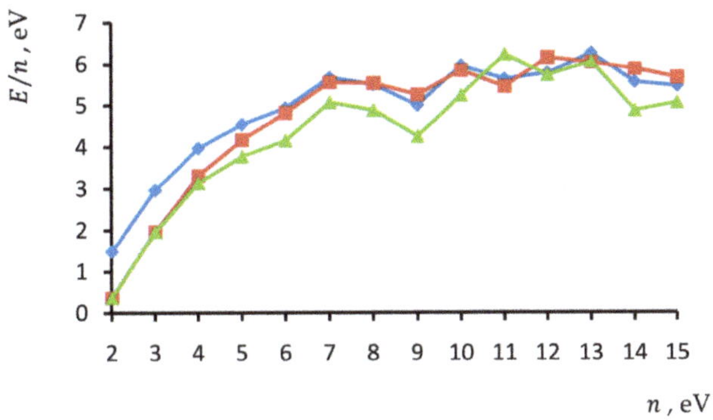

Figure 2. Specific binding energy of neutral (◆), positively (■) and negatively (▲) charged boron small planar clusters in dependence on number of atoms calculated in diatomic model.

For comparison, available experimental data are presented in Figure 3. In Reference [130], clusters in boron vapor were generated by the thermal decomposition of electrode made of a boron-rich metal boride. Note that, in this way, initially clusters are formed in neutral state but then ionized by collisions with released in the chamber energetic electrons and accelerated toward the mass-spectrometer. Boron cluster cations were generated [29] by laser evaporation of target compact made of boron dust with gold added to increase its stability. Here, a disproportionately intense peak related to traces of gold is erased because it masks that of $B^+{}_{18}$, as masses of gold Au atom and boron 18-atom cluster are almost undistinguishable. As for the boron cluster anions, they were produced [28] by laser vaporization from homogeneous pure boron target rods.

Again, for comparison, Figure 4 represents the theoretical specific binding energies of small boron clusters (in different, not only in planar, structures) depending on their size calculated by using QC methods [70]. One can note that such a typical curve looks similar to curves obtained in frames of diatomic model.

Thus, from the diatomic model, the theoretical equilibrium binding energies per B atom and B–B bond lengths are expected within ranges of 0.37–6.26 eV and 1.58–1.65 Å, respectively. For the most stabile neutral, positively and negatively charged species, $B_{13}{}^0$, $B_{12}{}^+$ and $B_{11}{}^-$ are predicted; their characteristics are 6.26, 6.15 and 6.23 eV and 1.59, 1.59 and 1.58 Å, respectively.

From experimental reports, the diboron molecule B_2 dissociation energy is expected within the ranges of 2.58–3.06 [15] and 2.26–3.12 eV [34]. Theoretical values obtained by HF [37], MO [38], CI [40], quasi-classical [41,42] and PES [53] methods are 2.86, 2.71, 2.70–2.78, 2.80 and 2.70 eV, respectively. As diboron molecule contains two atoms, the corresponding specific binding energies equal to (2.58–3.06)/2 = 1.29–1.53 and (2.26–3.12)/2 = 1.13–1.56 for measured and 2.86/2 = 1.43, 2.71/2 ≈ 1.36, (2.70–2.78)/2 = 1.35–1.39, 2.80/2 = 1.40 and 2.70/2 = 1.35 eV for calculated dissociation energies. Calculations based on DFT [30] and QC [70] yielded 1.36 and 1.39 eV, respectively. The value of 1.48 eV obtained for cluster $B_2{}^0$ from the diatomic model falls in both experimental ranges and seems only slightly overestimated if compared with previous theoretical ones.

The ground-state dissociation energy of diboron cation $B_2{}^+$ calculated with the CI approach [46] is 1.94 eV, which corresponds to 1.94/2 = 0.97 eV for specific binding energy, significantly exceeding 0.37 eV yielded by diatomic model for $B_2{}^+$. The same is true for the QC specific binding energy for $B_2{}^+$: 1.16 eV [70]. This discrepancy should be related not only to the diatomic model itself but mainly phenomenological estimation of static atomic charges used. The point is that, when calculating static atomic charges in diatomic species, the equal dividing of a single elemental charge between constituent atoms is too crude of an approximation, leading to the overestimated Coulomb repulsive energy. SCF CI [69] and calibrated hybrid DFT [82] approaches' $B_2{}^+ \rightarrow B_1{}^+ + B_1{}^0$ fragmentation energies are 1.47

and 1.96 eV, respectively. The corresponding specific binding energies are $1.47/2 \approx 0.74$ and $1.96/2 = 0.98$ eV.

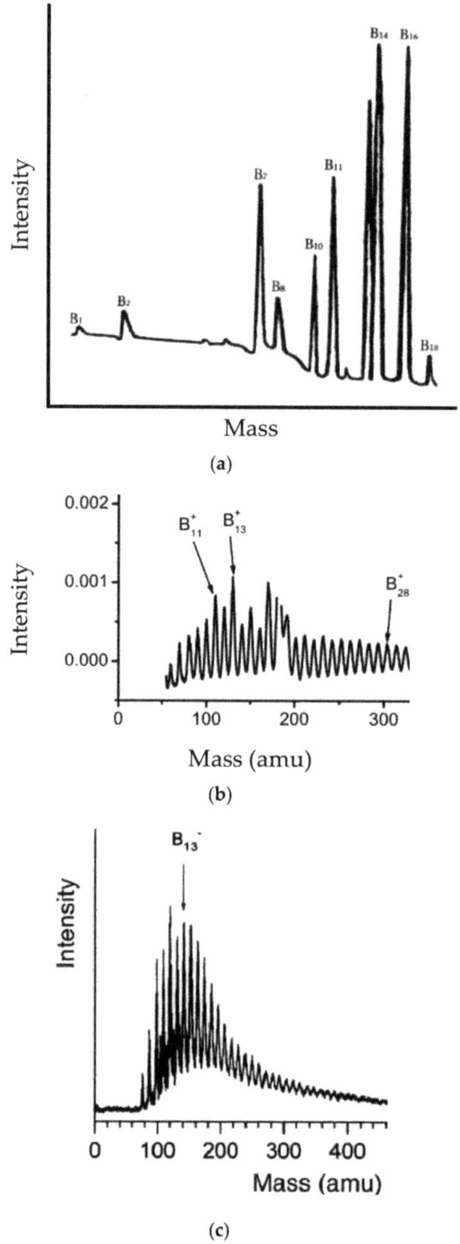

Figure 3. Experimentally recorded mass spectra of (**a**) neutral [130], Copyright (2015), with permission from Authors; (**b**) positively [29], Copyright (2010), with permission from KIT Scientific Publishing; and (**c**) negatively charged [28] boron clusters, Copyright (2003), with permission from Elsevier.

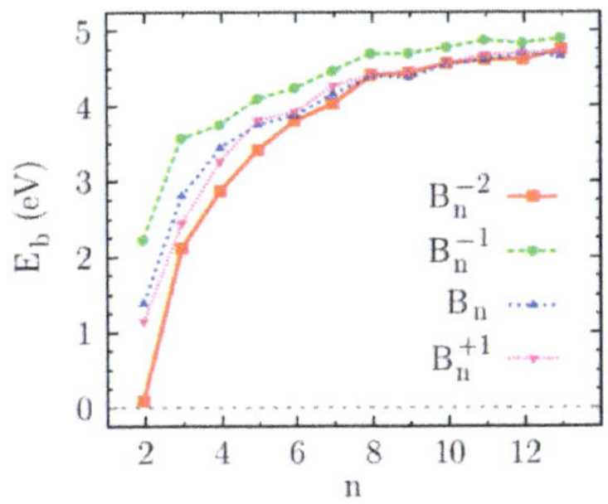

Figure 4. Binding energy per atom of neutral and ionized boron clusters as function of size [70], Copyright (2011), with permission from American Physical Society.

The QC specific binding energy of anionic cluster B_2^- is 2.24 eV [70]. As for the cationic cluster, this value significantly exceeds 0.37 eV, which is suggested by the diatomic model. The reason should be the same as for cationic isomer.

For the B_3 cluster, the HF scaled dissociation energy is 8.59 eV [37], while the PES study showed that atomization energy for B_3 would be in the range of 8.21–8.36 eV [53]. As the molecule contains three atoms, the corresponding specific binding energies are equal to $8.59/3 \approx 2.86$ and $(8.21–8.36)/3 \approx 2.74–2.79$ eV, respectively. Calculations based on DFT [30] and QC [70] approaches yielded 2.76 and 2.82 eV, respectively. The diatomic model'svalue of 2.96 eV for cluster B_3^0 again seems to be slightly overestimated if compared with other theoretical results.

The QC specific binding energy of B_3^+ and B_3^- clusters is 2.46 and 3.58 eV, respectively [70], while the diatomic model gives 1.96 eV for both clusterions.

A SCF calculation [51] predicted that the energy of fragmentation of the B_3 cluster to produce diatomic B_2 and atomic B is 4.96 eV. From the diatomic model, one obtains $3 \times 2.96 - 2 \times 1.48 = 5.92$ eV.

The SCF CI $B_3^+ \rightarrow B_1^+ + B_2^0$ and $B_3^+ \rightarrow B_2^+ + B_1^0$ fragmentation energies are 1.45 and 2.00 eV, respectively [68], while the $B_3^+ \rightarrow B_1^+ + B_2^0$ fragmentation energy calibrated in a hybrid DFT approach is 4.33 eV [82]. The diatomic model gives $3 \times 1.96 - 2 \times 0.37 = 5.14$ and $3 \times 1.96 - 2 \times 1.48 = 2.92$ eV for $B_3^+ \rightarrow B_1^+ + B_2^0$ and $B_3^+ \rightarrow B_2^+ + B_1^0$ reactions, respectively.

According to the study of the B_4 cluster PES, its total atomization energy is expected in the range of 13.46–13.64 eV [53]. As this cluster consists of four atoms, the corresponding specific binding energy is $(13.46–13.64)/4 \approx 3.37–3.41$ eV. The calculations based on DFT [30] and QC [70] approaches yielded 3.37 and 3.45 eV, respectively. The value of 3.97 eV obtained in the diatomic model for the neutral B_4^0 cluster is higher but comparable. However, the CI analysis of B_4 tetramer in rhombus geometry significantly underestimated its specific binding energy: 2.42 eV [54].

The QC specific binding energies of B_4^+ and B_4^- clusters are 3.28 and 3.76 eV, respectively [70]. The diatomic model gives quite close results of 3.31 and 3.13 eV for these cluster-ions.

SCF CI $B_4^+ \rightarrow B_1^+ + B_3^0$, $B_4^+ \rightarrow B_3^+ + B_1^0$ and $B_4^+ \rightarrow B_2^+ + B_2^0$ fragmentation energies are 1.75, 2.80 and 3.20 eV, respectively [68]. Moreover, $B_4^+ \rightarrow B_1^+ + B_3^0$ fragmentation energy calibrated in a hybrid DFT approach equals to 4.23 eV [82]. In the diatomic approach, their values are $4 \times 3.31 - 3 \times 2.96 = 4.36$, $4 \times 3.31 - 3 \times 1.96 = 7.36$ and $4 \times 3.31 - 2 \times 0.37 - 2 \times 1.48 = 9.54$, respectively.

The specific binding energies of the B_5^0 cluster calculated in DFT [30] and two versions of QC approach [31,70] are 3.67, 3.68 and 3.76 eV, respectively. Our result is 4.54 eV.

The QC specific binding energy of B_5^+ is 3.82 eV [70], while we obtained 4.17 eV.

The specific binding energies of the B_5^- cluster calculated in two versions of the QC approach are 4.04 [31] and 4.11 eV [70], respectively, while the diatomic model yields the lower value of 3.77 eV.

The SCF CI $B_5^+ \rightarrow B_1^+ + B_4^0$ fragmentation energy is 2.00 eV [68]. By calibrating this energy in a hybrid DFT approach, it was found 4.95 eV [82] to be in almost perfect agreement with the diatomic model result of $5 \times 4.17 - 4 \times 3.97 = 4.97$ eV.

By using DFT [30], QC [31,70] and diatomic model (this work) methods, we calculated the specific binding energy for the B_6^0 cluster as 3.79, 3.84, 3.88 and 4.94 eV, respectively. The QC approach [70] and diatomic model have yielded specific binding energy for B_6^+ cluster as 3.93 and 4.83 eV, respectively. According to two QC [31,70] and diatomic model calculations, the specific binding energy for B_6^- cluster is 4.16, 4.25 and 4.15 eV, respectively.

The SCF CI $B_6^+ \rightarrow B_5^+ + B_1^0$ fragmentation energy is 2.20 eV [68], while, in a hybrid DFT approach, it is found to be 4.11 eV [82]. The diatomic value of $6 \times 4.82 - 5 \times 4.17 = 8.07$ eV exceeds both of them.

DFT [30], QC [31,70], LMTO MD [71] and diatomic model specific binding energies for the B_7^0 cluster are 4.07, 4.11, 4.17, 5.24 and 5.67 eV, respectively. For the B_7^+ cluster, the QC [70] and diatomic-model specific binding energies are 4.28 and 5.55 eV, respectively. Moreover, the QC [31,70] and diatomic-model specific binding energies for B_7^- cluster are 4.49, 4.48 and 5.07 eV, respectively.

The $B_7^+ \rightarrow B_6^+ + B_1^0$ fragmentation energy calibrated in a hybrid DFT approach is 6.05 eV [82], i.e., lower than that predicted by the diatomic model: $7 \times 5.55 - 6 \times 4.82 = 9.93$ eV.

For the B_8^0 cluster, the DFT [30], QC [31,70] and diatomic-model specific binding energies are 4.31, 4.33, 4.41 and 5.50 eV, respectively. As for the B_8^+ cluster, its QC [70] and diatomic-model specific binding energies are 4.42 and 5.53 eV, respectively. Moreover, the B_8^- cluster's QC [31,70] and diatomic-model specific binding energies equal to 4.71, 4.71 and 4.88 eV, respectively.

The $B_8^+ \rightarrow B_7^+ + B_1^0$ fragmentation energy calibrated in a hybrid DFT approach is 5.04 eV [82], which is in good agreement with this work's calculation: $8 \times 5.53 - 7 \times 5.55 = 5.39$ eV.

The DFT [30], QC [70] and diatomic-model specific binding energies of B_9^0 cluster are 4.30, 4.39 and 5.00 eV, respectively. The QC [70] and diatomic-model specific binding energies of B_9^+ cluster are 4.46 and 5.25 eV, respectively. The QC [31,70] and diatomic-model specific binding energies of B_9^- cluster are 4.55, 4.71 and 4.25 eV, respectively.

The $B_9^+ \rightarrow B_8^+ + B_1^0$ fragmentation energy calibrated in a hybrid DFT approach is 4.29 eV [82]. The diatomic model yields the fragmentation energy of $9 \times 5.25 - 8 \times 5.53 = 3.01$ eV, which is surprisingly less than previous theoretical results.

The DFT [30], QC [70], LMTO MD [71] and diatomic-model specific binding energies of B_{10}^0 cluster are 4.45, 4.56, 5.78 and 5.96 eV, respectively. The QC [70] and diatomic-model specific binding energies of the B_{10}^+ cluster are 4.57 and 5.84 eV, respectively. The QC [70] and diatomic-model specific binding energies of the B_{10}^- cluster are 4.79 and 5.24 eV, respectively.

As for the $B_{10}^+ \rightarrow B_9^+ + B_1^0$ fragmentation energy calibrated in a hybrid DFT approach, it equals 5.58 eV [82], while the diatomic model gives $10 \times 5.84 - 9 \times 5.25 = 11.15$ eV.

The DFT [30], QC [70] and diatomic-model specific binding energies of the B_{11}^0 cluster are 4.52, 4.63 and 5.63 eV, respectively. QC [70] and diatomic-model specific binding energies of B_{11}^+ cluster are 4.69 and 5.45 eV, respectively. The QC [70] and diatomic-model specific binding energies of B_{11}^- cluster are 4.89 and 6.23 eV, respectively.

The $B_{11}^+ \rightarrow B_{10}^+ + B_1^0$ fragmentation energy calibrated in a hybrid DFT approach is 5.18 eV [82]. In this case, the diatomic model value is significantly less: $11 \times 5.45 - 10 \times 5.84 = 1.55$ eV.

In Reference [67], the Jahn–Teller distortion mechanism, which transforms the high-lysymmetric icosahedral structure B_{12}^0 into a quasi-planar disc-like structure with binding energy per atom as 4.60 eV, was proposed. The same value was suggested based on DFT calculations: 4.60 eV [30]. The QC specific binding energy of B_{12}^0 is 4.71 eV [70], while our diatomic model gives 5.77 eV. The QC [70] and diatomic-model specific binding energies of the B_{12}^+ and B_{12}^- charged cluster are 4.72 and 4.85, and 6.15 and 5.72 eV, respectively.

The $B_{12}^{+} \rightarrow B_{11}^{+} + B_1^{0}$ fragmentation energy calibrated in a hybrid DFT approach is 4.64 eV [82], while $12 \times 6.15 - 11 \times 5.45 = 13.85$ eV is obtained based on diatomic model.

The QC [70], LMTO MD [71] and diatomic-model specific binding energies of the B_{13}^{0} cluster are 4.67, 5.94 and 6.26 eV, respectively. The QC [70] and diatomic-model specific binding energies of the B_{13}^{+} and B_{13}^{-} charged cluster are 4.73 and 4.91, and 6.03 and 6.05 eV, respectively.

The $B_{13}^{+} \rightarrow B_{12}^{+} + B_1^{0}$ fragmentation energy calibrated in a hybrid DFT approach is 5.68 eV [82]. The diatomic model gives $13 \times 6.03 - 12 \times 6.15 = 4.59$ eV.

Finally, the $B_{14}^{+} \rightarrow B_{13}^{+} + B_1^{0}$ fragmentation energy calibrated in a hybrid DFT approach is 3.96 eV [82], which is in satisfactory agreement with diatomic model value of $14 \times 5.87 - 13 \times 6.03 = 3.79$ eV.

The above comparisons between the diatomic model and the previous results on specific binding and fragmentation energies of small boron clusters are summarized in Tables 2 and 3, respectively. Here, the term "specific fragmentation energy" implies the fragmentation energy per doubled difference between B–B bonds' numbers in the initial cluster (i.e., before fragmenting) and its fragments. For example, specific binding energies (in diatomic model), numbers of B atoms and B–B bonds of clusters B_4^{+}, B_2^{+} and B_2^{0} are 3.31, 0.37 and 1.48 eV; 4, 2 and 2; and 5, 1 and 1, respectively. Then, specific $B_4^{+} \rightarrow B_2^{+} + B_2^{0}$ fragmentation energy is $(4 \times 3.31 - 2 \times 0.37 - 2 \times 1.48)/(2 \times (5 - 1 - 1)) = 1.59$ eV.

Theoretical $E/n - n$ curves obtained in the diatomic model in general features coincide with experimental ones. In particular, we can see maxima in the range B_{11}–B_{13}. At a number of atoms ≤ 8, neutral clusters are predicted to be more stable than their charged isomers. Moreover, at a number of ≤ 9, cations seem to be more stable than anions. However, at a higher number of atoms in boron planar clusters, there is no common trend of relative stability in the dependence of the charge state. As for the specific bind energy averaged by charge states, it saturates. Discrepancies in details seem to be related to the different kinetics of experimental generation processes of small boron clusters, as well as assumptions of the diatomic method used in theoretical calculations.

As for quantitative agreement in specific binding and fragmentation energies and their dependences on charge states and number of atoms in small boron clusters with previously reported (in the most part theoretical) studies, it also seems satisfactory. However, the diatomic model frequently overestimates these energy characteristics. This could be related to the energy phenomenological parameter—dissociation energy of diboron molecule B_2—used in constructing the diatomic binding energy curve, as it is measured with too significant an error. Of course, such an almost systematic deviation from previous studies partially should be related to their assumptions on cluster structures and applied methods of calculation or measurement errors.

Table 2. Comparison of the literature data on specific binding energies (in eV) of boron small clusters with values calculated in diatomic model.

Cluster	Literature Data	Diatomic Model
B_2^{0}	1.29–1.53 [15] 1.36 [30] 1.13–1.56 [34] 1.43 [37] 1.36 [38] 1.35–1.39 [40] 1.40 [41,42] 1.35 [53] 1.39 [70]	1.48
B_2^{+}	0.97 [46] 0.74 [69] 1.16 [70] 0.98 [82]	0.37

Table 2. *Cont.*

Cluster	Literature Data	Diatomic Model
B_2^-	2.24 [70]	0.37
B_3	2.76 [30] 2.86 [37] 2.74–2.79 [53] 2.82 [70]	2.96
B_3^+	2.46 [70]	1.96
B_3^-	3.58 [70]	1.96
B_4	3.37 [30] 3.37–3.41 [53] 2.42 [54] 3.45 [70]	3.97
B_4^+	3.28 [70]	3.31
B_4^-	3.76 [70]	3.13
B_5^0	3.67 [30] 3.68 [31] 3.76 [70]	4.54
B_5^+	3.82 [70]	4.17
B_5^-	4.04 [31] 4.11 [70]	3.77
B_6^0	3.79 [30] 3.84 [31] 3.88 [70]	4.94
B_6^+	3.93 [70]	4.83
B_6^-	4.16 [31] 4.25 [70]	4.15
B_7^0	4.07 [30] 4.11 [31] 4.17 [70] 5.24 [71]	5.67
B_7^+	4.28 [70]	5.55
B_7^-	4.49 [31] 4.48 [70]	5.07
B_8^0	4.31 [30] 4.33 [31] 4.41 [70]	5.50
B_8^+	4.42 [70]	5.53
B_8^-	4.71 [31] 4.71 [70]	4.88
B_9^0	4.30 [30] 4.39 [70]	5.00
B_9^+	4.46 [70]	5.25
B_9^-	4.55 [31] 4.71 [70]	4.25
B_{10}^0	4.45 [30] 4.56 [70] 5.78 [71]	5.96
B_{10}^+	4.57 [70]	5.84
B_{10}^-	4.79 [70]	5.24

Table 2. *Cont.*

Cluster	Literature Data	Diatomic Model
B_{11}^{0}	4.52 [30] 4.63 [70]	5.63
B_{11}^{+}	4.69 [70]	5.45
B_{11}^{-}	4.89 [70]	6.23
B_{12}^{0}	4.60 [30] 4.60 [67] 4.71 [70]	5.77
B_{12}^{+}	4.72 [70]	6.15
B_{12}^{-}	4.85 [70]	5.72
B_{13}^{0}	4.67 [70] 5.94 [71]	6.26
B_{13}^{+}	4.73 [70]	6.03
B_{13}^{-}	4.91 [70]	6.05

Table 3. Comparison of the literature data on specific fragmentation energies (in eV) of boron small clusters with values calculated in diatomic model.

Chanel	Literature Data	Diatomic Model
$B_3^{0} \rightarrow B_2^{0} + B_1^{0}$	1.24 [51]	1.48
$B_3^{+} \rightarrow B_2^{0} + B_1^{+}$	0.36 [68] 1.08 [82]	1.29
$B_3^{+} \rightarrow B_2^{+} + B_1^{0}$	0.50 [68]	0.73
$B_4^{+} \rightarrow B_3^{0} + B_1^{+}$	0.44 [68] 1.06 [82]	1.09
$B_4^{+} \rightarrow B_3^{+} + B_1^{0}$	0.70 [68]	1.84
$B_4^{+} \rightarrow B_2^{0} + B_2^{+}$	0.80 [68]	2.39
$B_5^{+} \rightarrow B_4^{0} + B_1^{+}$	0.50 [68] 1.24 [82]	1.24
$B_6^{+} \rightarrow B_5^{+} + B_1^{0}$	0.55 [68] 1.03 [82]	2.02
$B_7^{+} \rightarrow B_6^{+} + B_1^{0}$	1.01 [82]	1.66
$B_8^{+} \rightarrow B_7^{+} + B_1^{0}$	1.26 [82]	1.35
$B_9^{+} \rightarrow B_8^{+} + B_1^{0}$	1.07 [82]	0.75
$B_{10}^{+} \rightarrow B_9^{+} + B_1^{0}$	0.93 [82]	1.86
$B_{11}^{+} \rightarrow B_{10}^{+} + B_1^{0}$	1.30 [82]	0.39
$B_{12}^{+} \rightarrow B_{11}^{+} + B_1^{0}$	0.77 [82]	2.31
$B_{13}^{+} \rightarrow B_{12}^{+} + B_1^{0}$	1.42 [82]	1.15
$B_{14}^{+} \rightarrow B_{13}^{+} + B_1^{0}$	0.99 [82]	0.95

4. Conclusions

In summary, a previously developed diatomic-type model for calculating clusters' specific (per atom) binding energy was applied to obtain this key parameter for boron small planar clusters'(B_n, n = 2–15) ground-state isomers in different charge states. The main result of the conducted study was that, in any of the three considered (neutral, single-cationic and single-anionic) charge states, the formation of B_{11}–B_{13} clusters is preferable. In general features, it is in good agreement with the available experimental data and previous theoretical reports.

As for quantitative agreement in specific binding and fragmentation energies, the diatomic model predicts the values in magnitude comparable with previous results, but it usually overestimates them. These deviations could be explained by different sets of assumptions made in theoretical calculations and/or too significant errors in experimental determination of boron clusters' energy characteristics, one of which (B_2 molecule dissociation energy), in particular, serves for key phenomenological parameter in diatomic model approach to calculating cluster specific binding energy.

Specific binding energies and B–B bond lengths of most stabile neutral, positively and negatively charged species are B_{13}^{0}—6.26 eV and 1.59 Å, B_{12}^{+}—6.15 eV and 1.59 Å and B_{11}^{-}—6.23 eV and 1.58 Å, respectively.

The success of the diatomic approach can serve for the basis for more detailed calculations of boron small clusters, including not only ground-state, but all the possible planar structural isomers, as well as competitive ring-like clusters. In this way, one can determine not only the specific binding energy but also other important characteristics, such as the cluster dipole moment, ionization potential and electron affinity, vibration, atomization, fragmentation energies, etc.

These results would be not only academic, but also practical interests, as boron (quasi)planar clusters serve for building blocks of borophene and other boron-based nanomaterials perspective for variety of technological applications, such as thin super-hard coatings, radiation shielding, solid fuel production, nanoelectronics, etc.

Funding: This research received no external funding.

Institutional Review Board Statement: Not applicable.

Informed Consent Statement: Not applicable.

Data Availability Statement: Data available via personal communication with proper reasons.

Conflicts of Interest: The author declares no conflict of interest.

Sample Availability: Samples of the compounds are not available from the authors.

References

1. Boustani, I. Towards novel boron nanostructural materials. *Chem. Modell.* **2011**, *8*, 1–44.
2. Chkhartishvili, L. Micro- and nano-structured boron. In *Boron: Compounds, Production and Application*; Perkins, G.L., Ed.; Nova Sci. Publ.: New York, NY, USA, 2011; Chapter 6; pp. 221–294.
3. Chkhartishvili, L. Nanoboron (An overview). *Nano Studies* **2011**, *3*, 227–314.
4. Becker, R.; Chkhartishvili, L.; Martin, P. Boron, the new graphene? *Vac. Technol. Coat.* **2015**, *16*, 38–44.
5. Chkhartishvili, L. All-boron nanostructures. In *CRC Concise Encyclopedia of Nanotechnology*; Kharisov, B.I., Kharissova, O.V., Ortiz–Mendez, U., Eds.; CRC Press: Boca Raton, FL, USA, 2016; Chapter 7; pp. 53–69.
6. Li, D.; Gao, J.; Cheng, P.; He, J.; Yin, Y.; Hu, Y.; Chen, L.; Cheng, Y.; Zhao, J. 2D boron sheets: Structure, growth, and electronic and thermal transport properties. *Adv. Funct. Mater.* **2019**, *30*, 1904349. [CrossRef]
7. Tian, Y.; Guo, Z.; Zhang, T.; Lin, H.; Li, Z.; Chen, J.; Deng, S.; Liu, F. Inorganic boron-based nanostructures: Synthesis, optoelectronic properties, and prospective applications. *Nanomaterials* **2019**, *9*, 538. [CrossRef] [PubMed]
8. Boustani, I. *Molecular Modelling and Synthesis of Nanomaterials. Applications in Carbon- and Boron-Based Nanotechnology*; Springer Nature: Cham, Switzerland, 2020.
9. Matsuda, I.; Wu, K. (Eds.) *2D Boron: Boraphene, Borophene, Boronene*; Springer Nature: Cham, Switzerland, 2021.
10. Alexandrova, A.N.; Boldyrev, A.I.; Zhai, H.-J.; Wang, L.-S. All-boron aromatic clusters as potential new inorganic ligands and building blocks in chemistry. *Coord. Chem. Rev.* **2006**, *250*, 2811–2866. [CrossRef]
11. Li, W.-L.; Chen, Q.; Tian, W.-J.; Bai, H.; Zhao, Y.-F.; Hu, H.-S.; Li, J.; Zhai, H.-J.; Li, S.-D.; Wang, L.-S. The B_{35} cluster with a double-hexagonal vacancy: A new and more flexible structural motif for borophene. *J. Am. Chem. Soc.* **2014**, *136*, 12257–12260. [CrossRef]
12. Kiran, B.; Bulusu, S.; Zhai, H.-J.; Yoo, S.; Zeng, X.C.; Wang, L.-S. Planar-to-tubular structural transition in boron clusters: B_{20} as the embryo of single-walled boron nanotubes. *Proc. Natl. Acad. Sci. USA* **2005**, *102*, 961–964. [CrossRef]
13. Boustani, I. Structural transitions and properties of boron nanoclusters. In *17th International Symposium on Boron, Borides and Related Materials*; Istanbul Tech. Univ.: Istanbul, Turkey, 2011; p. 49.

14. Arvanitidis, A.G.; Tai, T.B.; Nguyen, M.T.; Ceulemans, A. Quantum rules for planar boron nanoclusters. *Phys. Chem. Chem. Phys.* **2014**, *16*, 18311–18318. [CrossRef]
15. Verhaegen, G.; Drowart, J. Mass spectrometric determination of the heat of sublimation of boron and the dissociation energy of B_2. *J. Chem. Phys.* **1962**, *37*, 1367–1368. [CrossRef]
16. Mar, R.W.; Bedford, R.G. The sublimation of boron. *High. Temp. Sci.* **1976**, *8*, 365–376.
17. Goltz, D.M.; Chakrabarti, C.L.; Sturgeon, R.E.; Hughes, D.M.; Gregoire, D.C. Investigation of the vaporization and atomization of boron in a graphite furnace using digital imaging techniques. *Appl. Spectrosc.* **1995**, *49*, 1006–1016. [CrossRef]
18. Becker, R.C. Method for Generating a Boron Vapor. U.S. Patent # 5861630, 19 January 1999.
19. Boustani, I.; Becker, R. Boron clusters, single- and multiwalled nanotubes: Theoretical prediction and experimental observation. In *Proc. 9th Ann. Nanotechnol. Conf. & Trade Show*; Nano Sci. & Technol. Inst.: Boston, MA, USA, 2006; MO 60.802.
20. Medvedovski, E. Preparation of boron nitride-based coatings through thermal diffusion process. *Adv. Appl. Ceram.* **2018**, *117*, 221–230. [CrossRef]
21. Unsal, H.; Grasso, S.; Kovalcikova, A.; Hanzel, O.; Tatarkova, M.; Dlouhy, I.; Tatarko, P. In-situ graphene platelets formation and its suppression during reactive spark plasma sintering of boron carbide/titanium diboride composites. *J. Eur. Ceram. Soc.* **2021**, *41*, 6281–6289. [CrossRef]
22. Feldman, C.; Moorjani, K.; Blum, N.A. Mass-spectrometry, optical absorption, and electrical properties of boron amorphous films. In *Boron–Obtaining, Structure, and Properties*; Tsagareishvili, G.V., Tavadze, F.N., Eds.; Nauka: Moscow, Russia, 1974; pp. 130–138. (In Russian)
23. Kutelia, E.R.; Dekanosidze, R.N.; Maisuradze, N.I.; Dzigrashvili, T.A.; Petrov, V.I. Electron microscopic investigation of the structure of ultrafine elementary boron. In *Abs. 8th Int. Symp. Boron, Borides, Carbides, Nitrides & Rel. Comp.*; Tsagareishvili, G.V., Ed.; Metsniereba: Tbilisi, Georgia, 1984; pp. 12–13.
24. Kervalishvili, P.J.; Kutelia, E.R.; Dzigrashvili, T.A.; Dekanosidze, R.N.; Petrov, V.I. Electron-microscopic study of amorphous boron structure. *Phys. Solid State* **1985**, *27*, 1414–1418. (In Russian)
25. Dzigrashvili, T.A.; Kutelia, E.R. On structural relaxation of small particles. *Bull. Acad. Sci. Georgian SSR* **1988**, *129*, 53–56. (In Russian)
26. La Placa, S.J.; Roland, P.A.; Wynne, J.J. Boron clusters (B_n, n = 2–52) produced by laser ablation of hexagonal boron nitride. *Chem. Phys. Lett.* **1992**, *190*, 163–168. [CrossRef]
27. Liu, C.H.; Peng, W.; Sheng, L.M. Carbon and boron nanoparticles by pulsed-laser vaporization of boron carbide in liquids. *Carbon* **2001**, *39*, 144–147. [CrossRef]
28. Xu, S.-J.; Nilles, J.M.; Radisic, D.; Zheng, W.-J.; Stokes, S.; Bowen, K.H.; Becker, R.C.; Boustani, I. Boron cluster anions containing multiple B_{12} icosahedra. *Chem. Phys. Lett.* **2003**, *379*, 282–286. [CrossRef]
29. Oger, E. *Strukturaufklärung durch Mobilitätsmessungen an Massenselektierten Clusterionen in der Gasphase (Dissertation)*; KIT Sci. Publ.: Karlsruher, Germany, 2010. (In German)
30. Atis, M.; Ozdogan, C.; Guvenc, Z.B. Structure and energetic of B_n (n = 2–12) clusters: Electronic structure calculations. *Int. J. Quantum Chem.* **2007**, *107*, 729–744. [CrossRef]
31. Tai, T.B.; Grant, D.J.; Nguyen, M.T.; Dixon, D.A. Thermochemistry and electronic structure of small boron clusters (B_n, $n = 5-13$) and their anions. *J. Phys. Chem. A* **2010**, *114*, 994–1007. [CrossRef] [PubMed]
32. Tai, T.B.; Nguyen, M.T. Electronic structure and photoelectron spectra of B_n with n = 26–29: An overview of structural characteristics and growth mechanism of boron clusters. *Phys. Chem. Chem. Phys.* **2015**, *17*, 13672–13679. [CrossRef] [PubMed]
33. Wu, X.; Sai, L.; Zhou, S.; Zhou, P.; Chen, M.; Springborg, M.; Zhao, J. Competition between tubular, planar and cage geometries: A complete picture of structural evolution of B_n (n = 31–50) clusters. *Phys. Chem. Chem. Phys.* **2020**, *22*, 12959–12966. [CrossRef] [PubMed]
34. Vedeneev, V.I.; Gurvich, L.V.; Kondrat'ev, V.N.; Medvedev, V.L.; Frankevich, E.L. *Chemical Bonds Breaking Energy. Ionization Potentials and Electron Affinity (Handbook)*; Acad. Sci. USSR: Moscow, Russia, 1962. (In Russian)
35. Graham, W.R.M.; Weltner, W. B atoms, B_2 and H_2BO molecules: ESR and optical spectra at 4 K. *J. Chem. Phys.* **1976**, *65*, 1516–1521. [CrossRef]
36. Dupuis, M.; Liu, B. The ground electronic state of B_2. *J. Chem. Phys.* **1978**, *68*, 2902–2910. [CrossRef]
37. Martin, J.M.L.; Francois, J.P.; Gijbels, R. Ab initio study of boron, nitrogen, and boron–nitrogen clusters. I. Isomers and thermochemistry of B_3, B_2N, BN_2, and N_3. *J. Chem. Phys.* **1989**, *90*, 6469–6485. [CrossRef]
38. Deutsch, P.W.; Curtiss, L.A.; Pople, J.A. Boron dimer: Dissociation energy and ionization potentials. *Chem. Phys. Lett.* **1990**, *174*, 33–36. [CrossRef]
39. Huber, K.P.; Herzberg, H. *Molecular Spectra and Molecular Structure. IV. Constants of Diatomic Molecules*; van Nostrand Reinhold Co.: New York, NY, USA, 1979.
40. Langhoff, S.R.; Bauschlicher, C.W. Theoretical study of the spectroscopy of B_2. *J. Chem. Phys.* **1991**, *95*, 5882–5888. [CrossRef]
41. Chkhartishvili, L.; Lezhava, D.; Tsagareishvili, O.; Gulua, D. Ground-state parameters of diatomic molecules B_2, BC, BN, and BO. *Proc. Georgian Police Acad.* **1999**, *1*, 195–300. (In Russian)
42. Chkhartishvili, L.; Lezhava, D.; Tsagareishvili, O. Quasi-classical determination of electronic energies and vibration frequencies in boron compounds. *J. Solid State Chem.* **2000**, *154*, 148–152. [CrossRef]

43. Chkhartishvili, L. On quasi-classical estimations of boron nanotubes ground-state parameters. *J. Phys. Conf. Ser.* **2009**, *176*, 012013. [CrossRef]
44. Chkhartishvili, L. Molar binding energy of the boron nanosystems. In *Proceedings of the 4th International Boron Symposium*; Konuk, A., Kurama, H., Ak, H., Iphar, M., Eds.; Osmangazi Univ.–TMMOB: Ankara, Turkey, 2009; pp. 153–160.
45. Chkhartishvili, L. Nanotubular boron: Ground-state estimates. In *New Developments in Materials Science*; Chikoidze, E., Tchelidze, T., Eds.; Nova Sci. Publ.: New York, NY, USA, 2013; Chapter 8; pp. 67–80.
46. Bruna, P.J.; Wright, J.S. Strongly bound multiply excited states of B_2^+ and B_2. *J. Chem. Phys.* **1989**, *91*, 1126–1136. [CrossRef]
47. Mierzwa, G.; Gordon, A.J.; Berski, S. The nature of the triple B≡B, double B=B, single B–B, and one-electron B·B boron–boron bonds from the topological analysis of Electron Localization Function (ELF) perspective. *J. Mol. Str.* **2020**, *1221*, 128530. [CrossRef]
48. Boustani, I. New convex and spherical structures of bare boron clusters. *J. Solid State Chem.* **1997**, *133*, 182–189. [CrossRef]
49. Boustani, I. New quasi-planar surfaces of bare boron. *Surf. Sci.* **1997**, *370*, 355–363. [CrossRef]
50. Boustani, I.; Quandt, A. Boron in ab initio calculations. *Comput. Mater. Sci.* **1998**, *11*, 132–137. [CrossRef]
51. Hernandez, R.; Simons, J. Electronic energies, geometries, and vibrational frequencies of the ground and low-lying excited states of the boron trimer. *J. Chem. Phys.* **1991**, *94*, 2961–2967. [CrossRef]
52. Zhai, H.-J.; Wang, L.-S.; Alexandrova, A.N.; Boldyrev, A.I.; Zakrzewski, V.G. Photoelectron spectroscopy and ab initio study of B_3^- and B_4^- anions and their neutrals. *J. Chem. Phys. A* **2003**, *107*, 9319–9328. [CrossRef]
53. Martin, J.M.L.; Francois, J.P.; Gijbels, R. Potential energy surface of B_4 and total atomization energies of B_2, B_3, and B_4. *Chem. Phys. Lett.* **1992**, *189*, 529–536. [CrossRef]
54. Koutecky, J.; Pacchioni, G.; Jeung, G.H.; Hass, E.C. Comparative study of tetramers built from Ia, IIa, IIIa, and IVa atoms. *Surf. Sci.* **1985**, *156*, 650–669. [CrossRef]
55. Li, Q.S.; Jin, H.W. Structure and stability of B_5, B_5^+, and B_5^- clusters. *J. Phys. Chem. A* **2002**, *106*, 7042–7047. [CrossRef]
56. Zhai, H.-J.; Wang, L.-S.; Alexandrova, A.N.; Boldyrev, A.I. Electronic structure and chemical bonding of B_5^- and B_5 by photoelectron spectroscopy and ab initio calculations. *J. Chem. Phys.* **2002**, *117*, 7917–7924. [CrossRef]
57. Alexandrova, A.N.; Boldyrev, A.I.; Zhai, H.-J.; Wang, L.-S.; Steiner, E.; Fowler, P.W. Structure and bonding in B_6^- and B_6: Planarity and antiaromaticity. *J. Phys. Chem. A* **2003**, *107*, 1359–1369. [CrossRef]
58. Alexandrova, A.N.; Boldyrev, A.I.; Zhai, H.-J.; Wang, L.-S. Electronic structure, isomerism, and chemical bonding in B_7^- and B_7. *J. Phys. Chem. A* **2004**, *108*, 3509–3517. [CrossRef]
59. Hanley, L.; Anderson, S.L. Production and collision-induced dissociation of small boron cluster ions. *J. Phys. Chem.* **1987**, *91*, 5161–5163. [CrossRef]
60. Ray, A.K.; Howard, I.A.; Kanal, K.M. Structure and binding in small neutral and cationic boron clusters. *Phys. Rev. B* **1992**, *45*, 14247–14255. [CrossRef]
61. Kato, H.; Tanaka, E. Stabilities of small Be_n and B_n clusters ($4 \leq n \leq 8$) by vibrational analysis. *J. Comput. Chem.* **1991**, *12*, 1097–1109. [CrossRef]
62. Zhai, H.-J.; Alexandrova, A.N.; Birch, K.A.; Boldyrev, A.I.; Wang, L.-S. Hepta- and octacoordinate boron in molecular wheels of eight- and nine-atom boron clusters: Observation and confirmation. *Angew. Chem. Int. Ed.* **2003**, *42*, 6004–6008. [CrossRef]
63. Drummond, M.L.; Meunier, V.; Sumpter, B.G. Structure and stability of small boron and boron oxide clusters. *J. Phys. Chem. A* **2007**, *111*, 6539–6551. [CrossRef]
64. Kato, H.; Yamashita, K.; Morokuma, K. Ab initio MO study of neutral and cationic boron clusters. *Chem. Phys. Lett.* **1992**, *190*, 361–366. [CrossRef]
65. Bambakidis, G.; Wagner, R.P. Electronic structure and binding energy of the icosahedral boron cluster B_{12}. *J. Phys. Chem. Solids* **1981**, *42*, 1023–1025. [CrossRef]
66. Kawai, R.; Weare, J.H. Instability of the B_{12} icosahedral cluster: Rearrangement to a lower energy structure. *J. Chem. Phys.* **1991**, *95*, 1151–1159. [CrossRef]
67. Bhattacharyya, P.; Boustani, I.; Shukla, A. First principles study of structural and optical properties of B_{12} isomers. *arXiv* **2018**, arXiv:1802.01072. [CrossRef]
68. Hanley, L.; Whitten, J.L.; Anderson, S.L. Collision-induced dissociation and ab initio studies of boron cluster ions: Determination of structures and stabilities. *J. Phys. Chem.* **1988**, *92*, 5803–5812. [CrossRef]
69. Hanley, L.; Whitten, J.L.; Anderson, S.L. Collision-induced dissociation and ab initio studies of boron cluster ions: Determination of structures and stabilities [Erratum to document cited in CA109(18):156723t]. *J. Phys. Chem.* **1990**, *94*, 2218. [CrossRef]
70. Akman, N.; Tas, M.; Ozdogan, C.; Boustani, I. Ionization energies, Coulomb explosion, fragmentation, geometric, and electronic structures of multicharged boron clusters B_n (n = 2–13). *Phys. Rev. B* **2011**, *84*, 075463. [CrossRef]
71. Cao, P.-l.; Zhao, W.; Li, B.-X.; Song, B.; Zhou, X.-Y. A full-potential linear-muffin-tin-orbital molecular-dynamics study of B_7, B_{10} and B_{13} clusters. *J. Phys. Cond. Matter* **2001**, *13*, 5065–5076. [CrossRef]
72. Kato, H.; Yamashita, K.; Morokuma, K. Ab initio study of neutral and cationic B_{12} and B_{13} clusters. *Bull. Chem. Soc. Jpn.* **1993**, *66*, 3358–3361. [CrossRef]
73. Gu, F.L.; Yang, X.; Tang, A.-C.; Jiao, H.; von Schleyer, P.R. Structure and stability of B_{13}^+ clusters. *J. Comput. Chem.* **1998**, *19*, 203–214. [CrossRef]
74. Kiran, B.; Kumar, G.G.; Nguyen, M.T.; Kandalam, A.K.; Jena, P. Origin of the unusual stability of B_{12} and B_{13}^+ clusters. *Inorg. Chem.* **2009**, *48*, 9965–9967. [CrossRef]

75. Kawai, R.; Weare, J.H. Anomalous stability of B_{13}^+ clusters. *Chem. Phys. Lett.* **1992**, *191*, 311–314. [CrossRef]
76. Fowler, J.E.; Ugalde, J.M. The curiously stable B_{13}^+ cluster and its neutral and anionic counterparts: The advantages of planarity. *J. Phys. Chem. A* **2000**, *104*, 397–403. [CrossRef]
77. Aihara, J.-I. B_{13}^+ is highly aromatic. *J. Phys. Chem. A* **2001**, *105*, 5486–5489. [CrossRef]
78. Shinde, R.; Tayade, M. Optical absorption in B_{13}cluster: A Time-dependent density functional approach. *AIP Conf. Proc.* **2012**, *G–163*, 1–2.
79. Boustani, I. Systematic LSD investigation on cationic boron clusters: B_n^+ (n = 2–14). *Int. J. Quant. Chem.* **1994**, *52*, 1081–1111. [CrossRef]
80. Boustani, I. A comparative study of ab initio SCF–CI and DFT. Example of small boron clusters. *Chem. Phys. Lett.* **1995**, *233*, 273–278. [CrossRef]
81. Boustani, I. Structure and stability of small boron clusters. A density functional theoretical study. *Chem. Phys. Lett.* **1995**, *240*, 135–140. [CrossRef]
82. Ricca, A.; Bauschlicher, C.W., Jr. The structure and stability of B_n^+ clusters. *Chem. Phys.* **1996**, *208*, 233–242. [CrossRef]
83. Boustani, I. Systematic ab initio investigation of bare boron clusters: Determination of the geometry and electronic structures of B_n (n = 2–14). *Phys. Rev. B* **1997**, *55*, 16426–16438. [CrossRef]
84. Zhai, H.-J.; Kiran, B.; Li, J.; Wang, L.-S. Hydrocarbon analogues of boron clusters–Planarity, aromaticity and antiaromaticity. *Nat. Mater.* **2003**, *2*, 827–833. [CrossRef]
85. Sergeeva, A.P.; Zubarev, D.Y.; Zhai, H.-J.; Boldyrev, A.I.; Wang, L.-S. A photoelectron spectroscopic and theoretical study of B_{16}^- and B_{16}^{2-}: An all-boron naphthalene. *J. Am. Chem. Soc.* **2008**, *130*, 7244–7246. [CrossRef] [PubMed]
86. Huang, W.; Sergeeva, A.P.; Zhai, H.-J.; Averkiev, B.B.; Wang, L.-S.; Boldyrev, A.I. A concentric planar doubly π-aromatic B_{19}^- cluster. *Nat. Chem.* **2010**, *2*, 202–206. [CrossRef] [PubMed]
87. An, W.; Bulusu, S.; Gao, Y.; Zeng, X.C. Relative stability of planar versus double-ring tubular isomers of neutral and anionic boron cluster B_{20} and B_{20}^-. *J. Chem. Phys.* **2006**, *124*, 154310. [CrossRef] [PubMed]
88. Oger, E.; Crawford, N.R.M.; Kelting, R.; Weis, P.; Kappes, M.M.; Ahlrichs, R. Boron cluster cations: Transition from planar to cylindrical structures. *Angew. Chem. Int. Ed.* **2007**, *46*, 8503–8506. [CrossRef]
89. Chacko, S.; Kanhere, D.G.; Boustani, I. Ab initio density functional investigation of B_{24} clusters: Rings, tubes, planes, and cages. *Phys. Rev. B* **2003**, *68*, 035414. [CrossRef]
90. Piazza, Z.A.; Popov, I.A.; Li, W.-L.; Pal, R.; Zeng, X.C.; Boldyrev, A.I.; Wang, L.-S. A photoelectron spectroscopy and ab initio study of the structures and chemical bonding of the B_{25}^- cluster. *J. Chem. Phys.* **2014**, *141*, 034303. [CrossRef]
91. Li, W.-L.; Pal, R.; Piazza, Z.A.; Zeng, X.C.; Wang, L.-S. B_{27}^-: Appearance of the smallest planar boron cluster containing a hexagonal vacancy. *J. Chem. Phys.* **2015**, *142*, 204305. [CrossRef]
92. Boustani, I.; Rubio, A.; Alonso, J.A. Ab initio study of B_{32} clusters: Competition between spherical, quasiplanar and tubular isomers. *Chem. Phys. Lett.* **1999**, *311*, 21–28. [CrossRef]
93. Chen, Q.; Wei, G.-F.; Tian, W.-J.; Bai, H.; Liu, Z.-P.; Zhai, H.-J.; Li, S.-D. Quasi-planar aromatic B_{36} and B_{36}^- clusters: All-boron analogues of coronene. *Phys. Chem. Chem. Phys.* **2014**, *16*, 18282–18287. [CrossRef]
94. Piazza, Z.A.; Hu, H.-S.; Li, W.-L.; Zhao, Y.-F.; Li, J.; Wang, L.-S. Planar hexagonal B_{36} as a potential basis for extended single-atom layer boron sheets. *Nat. Commun.* **2014**, *5*, 3113. [CrossRef]
95. Bai, H.; Chen, T.-T.; Chen, Q.; Zhao, X.-Y.; Zhang, Y.-Y.; Chen, W.-J.; Li, W.-L.; Cheung, L.F.; Bai, B.; Cavanagh, J.; et al. Planar B_{41}^- and B_{42}^- clusters with double-hexagonal vacancies. *Nanoscale* **2019**, *11*, 23286–23295. [CrossRef] [PubMed]
96. Mukhopadhyay, S.; He, H.; Pandey, R.; Yap, Y.K.; Boustani, I. Novel spherical boron clusters and structural transition from 2D quasi-planar structures to 3D double-rings. *J. Phys. Conf. Ser.* **2009**, *176*, 012028. [CrossRef]
97. Oyaidzu, M.; Yoshikawa, A.; Kodama, H.; Oya, Y.; Sagara, A.; Noda, N.; Okuno, K. Preparation of pure boron coating film and its characterization by XPS and TDS. *Appl. Surf. Sci.* **2005**, *244*, 240–243. [CrossRef]
98. Becker, R.C.; Bunker, S.N. Method for depositing boron-rich coatings. Patent Appl. # US 2005/0208218 A1, 22 September 2005.
99. Becker, R.; Chkhartishvili, L.; Martin, P. Tribological applications for boron. *Vac. Technol. Coat.* **2015**, *16*, 36–41.
100. Korbut, E.V.; Labunets, V.F.; Radko, O.V.; Zagrebelnyi, V.V.; Yakobchuk, O.E. Improve of the high-speed steel wear resistance using boron coating. In *Proc. 5th Int. Conf. High Mat Tech*; IPMS: Kyiv, Ukraine, 2015; p. 189.
101. Zhu, Z.; Kwon, D.-G.; Kwon, Y.-K.; Tomanek, D. Enhancing mechanical toughness of aluminum surfaces by nano-boron implantation: An ab initio study. *Chem. Phys. Lett.* **2015**, *620*, 25–28. [CrossRef]
102. Demirbas, A. Energy from boron and non-nuclear metallic fuels. *Energy Sources A* **2008**, *30*, 1108–1113. [CrossRef]
103. Young, G.; Sullivan, K.; Zachariah, M.R.; Yu, K. Combustion characteristics of boron nanoparticles. *Combust. Flame* **2009**, *156*, 322–333. [CrossRef]
104. Hussmann, B.; Pfitzner, M. Extended combustion model for single boron particles–Part I: Theory. *Combust. Flame* **2010**, *157*, 803–821. [CrossRef]
105. Hussmann, B.; Pfitzner, M. Extended combustion model for single boron particles–Part II: Validation. *Combust. Flame* **2010**, *157*, 822–833. [CrossRef]
106. Katoshevski, D.; Chkhartishvili, L. Manipulating grouping dynamics of nanoscale boron particles as basis for environmentally friendlier combustion and efficient filtration. In *Nanotechnology in Environmental Science*; Hussain, C.M., Mishra, A.K., Eds.; Wiley–VCH Verlag GmbH & Co. KGaA: Weinheim, Germany, 2018; Chapter 13; pp. 413–442.

107. Ojha, P.K.; Karmakar, S. Boron for liquid fuel engines–A review on synthesis, dispersion stability in liquid fuel, and combustion aspects. *Prog. Aerospace Sci.* **2018**, *100*, 18–45. [CrossRef]
108. Gan, Y.; Lim, Y.S.; Qiao, L. Combustion of nanofluid fuels with the addition of boron and iron particles at dilute and dense concentrations. *Combust. Flame* **2012**, *159*, 1732–1740. [CrossRef]
109. Chintersingh, K.-L.; Schoenitz, M.; Dreizin, E.L. Combustion of boron and boron–iron composite particles in different oxidizers. *Combust. Flame* **2018**, *192*, 44–58. [CrossRef]
110. Epshteyn, A.; Weismiller, M.R.; Huba, Z.J.; Maling, E.L.; Chaimowitz, A.S. Optimization of a high-energy Ti–Al–B nanopowder fuel. *Energy Fuels* **2017**, *31*, 1811–1819. [CrossRef]
111. Hanley, L.; Anderson, S.L. Oxidation of small boron cluster ions (B_{1-13}^+) by oxygen. *J. Chem. Phys.* **1988**, *89*, 2848–2860. [CrossRef]
112. Ruatta, S.A.; Hintz, P.A.; Anderson, S.L. Boron cluster ion oxidation: Reactions with CO_2, dissociation of boron cluster oxide (B_nO^+) ions, and sequential oxidation. *J. Chem. Phys.* **1991**, *94*, 2833–2847. [CrossRef]
113. Chkhartishvili, L.; Tsagareishvili, O.; Gabunia, D. ^{10}B-based materials for neutron-shielding. In *Proc. 1st Int. Conf. "Modern Technologies and Methods of Inorganic Materials Science"*; Khantadze, J., Chkhartishvili, L., Gabunia, D., Ramazashvili, D., Eds.; Meridian: Tbilisi, Georgia, 2012; pp. 188–202.
114. Chkhartishvili, L. Interaction between neutron-radiation and boron-containing materials. In *Radiation Synthesis of Materials and Compounds*; Kharisov, B.I., Kharissova, O.V., Mendez, U.O., Eds.; CRC Press–Taylor & Francis Group: Boca Raton, FL, USA, 2013; Chapter 3; pp. 43–80.
115. Chkhartishvili, L. Boron-contained nanostructured materials for neutron-shields. In *Nanostructured Materials for the Detection of CBRN*; Bonca, J., Kruchinin, S., Eds.; Springer Science: Dordrecht, The Netherlands, 2018; Chapter 11; pp. 133–154.
116. Chkhartishvili, L. Neutron-fluence nanosensors based on boron-containing materials. In *Nanomaterials for Environmental Protection*; Kharisov, B.I., Kharissova, O.V., Rasika Dias, H.V., Eds.; John Wiley & Sons Inc.: Hoboken, NJ, USA, 2014; Chapter 26; pp. 445–449.
117. Chkhartishvili, L.; Tsagareishvili, O.; Tavadze, G. Neutron detectors based on ^{10}B-containing nanomaterials. In *Nuclear Radiation Nanosensors and Nanosensory Systems*; Kervalishvili, P.J., Yannakopoulos, P.H., Eds.; Springer Science: Dordrecht, The Netherlands, 2016; Chapter 12; pp. 187–196.
118. Chkhartishvili, L.; Murusidze, I. Band structure of all-boron 2D metallic crystals as a prospective electromagnetic shielding material. In *Proc. Int. Conf. "Fundamental and Applied Nano Electromagnetics"*; Belarusian State Univ.: Minsk, Belarus, 2012; p. 11.
119. Wang, Y.-J.; Zhao, X.-Y.; Chen, Q.; Zhai, H.-J.; Li, S.-D. B_{11}^-: A moving subnanoscale tank tread. *Nanoscale* **2015**, *7*, 16054–16060. [CrossRef]
120. He, C.-C.; Xu, S.-G.; Zhao, Y.-J.; Xu, H.; Yang, X.-B. All-boron planar ferromagnetic structures: From clusters to monolayers. *Nanoscale* **2021**, *13*, 9881–9887. [CrossRef]
121. Wang, Z.-Q.; Lu, T.-Y.; Wang, H.-Q.; Feng, Y.P.; Zheng, J.-C. Review of borophene and its potential applications. *Front. Phys.* **2019**, *14*, 33403. [CrossRef]
122. Fermi, E. *Molecules and Crystals*; Barth: Leipzig, Germany, 1938.
123. Novikova, S.I. *Thermal Expansion of Solids*; Nauka: Moscow, Russia, 1974. (In Russian)
124. Slutsker, A.I.; Gilyarov, V.L.; Luk'yanenko, A.S. Energy features of an adiabatically loaded anharmonic oscillator. *Phys. Solid State* **2006**, *48*, 1947–1953. [CrossRef]
125. Chkhartishvili, L.; Becker, R.; Avci, R. Relative stability of boron quasi-planar clusters. In *Proc. Int. Conf. "Advanced Materials and Technologies"*; Darsavelidze, G., Guldamashvili, A., Chedia, R., Sichinava, A., Kadaria, M., Eds.; Universal: Tbilisi, Georgia, 2015; pp. 42–46.
126. Chkhartishvili, L. Small elemental clusters in pair interaction approximation. In *Proc. 4th Int. Conf. Exh. Adv. Nano Mater.*; IAEMM: Montreal, QC, Canada, 2016; pp. 128–132.
127. Chkhartishvili, L. Quasi-planar elemental clusters in pair interactions approximation. *Open Phys.* **2016**, *14*, 617–620. [CrossRef]
128. Chkhartishvili, L. Planar clusters of identical atoms in equilibrium: 1. Diatomic model approach. *Am. J. Nano Res. Appl.* **2017**, *5*, 1–4.
129. Chkhartishvili, L.; Becker, R. Effective atomic charges and dipole moment of small boron clusters. In *Proc. 3rd Int. Conf. Exh. Adv. Nano Mater.*; IAEMM: Ottawa, ON, Canada, 2015; pp. 130–147.
130. Becker, R.; Chkhartishvili, L. Dipole moment of quasi-planar boron clusters. *Nano Studies* **2015**, *11*, 29–48.
131. Chkhartishvili, L. Boron quasi-planar clusters. A mini-review on diatomic approach. In *Proc. IEEE 7th Int. Conf. Nanomaterials: Applications & Properties (NAP–2017), Track: Nanomaterials for Electronics, Spintronics and Photonics*; Sumy State Univ.: Sumy, Ukraine, 2017; Part 4; pp. 04NESP10-1–04NESP10-5.
132. Chkhartishvili, L. Boron triangular sheet: Calculation of ground-state and electronic-structure parameters. In *Proc. 7th Int. Conf. Exh. Adv. Nano Mater.*; IAEMM: Montreal, QC, Canada, 2019; pp. 11–16.
133. Chkhartishvili, L. Relative stability of planar clusters B_{11}, B_{12}, and B_{13} in neutral- and charged-states. *Char. Appl. Nanomater.* **2019**, *2*, 761-1–761-7. [CrossRef]
134. Chkhartishvili, L.; Murusidze, I.; Becker, R. Electronic structure of boron flat holeless sheet. *Condensed Matter* **2019**, *4*, 28. [CrossRef]
135. Chkhartishvili, L. Nanoclusters binding energy in diatomic model. *Int. J. Adv. Nano Comput. Anal.* **2021**, *1*, 80–83.

molecules

Communication

Structure of χ_3-Borophene Studied by Total-Reflection High-Energy Positron Diffraction (TRHEPD)

Yuki Tsujikawa [1,*], Makoto Shoji [2], Masashi Hamada [2], Tomoya Takeda [2], Izumi Mochizuki [3], Toshio Hyodo [3], Iwao Matsuda [1] and Akari Takayama [2,*]

[1] Institute for Solid State Physics, The University of Tokyo, Chiba 277-8581, Japan; imatsuda@issp.u-tokyo.ac.jp
[2] Department of Physics and Applied Physics, Waseda University, Tokyo 169-8555, Japan; m.shoji-bmae2358@ruri.waseda.jp (M.S.); hamcchi.8@akane.waseda.jp (M.H.); zuratantakeabc@fuji.waseda.jp (T.T.)
[3] Institute of Materials Structure Science, High Energy Accelerator Research Organization (KEK), Ibaraki 305-0801, Japan; mochizu@post.kek.jp (I.M.); hyodot@post.kek.jp (T.H.)
* Correspondence: ytsujika@issp.u-tokyo.ac.jp (Y.T.); a.takayama@waseda.jp (A.T.)

Abstract: We have investigated the structure of χ_3-borophene on Ag(111), a monolayer material of boron atoms, via total-reflection high-energy positron diffraction (TRHEPD). By comparing the experimental rocking-curves with ones for several structures calculated by using dynamical diffraction theory, we confirmed that the χ_3-borophene layer has a flat structure. The distance from the topmost layer of the metal crystal is 2.4 Å, which is consistent with results reported by X-ray standing wave-excited X-ray photoelectron spectroscopy. We also demonstrated that the in-plane structure of χ_3-borophene is compatible with the theoretical predictions. These structural properties indicate that χ_3-borophene belongs to a group of epitaxial monolayer sheets, such as graphene, which have weak interactions with the substrates.

Keywords: monolayer material; borophene; diffraction; TRHEPD

Citation: Tsujikawa, Y.; Shoji, M.; Hamada, M.; Takeda, T.; Mochizuki, I.; Hyodo, T.; Matsuda, I.; Takayama, A. Structure of χ_3-Borophene Studied by Total-Reflection High-Energy Positron Diffraction (TRHEPD). *Molecules* **2022**, *27*, 4219. https://doi.org/10.3390/molecules27134219

Academic Editor: Mohammed Baalousha

Received: 31 May 2022
Accepted: 24 June 2022
Published: 30 June 2022

1. Introduction

Monolayer (ML) materials that are composed of a single element, which are referred to by the term "Xene", have attracted great interest. It is because they have intriguing electronic states, such as Dirac Fermions [1], which are derived by a structural arrangement of a two-dimensional (2D) honeycomb lattice. Today, Xenes have been investigated by a large number of researchers in both the academic and technological fields. Typical examples of Xenes are composed of group 14 elements and include materials such as graphene, silicene, and germane [2,3]. Recently, a 2D material of boron (a group 13 element) called "borophene" has been attracting attention, especially after the success of its fabrication on metal substrates [4–8]. Similar to 3D boron, borophene layers are predicted to have various types of atomic structures due to the multi-center bonding scheme of boron atoms [9]. Theoretical works have found that a flat layer of boron is stable when it is composed of triangular lattices and hexagonal hollows [9–11]. Experimental observations of epitaxial borophene on Ag(111) have been consistent with the theoretical structural models [6,12], but the atomic structure has not been examined directly. Thus, there is a strong need to conduct structural analysis by an appropriate diffraction experiment on the surface.

In the present research, we studied the atomic structure of χ_3-borophene on Ag(111) by means of total-reflection high-energy positron diffraction (TRHEPD). This structural analysis method is exceedingly surface-sensitive, and it has determined varieties of atomic layers precisely [1,13]. The present experiment has revealed that χ_3-borophene is flat in the 2D layer with a distance of 2.4 Å from the Ag(111) surface. This result agrees with the structural models proposed in previous experimental and theoretical studies [12,14]. Notably, its structure is more like graphene, which interacts weakly with the substrate, and unlike silicene and germane, which are epitaxial on the metal substrate.

2. Results and Discussion

2.1. Calculated Rocking Curves

Since this work is the first case of applying TRHEPD for structural analysis of a 2D boron material, here we briefly introduce the principles and the methods (see Section 3.2 for details). The TRHEPD experiment consists of measuring a series of the diffraction patterns for a fixed incident azimuthal direction at various glancing angles. Following that, the diffraction intensity of the specular (00) spot is plotted as a function of θ. The plot is called a rocking curve (RC). In the structural analysis, the experimental RCs are compared with those calculated for various structural models by using dynamical diffraction theory. A reliability factor R [15,16] is used as a criterion to judge goodness of the agreement. Typical measurements are made under one-beam (OB) and many-beam (MB) conditions. In the OB condition the beam is incident along an off-symmetric direction, while in the MB condition it is incident along a symmetric direction. The RC in the OB condition essentially gives the information on the atomic positions in the out-of-plane direction only, while the RC in the MB condition includes information on the in-plane structure as well [17,18]. Before showing the experimental data of the TRHEPD and the optimized results, we demonstrate the sensitivity of the RC of TRHEPD to the details of structural models. Figure 1a shows the top view of a structural model of χ_3-borophene on Ag(111), with arrows indicating the beam incident directions for the OB (red) and MB (blue) conditions. The positron beam is incident with a 17° deviation from the $[1\bar{1}0]$ direction for the OB condition, while it is along the $[1\bar{1}0]$ direction for the MB condition. Usually, the TRHEPD analysis of the data under the OB condition is performed first to extract the out-of-plane structure only, i.e., in the present case, the interlayer distance between the Ag substrate and borophene and the possible buckling of the borophene sheet.

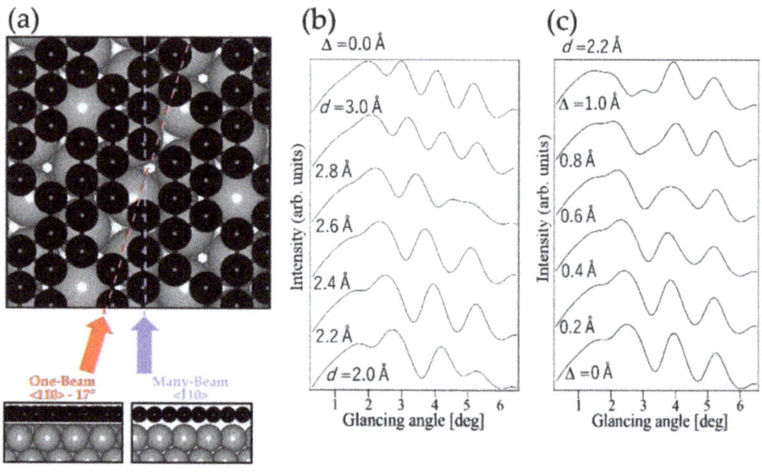

Figure 1. (**a**) Structural model of χ_3-borophene. Beam directions of the one- and many-beam condition are depicted by the red and blue arrows, respectively. The bottom pictures show the side view along each condition. (**b**) Calculated rocking curves of χ_3 -borophene in one-beam condition with different interlayer distance d, with no buckling (Δ = 0.0 Å). (**c**) Calculated rocking curves of χ_3 -borophene with different buckling Δ, with fixed interlayer distance (d = 2.2 Å) which is the height of the lowest boron atoms from the Ag substrate.

2.1.1. One-Beam (OB) Condition

Figure 2b,c show the calculated RCs under the OB condition using the structural model of the χ_3-borophene [12]. The interlayer distance d from the substrate and the magnitude of the buckling Δ in the layer were changed systematically. It is of note that the proposed model [12] has the parameters d = 2.4 Å and Δ = 0.04 Å. The calculations here assume that the 1 ML borophene sheet completely covers the Ag substrate. Figure 1b shows RCs calculated for various values of d with Δ fixed to 0.0 Å (i.e., flat or no buckling). One can find that the shape of the RC apparently depends on d. For example, a peak near 4° at d = 2.0 Å approaches 3° at d = 3.0 Å. Figure 1c shows a series of RCs calculated for various Δ with d fixed at 2.2 Å. Furthermore, it is clearly seen that the shape of the RC changes as Δ increases. These results in Figure 1b,c demonstrate that RCs measured by TRHEPD are sensitive enough to accurately examine the d and Δ values of borophene.

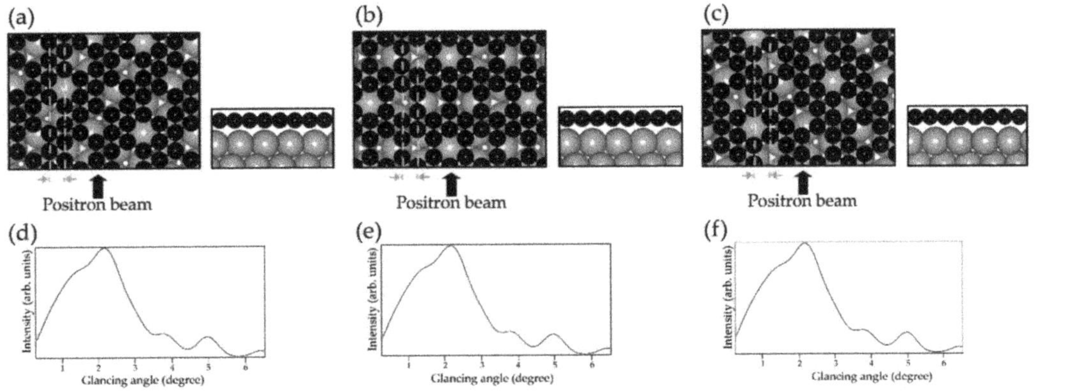

Figure 2. **(a–c)** Structure of each of the three domains of χ_3-borophene grown on different directions to Ag(111), viewed from the top and side. **(d–f)** The RCs of **(a–c)**, respectively. The shape of the RCs was the same for all.

2.1.2. Many-Beam (MB) Condition

The in-plane structure is determined from the RCs under MB condition; for χ_3-borophene on the Ag(111), the beam is incident along the [11$\bar{2}$] and [1$\bar{1}$0] directions. It is reported that the χ_3-borophene on the Ag(111) has three domains that are 120°-rotated from each other [14]. The structures in the different domains are illustrated in Figure 2a–c. Essentially, we must pay attention to the domain direction relative to the beam incidence direction. Interestingly, the calculated results of RCs of the three domains along the [1$\bar{1}$0] direction are identical to one another, as presented in Figure 2d–f. This is because the in-plane structure perpendicular to the beam incidence direction is reflected in the RC. It means that existence of domains is not an issue when the incident positron beam is along the [1$\bar{1}$0] direction. In this study, we focus on data along the [1$\bar{1}$0] direction for a sample without domain control.

2.2. TRHEPD Measurements

2.2.1. OB Measurement

Figure 3 shows the RCs of (a) a pristine Ag(111) surface and (b) the borophene grown on it, measured under the OB condition. In Figure 3a, the experimental curve is superimposed on the calculated one, based on a previous TRHEPD report of Ag(111) [2]. The profiles show good agreement with each other except for a small deviation below 2°. The suppressed intensity at the low glancing angles is likely due to the surface roughness, or terraces with various heights on the Ag substrate. It has been made from the multiple sputtering and annealing processes breaking the ideal total-reflection condition. Such an effect is observed only in TRHEPD measurement whose surface selectivity is exceedingly high and is not a concern in other techniques of surface analysis such as electron diffraction. Considering this, we only used data above 1.6° for accurate analysis.

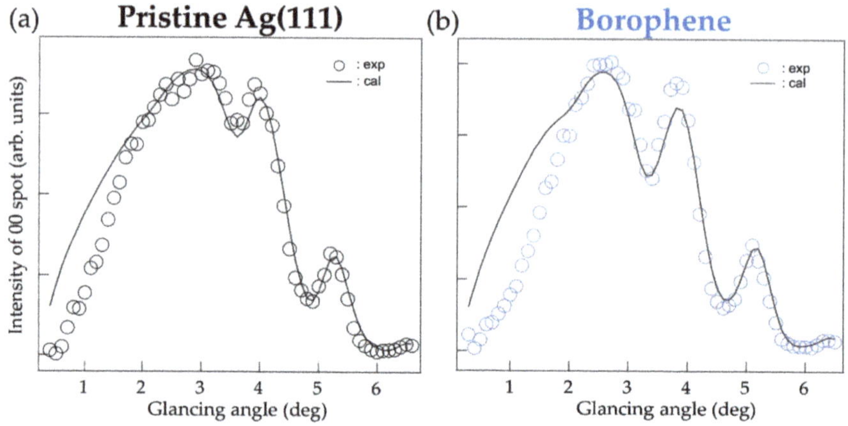

Figure 3. Rocking curves under one-beam (OB) condition with calculated curves for (**a**) pristine Ag(111) and (**b**) χ_3-borophene on Ag(111).

In Figure 3b, a result for the χ_3-borophene layer grown on the Ag(111) substrate obtained in the OB condition was compared with the corresponding calculation. The experimental profile was reproduced by the proposed model with d = 2.4 Å and Δ = 0.04 Å [12]. In addition, it resulted in a surface occupancy of 56% on the Ag(111) substrate with a minimum R of 1.38%. These agree with the previous report of the microscopic imaging by scanning tunneling microscopy that observed a surface covered partially by domains of the borophene [6]. We found that the agreements between the experiment and the calculation were never possible with a structure with appreciable buckling. This indicates that there is no buckling considering by the sensitivity of the shape of the RC to the amount of buckling as shown in Figure 1c. Using the diffraction method, we have directly confirmed that the χ_3-borophene layer is a flat sheet and found that the structural model determined in this study is completely consistent with that previously proposed by the X-ray standing wave photoemission experiment [12].

2.2.2. MB Measurement

Figure 4 shows the RCs acquired by the THREPD measurements along the $[1\bar{1}0]$ direction, i.e., under the MB condition, for (a) the pristine Ag(111) surface and (b) epitaxial χ_3-borophene. As in the case of the data under the OB condition, the intensity at low glancing angles (<~2°) was suppressed due to surface roughness that partly hindered the total reflection of the positron beam. RCs measured under MB conditions include information on the in-plane and out-of-plane structure, i.e., the analysis must be conducted involving many parameters that are not independent. In this study, we have assumed the in-plane structural model proposed in previous theoretical studies as shown in Figure 1a [14], and adjusted d and Δ so that R for the glancing angles above 2.0° is smallest. We found that the experimental curves were well reproduced with the same parameters optimized for the OB condition, d = 2.4 Å, Δ = 0.0 Å, and 54% coverage. Other possible structural models would not reproduce the experimental RC well. Thus, the result supports the structural model proposed in theoretical studies, which state that χ_3-borophene on Ag(111) is a structure with a triangular lattice and hexagonal hollows. In the present study we did not perform a full analysis of the MB RCs, where all the in-plane atomic coordinates are optimized. For a more accurate structure determination, we plan to prepare a sample with a single-domain and analyze the RC under MB conditions along the $[11\bar{2}]$ direction.

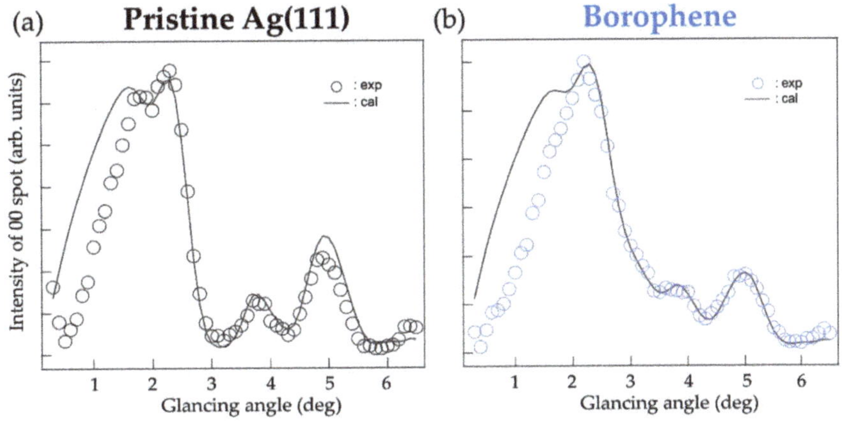

Figure 4. Rocking curve under many-beam (MB) condition with calculated curves for (**a**) pristine Ag(111) and (**b**) χ_3-borophene on Ag(111).

It would be worth while to discuss if or how TRHEPD is able to observe the concentric superlattice structure in borophene via self-assembly of twin boundaries. Previous STM works [12] have established the very interesting defect-mediated self-assembly as a pathway to unique borophene structures and properties. Unfortunately, considering the positron beam diameter of the current TRHPED measurement, it is difficult to observe and analyze such structures, which depend on local domains or boundaries. Structural analysis for these concentric superlattice structures may become possible if diffraction patterns originating from the structural periodicity of such domains could be observed, or by achieving microscopic measurement comparable to the domain size.

3. Materials and Methods

3.1. Sample

A clean surface of Ag(111) crystal was prepared by several cycles of Ar+ sputtering at 0.5 keV for 10 min and annealing processes at 450 °C, followed by confirmation through clear 1 × 1 pattern of reflection high-energy electron diffraction (RHEED). Subsequently, the epitaxial χ_3-borophene was prepared by boron deposition at 300 °C. We checked the formation of the borophene layer with the 3 × 3 pattern by RHEED and TRHEPD, as shown in Figure 5a,b, respectively.

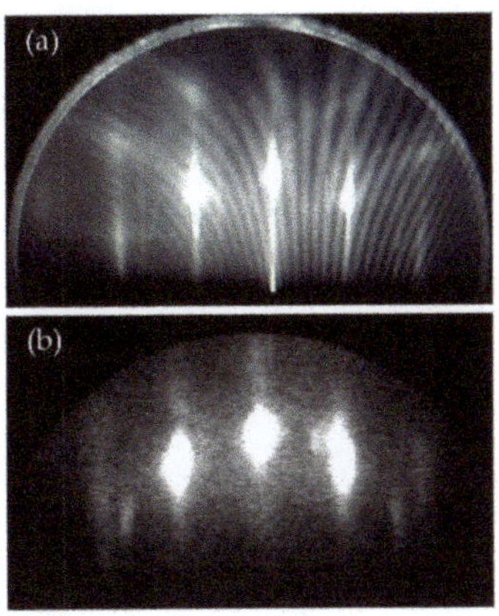

Figure 5. (**a**) The RHEED and (**b**) TRHEPD pattern of the epitaxial χ_3-borophene on Ag(111).

3.2. TRHEPD Method

3.2.1. Features of TRHEPD

Figure 6 schematically shows a measurement configuration of the TRHEPD method. The set-up resembles that of RHEED that uses the electron, its antiparticle. The surface sensitivity of TRHEPD and RHEED results from the fact that elastic scattering or diffraction events can only be available from a surface within a depth of the inelastic mean free path. In addition, compared to electrons, a positron beam has the advantage of much higher surface sensitivity because it senses positive electrostatic potential in every material [15,19,20]; when a positron of kinetic energy E_0 is incident on a solid whose mean internal potential is V_0, the conservation of total energy leads to the mean kinetic energy (E) inside to be $E = E_0 - eV_0$, while it is $E = E_0 + eV_0$ in the case of an electron. Here, for incidence at a glancing angle (θ), the component of E related to the perpendicular momentum component inside materials ($E_\perp$) is described as $E_\perp = E_0 \sin^2\theta + eV_0$ due to conservation of the parallel component of the momentum. Since no state is allowed for $E_\perp < 0$, positrons are totally reflected for θ below the critical angle (θ_C)[$\theta < \sin^{-1}\sqrt{eV_0/E_0} = \theta_c$], while the positron beam penetrates into the crystal at the glancing angle above θ_c.

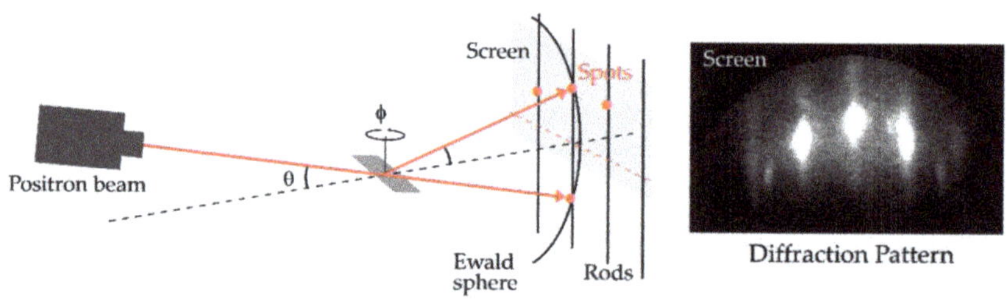

Figure 6. Schematic diagram of TRHEPD. A positron beam is incident on a sample surface with a glancing angle (θ) and an orientation fixed at a specific azimuthal angle (ϕ). The diffraction spots appear at the intersection of the Ewald sphere and reciprocal-lattice rods and the spots are projected onto the screen as the diffraction pattern on the right.

3.2.2. TRHEPD Analysis

The R-factor is defined as $R = \sqrt{\sum_i \left[I_{\text{exp}}(\theta_i) - I_{\text{cal}}(\theta_i)\right]^2} \times 100$ (%) where $I_{\text{exp}}(\theta_i)$ and $I_{\text{cal}}(\theta_i)$ are the experimental and calculated intensities of a diffraction spot at the glancing angle θ_i, respectively, normalized under the following condition: $\sum_i I_{\text{exp}}(\theta_i) = \sum_i I_{\text{cal}}(\theta_i) = 1$. The most appropriate structure should be the one with the minimum R, as reduction in R corresponds to a smaller difference between the measured and the calculated RCs with structural optimization.

3.2.3. Experimental Condition for TRHEPD

The experiment was performed at the Slow Positron Facility (SPF) in IMSS, KEK. The incident energy and the spot size of the positron beam were 10 keV and ϕ 1 mm, respectively. The TRHED rocking curves were measured in a glancing angle range of $0.3^\circ \sim 6.5^\circ$ with 0.1° steps. All of the measurements were performed at room temperature. At incident energy E = 10 keV, the critical angle for an Ag crystal with $eV_0 = 23$ eV is $\theta_C = 2.8^\circ$ [2]. Calculations of the rocking curves for TRHEPD were performed by the structure-analysis program, "2DMAT" [21], as well as the one build by Hanada et al. [22,23].

4. Conclusions

We performed a THRPED study to examine the atomic structure of the a χ_3-borophene layer on Ag(111). The structural parameters with d = 2.4 Å and Δ = 0.04 Å, indicate that the χ_3-borophene belongs to a group of the monolayer sheets that have weak interactions with substrates, such as graphene. The analysis of the RC under the MB condition along the [1$\bar{1}$0] direction was also compatible with the in-plane structural model proposed by theoretical structural model for χ_3-borophene. The present study is the first application of the diffraction method for the structural analysis of borophene. The method of THRPED is described in some detail because the techniques is novel, and it can be applied to the other 2D boron materials.

Author Contributions: Conceptualization, A.T. and I.M. (Iwao Matsuda); formal analysis, Y.T. and I.M. (Izumi Mochizuki); investigation, Y.T., M.S., T.T., M.H., I.M. (Izumi Mochizuki) and A.T.; resource, I.M. (Izumi Mochizuki) and T.H.; data curation, Y.T. and I.M. (Izumi Mochizuki); writing—original draft preparation, Y.T.; writing—review and editing, Y.T., I.M. (Iwao Matsuda), T.H. and A.T.; visualization, Y.T.; supervision, A.T.; project administration, A.T.; funding acquisition, I.M. (Iwao Matsuda) and A.T. All authors have read and agreed to the published version of the manuscript.

Funding: This research was funded by JSPS KAKENHI Grants No. JP19H04398, JP21H05012, JP18H03874 and JP21J21993. This study was performed with the approval of the Photon Factory Program Advisory Committee of the Institute of Materials Structure Science (Proposal No. 2019G123, 2020G625, 2021T002). Y. T. thanks JSPS for its financial support.

Institutional Review Board Statement: Not applicable.

Informed Consent Statement: Not applicable.

Data Availability Statement: Data is contained in article.

Acknowledgments: We thank F. Komori and T. Iimori for their help in the experiment and the insightful discussions.

Conflicts of Interest: The authors declare no conflict of interest.

Sample Availability: Samples of the compounds are not available from the authors.

References

1. Feng, B.; Sugino, O.; Liu, R.-Y.; Zhang, J.; Yukawa, R.; Kawamura, M.; Iimori, T.; Kim, H.; Hasegawa, Y.; Li, H.; et al. Dirac Fermions in Borophene. *Phys. Rev. Lett.* **2017**, *118*, 096401. [CrossRef] [PubMed]
2. Fukaya, Y.; Mochizuki, I.; Maekawa, M.; Wada, K.; Hyodo, T.; Matsuda, I.; Kawasuso, A. Structure of silicene on a Ag(111) surface studied by reflection high-energy positron diffraction. *Phys. Rev. B* **2013**, *88*, 205413. [CrossRef]
3. Yuhara, J.; Shimazu, H.; Ito, K.; Ohta, A.; Araidai, M.; Kurosawa, M.; Nakatake, M.; Le Lay, G. Germanene Epitaxial Growth by Segregation through Ag(111) Thin Films on Ge(111). *ACS Nano* **2018**, *12*, 11632–11637. [CrossRef] [PubMed]
4. Feng, B.; Zhang, J.; Zhong, Q.; Li, W.; Li, S.; Li, H.; Cheng, P.; Meng, S.; Chen, L.; Wu, K. Experimental Realization of Two-Dimensional Boron Sheets. *Nat. Chem.* **2016**, *8*, 563–568. [CrossRef] [PubMed]
5. Mannix, A.J.; Zhou, X.-F.; Kiraly, B.; Wood, J.D.; Alducin, D.; Myers, B.D.; Liu, X.; Fisher, B.L.; Santiago, U.; Guest, J.R.; et al. Synthesis of Borophenes: Anisotropic, Two-Dimensional Boron Polymorphs. *Science* **2015**, *350*, 1513–1516. [CrossRef]
6. Wu, R.; Drozdov, I.K.; Eltinge, S.; Zahl, P.; Ismail-Beigi, S.; Božović, I.; Gozar, A. Large-Area Single-Crystal Sheets of Borophene on Cu(111) Surfaces. *Nat. Nanotechnol.* **2019**, *14*, 44–49. [CrossRef]
7. Vinogradov, N.A.; Lyalin, A.; Taketsugu, T.; Vinogradov, A.S.; Preobrajenski, A. Single-Phase Borophene on Ir(111): Formation, Structure, and Decoupling from the Support. *ACS Nano* **2019**, *13*, 14511–14518. [CrossRef]
8. Li, W.; Kong, L.; Chen, C.; Gou, J.; Sheng, S.; Zhang, W.; Li, H.; Chen, L.; Cheng, P.; Wu, K. Experimental Realization of Honeycomb Borophene. *Sci. Bull.* **2018**, *63*, 282–286. [CrossRef]
9. Matsuda, I.; Wu, K. *2D Boron: Boraphene, Borophene, Boronene*, 1st ed.; Springer: Cham, Switzerland, 2021. [CrossRef]
10. Wang, Z.Q.; Lu, T.Y.; Wang, H.Q.; Feng, Y.P.; Zheng, J.C. Review of Borophene and Its Potential Applications. *Front. Phys.* **2019**, *14*, 33403. [CrossRef]
11. Zhang, Z.; Yang, Y.; Gao, G.; Yakobson, B.I. Two-Dimensional Boron Monolayers Mediated by Metal Substrates. *Angew. Chem. Int. Ed.* **2015**, *54*, 13022–13026. [CrossRef]
12. Campbell, G.P.; Mannix, A.J.; Emery, J.D.; Lee, T.-L.; Guisinger, N.P.; Hersam, M.C.; Bedzyk, M.J. Resolving the Chemically Discrete Structure of Synthetic Borophene Polymorphs. *Nano Lett.* **2018**, *18*, 2816–2821. [CrossRef]
13. Fukaya, Y.; Entani, S.; Sakai, S.; Mochizuki, I.; Wada, K.; Hyodo, T.; Shamoto, S. Spacing between Graphene and Metal Substrates Studied with Total-Reflection High-Energy Positron Diffraction. *Carbon* **2016**, *103*, 1–4. [CrossRef]
14. Liu, L.; Zhang, Z.; Liu, X.; Xuan, X.; Yakobson, B.I.; Hersam, M.C.; Guo, W. Borophene Concentric Superlattices via Self-Assembly of Twin Boundaries. *Nano Lett.* **2020**, *20*, 1315–1321. [CrossRef]
15. Fukaya, Y.; Kawasuso, A.; Ichimiya, A.; Hyodo, T. Total-Reflection High-Energy Positron Diffraction (TRHEPD) for Structure Determination of the Topmost and Immediate Sub-Surface Atomic Layers. *J. Phys. D Appl. Phys.* **2018**, *52*, 013002. [CrossRef]

16. Fukaya, Y.; Shigeta, Y.; Maki, K. Dynamic Change in the Surface and Layer Structures during Epitaxial Growth of Si on a Si(111) 7×7 Surface. *Phys. Rev. B* **2000**, *61*, 13000–13004. [CrossRef]
17. Ichimiya, A. Numerical Convergence of Dynamical Calculations of Reflection High-Energy Electron Diffraction Intensities. *Surf. Sci.* **1990**, *235*, 75–83. [CrossRef]
18. Ichimiya, A. Rheed Intensity Analysis of Si(111)7×7 at One-Beam Condition. *Surf. Sci.* **1987**, *192*, L893–L898. [CrossRef]
19. Ichimiya, A. Reflection High-Energy Positron Diffraction (RHEPD). *Solid State Phenom.* **1992**, *28–29*, 143–148. [CrossRef]
20. Hugenschmidt, C. Positrons in Surface Physics. *Surf. Sci. Rep.* **2016**, *71*, 547–594. [CrossRef]
21. Motoyama, Y.; Yoshimi, K.; Iwamoto, H.; Ichinose, H.; Hoshi, T. Data-analysis software framework 2DMAT and its application to experimental measurements for two-dimensional material structures. *arXiv* **2022**, arXiv:2204.04484. [CrossRef]
22. Hanada, T.; Daimon, H.; Ino, S. Rocking-curve analysis of reflection high-energy electron diffraction from the Si(111)-($\sqrt{3}\times\sqrt{3}$) $R30°$-Al, -Ga, and -In surfaces. *Phys. Rev. B* **1995**, *51*, 13320–13325. [CrossRef]
23. Hanada, T.; Motoyama, Y.; Yoshimi, K.; Hoshi, T. sim-trhepd-rheed—Open-source simulator of total-reflection high-energy positron diffraction (TRHEPD) and reflection high-energy electron diffraction (RHEED). *Comput. Phys. Commun.* **2022**, *277*, 108371. [CrossRef]

molecules

MDPI

Article

Vibrational Property of α-Borophene Determined by Tip-Enhanced Raman Spectroscopy

Ping Zhang [1,2,†], Xirui Tian [3,†], Shaoxiang Sheng [1,2,†], Chen Ma [1,2], Linjie Chen [4], Baojie Feng [1,2], Peng Cheng [1,2], Yiqi Zhang [1,2], Lan Chen [1,2], Jin Zhao [5,*] and Kehui Wu [1,2,6,*]

1 Institute of Physics, Chinese Academy of Sciences, Beijing 100190, China; zhangping@iphy.ac.cn (P.Z.); sxsheng@iphy.ac.cn (S.S.); chma@iphy.ac.cn (C.M.); bjfeng@iphy.ac.cn (B.F.); pcheng@iphy.ac.cn (P.C.); yiqi.zhang@iphy.ac.cn (Y.Z.); lchen@iphy.ac.cn (L.C.)
2 School of Physical Sciences, University of Chinese Academy of Sciences, Beijing 100190, China
3 Department of Physics, University of Science and Technology of China, Hefei 230026, China; txr@mail.ustc.edu.cn
4 Department of Chemical Physics, School of Chemistry, University of Science and Technology of China, Hefei 230026, China; ljchen9@mail.ustc.edu.cn
5 CAS Key Laboratory of Strongly-Coupled Quantum Matter Physics, Department of Physics, University of Science and Technology of China, Hefei 230026, China
6 Songshan Lake Materials Laboratory, Dongguan 523808, China
* Correspondence: zhaojin@ustc.edu.cn (J.Z.); khwu@iphy.ac.cn (K.W.)
† These authors contributed equally to this work.

Abstract: We report a Raman characterization of the α borophene polymorph by scanning tunneling microscopy combined with tip-enhanced Raman spectroscopy. A series of Raman peaks were discovered, which can be well related with the phonon modes calculated based on an asymmetric buckled α structure. The unusual enhancement of high-frequency Raman peaks in TERS spectra of α borophene is found and associated with its unique buckling when landed on the Ag(111) surface. Our paper demonstrates the advantages of TERS, namely high spatial resolution and selective enhancement rule, in studying the local vibrational properties of materials in nanoscale.

Keywords: borophene; vibrational modes; tip-enhanced Raman spectroscopy; scanning tunneling microscopy; density functional theory calculations

Citation: Zhang, P.; Tian, X.; Sheng, S.; Ma, C.; Chen, L.; Feng, B.; Cheng, P.; Zhang, Y.; Chen, L.; Zhao, J.; et al. Vibrational Property of α-Borophene Determined by Tip-Enhanced Raman Spectroscopy. *Molecules* **2022**, *27*, 834. https://doi.org/10.3390/molecules27030834

Academic Editors: Takahiro Kondo, Iwao Matsuda and Josep M. Oliva-Enrich

Received: 26 December 2021
Accepted: 11 January 2022
Published: 27 January 2022

Publisher's Note: MDPI stays neutral with regard to jurisdictional claims in published maps and institutional affiliations.

1. Introduction

Borophene is an emerging two-dimensional (2D) material with novel properties, such as structural anisotropy [1], high thermal conductivity [2], metallicity [3,4], possible superconductivity [5] and polymorphism [4,6–8]. The unique polymorphism of borophene stems from the tremendous possible arrangements of hexagonal holes (HHs) in a planar triangular lattice. However, among the huge number of 2D borophene polymorphs that have been designed previously [7,9], only a few of them have been realized experimentally [10]. Using molecule beam epitaxial (MBE) to deposit boron on various metal substrates, different borophene polymorphs have been found on Ag(111) [4], Ag(100) [8], Cu(111) [11,12], Ag(110) [13], Au(111) [14], Al(111) [15] and Ir(111) [16,17]. Among them, the most studied phases are the β_{12} and χ_3 on Ag(111), and both of them can form large-area single phases with appropriate growth conditions [4]. Their structures and properties have been well established with different methods, such as *in-situ* Raman [18], angle-resolved photoemission spectroscopy (ARPES) [19,20] and high-resolution electron energy loss spectroscopy (HREELS) [21].

Among various 2D borophene polymorphs, the α phase with three-fold symmetry and 1/9 HH density is particularly interesting, as it was predicted to be one of the most stable and fundamental borophene structures [9,10]. Previously, Zhong et al. reported the observation of small α-phase domains coexisting with β_{12} and χ_3 phases in Ag(111) substrate [6]. Recently, Liu et al. reported the observation of bi-layer borophene on Ag(111), which was assigned to two covalently bonded α-layers [22]. In both cases, the α phase only exists in small islands of typically nanometer size, and a complete α-layer is still not available. This poses a great challenge to the understanding of the physical properties of α borophene, as even microscopic characterization techniques usually require samples of micrometer size. Thus, the properties of α borophene remain elusive so far.

In this paper, the vibrational properties of the α borophene were studied by combining scanning tunneling microscopy (STM) with tip-enhanced Raman spectroscopy (TERS). TERS allows one to detect the local vibrational properties with high spatial resolution (<0.5 nm) by the help of the strong localized electric field under the probe tip [23]. We obtained dramatically different Raman spectra from α borophene, as compared with those from other phases in the previous report [18]. DFT calculations reproduce the vibrational modes observed by the Raman spectra well, based on a buckled α-phase model on Ag(111). Our results provide a fundamental data set for further studies of borophene and demonstrate the capability of TERS in the study of local properties of 2D materials

2. Method

All STM and TERS measurements were performed at 77 K using a home-made STM-TERS system (located in Institute of Physics, CAS, Beijing, China), the base pressure being 10^{-8} Pa. The single crystalline Ag(111) surface was cleaned by standard cycles of Ar^+ ion sputtering and annealing at 800 K. Pure boron was evaporated from an e-beam evaporator to the Ag(111) substrate held at 570 K during deposition [4]. The TERS measurement was performed with side illumination and backscattering collection configuration [24]. A 532 nm laser was focused at the tunneling gap using aspheric lens attached to the side of the STM head in the ultrahigh vacuum chamber. The scattered Raman signals were dispersed by 1200 grooves/mm grating and collected by a liquid-nitrogen-cooled charge coupled device (CCD) (SP2300i, Princeton Instrument, Trenton, NJ, USA).

The first-principle calculations were performed within the framework of projector-augmented wave (PAW) method [25], as implemented in Vienna Ab-initio Simulation Package (VASP) [26,27]. The electronic exchange–correlation interaction was described by Perdew–Burke–Ernzerhof (PBE) functional [28], and the van der Waals (vdW) correction was included using DFT-D3 method with Becke-Jonson damping [29]. We set a 500 eV plane-wave cutoff and adopted a $12 \times 12 \times 1$ k-grid to sample the first Brillouin zone of the unit cell. All the atomic structures of borophene were fully relaxed on a two-layer Ag(111) surface until the changes in energy and force between each iteration step were respectively smaller than 10^{-8} eV and 0.001 eV/Å. To avoid the interlayer interaction, a 30Å vacuum interval was set up. With regard to phonon calculations, we employed the frozen-phonon method with $4 \times 4 \times 1$ supercell and $3 \times 3 \times 1$ k-grid. The phonon dispersion was obtained based on the frozen-phonon results with the help of Phonopy [30]. Finally, all the models were shown using VESTA [31].

3. Results and Discussion

After the deposition of about 0.8 ML boron atoms on the Ag(111) substrate, the Ag(111) surface was covered mainly with the β_{12} borophene islands, which exhibited parallel Moiré stripes in parallel with the high-symmetry orientations of the Ag(111) substrate. Meanwhile, small domains with hexagonal Moiré patterns are frequently found to coexist with the β_{12} phase, an example of which is shown in Figure 1a. The high resolution STM image in Figure 1b shows the hexagonal structure of this phase, with the lattice constant $a = 0.52 \pm 0.01$ nm. This structure is consistent with the previously reported α borophene on Ag(111) [6].

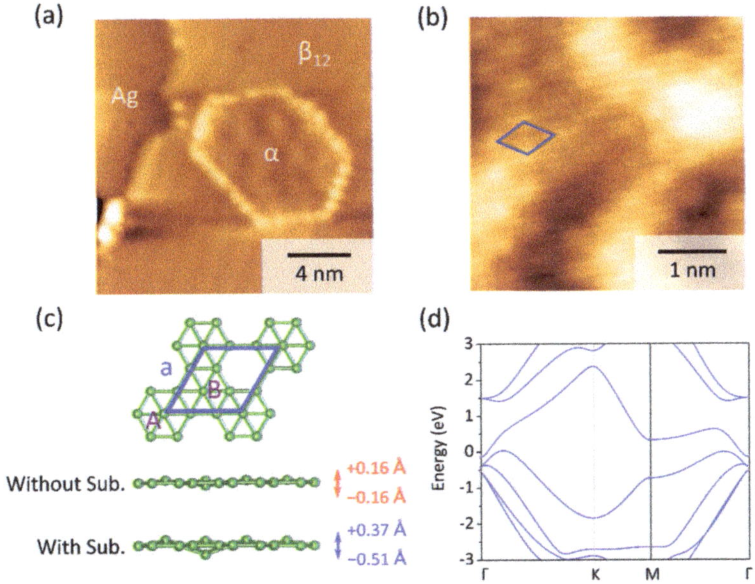

Figure 1. (**a**) STM image (1.3 V, 190 pA) of monolayer borophene sheet grown on Ag(111), showing an α domain between β_{12} islands. (**b**) High-resolution STM image (−0.8 V, 190 pA) of the surface of α borophene; the unit cell is marked by a blue rhombus. (**c**) The model of planar α phase, shown together with the symmetric buckled α phase (relaxed without Ag(111) substrate) and asymmetric buckled α phase (relaxed on Ag(111) substrate). (**d**) The simulated electronic band structure of asymmetric buckled α phase.

To understand the structure and properties of the α phase, it is worth noting that a completely flat α structure is unstable because of a large negative phonon frequency [9]. Instead, a symmetric and slightly buckled α phase with a vertical distance from the plane of about ±0.16 Å is found to be stable in the freestanding form [9]. The upward buckled boron atom in this model is marked by A, while the downward one is marked by B (Figure 1c). Furthermore, after relaxing the symmetric buckled α phase on the Ag(111) substrate, we found that the vertical distance of two buckled boron atoms from the plane further increased to 0.36 Å and −0.51 Å, exhibiting an asymmetric buckled structure. The electronic band structure of this asymmetric buckled phase is shown in Figure 1d. Its metallic properties are also consistent with previous STS result [6]. The asymmetric buckling is found to be critical for the vibrational properties of α borophene on Ag(111), as will be shown and discussed below.

TERS measurement was performed to obtain the vibrational information from the α borophene. As Figure 2a shows, when the STM tip is far from the surface, the far-field Raman signal is very weak due to the small Raman scattering cross-section of borophene [18]. When the STM probe tip is brought close to the surface of α borophene, a dramatic enhancement of Raman signal is observed, exhibiting a strong increment with the decrease in gap distance. The near-field TERS spectra clearly show a series of characteristic peaks, as illustrated in the background subtracted spectrum (the red curve in Figure 2b). Five strong peaks are found, located at 116.8 cm^{-1}, 157.3 cm^{-1}, 339.0 cm^{-1}, 702.6 cm^{-1} and 920.4 cm^{-1}, together with three weak peaks at 406.4 cm^{-1}, 446.8 cm^{-1} and 1230.0 cm^{-1}. For comparison, the TERS spectrum of α borophene is quite different from that of β_{12} phase (blue curve in Figure 2b), as well as from that of χ_3 phase reported in our previous study [18]. In particular, we observe significant enhancements of high-frequency peaks over 500 cm^{-1}, in contrast with the cases of β_{12} and χ_3 phases, where only the low-frequency peaks are enhanced [18]. In view of the selective enhancement mechanism of TERS [18,23], only vibration modes that contain out-of-plane components can be enhanced effectively. For completely flat 2D borophene phases, such as β_{12} and χ_3, their high-frequency vibrational modes contain only in-plane components, and thus cannot be enhanced in TERS [18]. Therefore, we speculate that the obvious enhancement of the high-frequency Raman modes in α borophene implies that these vibrational modes contain out-of-plane components. This perfectly agrees with the fact that the α borophene is significantly buckled on Ag(111), according to DFT calculations.

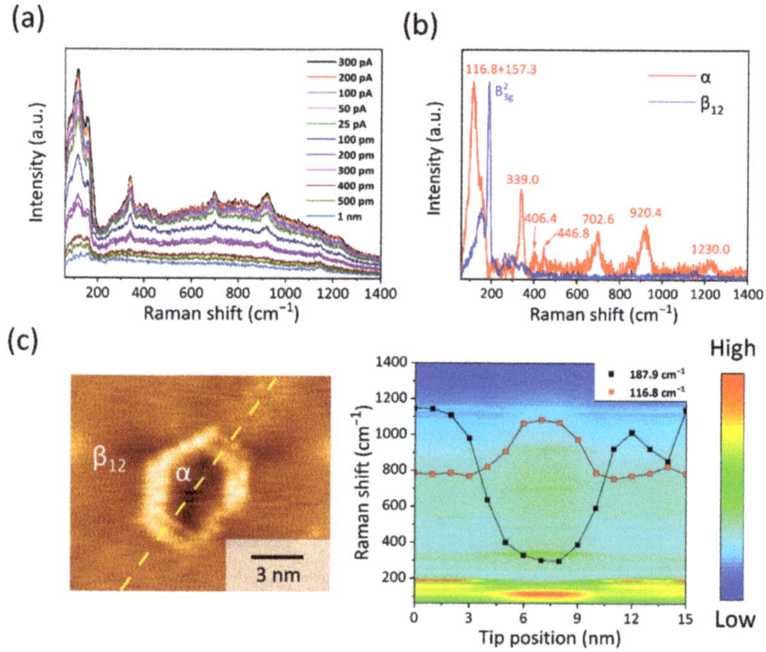

Figure 2. (**a**) Gap distance-dependent TERS spectra of α borophene (10 mW, 0.3 V, the accumulation time for each spectrum is 50 s). The tip-sample distance was controlled by first decreasing the tunneling current from 300 pA to 25 pA, and then the tip was retracted from the surface in 100 pm steps with the feedback loop off. (**b**) Comparison of the TERS spectra of α and β_{12} phases after background subtraction and normalization. (**c**) TERS spectra were taken along the yellow line, crossing the α borophene domain. The Raman intensity map was plotted in the right panel, where two dotted lines are the TERS intensity profiles of the two characteristic peaks of α borophene (red) and β_{12} borophene (black), respectively.

We also emphasize that our TERS measurement renders Raman spectrum with extremely high spatial resolution. A series of TERS spectra were taken along the yellow line in the STM image shown in Figure 2c, crossing β_{12}-α-β_{12} regions as the α domain is surrounded by β_{12} domains. One can see that when the STM tip moves from β_{12} to α phase, the intensity of the characteristic $B_{3g}{}^2$ peak from the β_{12} phase drop immediately, accompanied by the appearance of the characteristic 116.8 cm^{-1} peak from α borophene (the right panel of Figure 2c). Therefore, the high spatial resolution of TERS allows us to well separate the Raman signal of α borophene from that of surrounding β_{12} phase, even though the size of the α borophene domain is only a few nanometers.

To account for these TERS peaks, we performed DFT calculations. The phonon spectra of both symmetric buckled α phase and asymmetric buckled α phase were simulated by VASP, respectively. The phonon spectrum of symmetric buckled α phase was found to largely deviate from our experimental TERS spectra. The phonon modes are completely absent in the vicinity of 116.8 cm^{-1}, 157.3 cm^{-1}, 339.0 cm^{-1}, 406.4 cm^{-1}, 1230.0 cm^{-1} at the Γ point. In contrast, after relaxing the structure to the asymmetric buckled α phase, its symmetry changes from D_{3d} to C_{3v}, causing the change of phonon spectrum. As shown in Figure 3a, the phonon spectrum of asymmetric buckled α phase shows no obvious negative phonon frequencies, indicating a stable structure. Importantly, most peaks in experimental TERS spectra can be assigned to phonon modes at the Γ point, as shown in Figure 3b. A detailed comparison of experimental and simulated peaks is shown in Table 1. The five low-frequency peaks located at 116.8 cm^{-1}, 157.3 cm^{-1}, 339.0 cm^{-1}, 406.4 cm^{-1}, 446.8 cm^{-1} peak can be associated with E^7, E^6, A_1^4, A_1^3, A_2^3 and phonon modes, respectively, within a reasonable error range. The atomic displacements of these peaks, as shown in Figure 3c, are composed of nearly pure out-of-plane vibrational components, which accords with the selective enhancement rule in TERS. For the two high-frequency peaks located at 702.6 cm^{-1} and 920.4 cm^{-1}, we can assign them to two phonon modes E^4 and E^3. The schematics of atomic displacements show that these two modes are composed of nearly in-plane vibrational components; however, the out-of-plane vibrational components still exist because of the two buckled boron atoms. Therefore, the enhancement of these two peaks in TERS accords with our model.

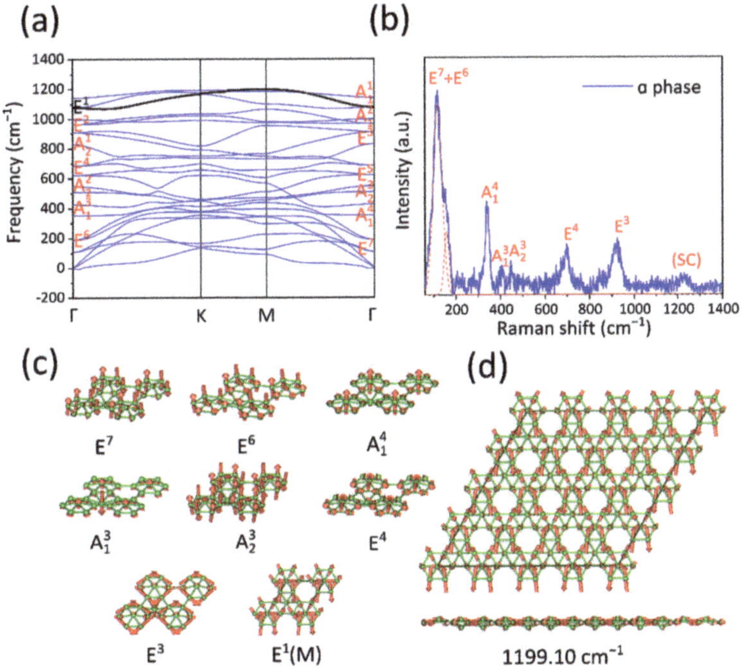

Figure 3. (**a**) The calculated phonon dispersion curves of asymmetric buckled α phase. (**b**) The near-field TERS signal of α phase, E^7 and E^6 peaks are fitted by red dashed curve. (**c**) Vibration modes of unit cell associated with TERS peaks. (**d**) The vibration mode of supercell associated with the 1230 cm^{-1} peak in TERS.

Table 1. TERS modes of the α phase, as compared with the calculation, cm^{-1}.

TERS	Simulation	Modes
116.8	107.85	E^7
157.3	188.6	E^6
339.0	352.94	A_1^4
406.4	425.2	A_1^3
446.8	507.83	A_2^3
702.6	684.32	E^4
920.4	907.16	E^3
1230.0	1199.10	Vibration of SC

Finally, the TERS peak located at 1230 cm^{-1} is about 90 cm^{-1} above the highest phonon mode at Γ in the phonon spectrum. To account for this peak, we attribute it to a vibration mode of a supercell. Here, a 4 × 4 × 1 supercell is considered because the periodicity of Moiré pattern of the α phase is about four times that of unit cell. Due to the Brillouin zone folding, the M point of the unit cell will be folded to the Γ point in the new Brillouin zone of the 4 × 4 × 1 supercell. The E^1 mode at the M point contributes to a supercell vibrational mode with a frequency of 1199.10 cm^{-1} (Figure 3d), which matches our measurement better than any other modes contained in such a supercell. In addition to the frequency consistency, this mode also contains the out-of-plane vibration component. Therefore, it very likely corresponds to the 1230 cm^{-1} TERS peak.

4. Conclusions

In conclusion, we determine the characteristic Raman spectrum of α borophene with the help of high spatial resolution of TERS, combining with DFT calculations. All Raman peaks can be well associated with the phonon modes calculated based on an asymmetric buckled α structure. The unusual enhancement of high-frequency Raman peaks in TERS spectra of α borophene is also related to its unique buckling when landed on the Ag(111) surface. Our work provides not only the basic Raman characterization of the highly interesting α borophene, but also demonstrates the high prospect of TERS in studying local vibrational properties of nanoscale structures.

Author Contributions: P.Z., X.T. and S.S. have contributed in an equal manner in manuscript writing. Experiment performing, P.Z., S.S. and B.F.; DFT simulation, X.T. and L.C. (Linjie Chen); data analyzing, P.Z., C.M., P.C., Y.Z. and L.C. (Lan Chen); editing and interpretation, J.Z. and K.W. All authors have read and agreed to the published version of the manuscript.

Funding: This research was funded by the National Key R&D Program of China (grant number 2021YFA1400500), National Natural Science Foundation of China (grands number 11825405, 12134019, 11974322, 12125408), and the Strategic Priority Research Program of the Chinese Academy of Sciences (grand number XDB30000000).

Informed Consent Statement: Not applicable.

Data Availability Statement: Data available via personal communication with proper reasons.

Conflicts of Interest: The authors declare no conflict of interest.

References

1. Feng, B.; Zhang, J.; Liu, R.-Y.; Iimori, T.; Lian, C.; Li, H.; Chen, L.; Wu, K.; Meng, S.; Komori, F.; et al. Direct evidence of metallic bands in a monolayer boron sheet. *Phys. Rev. B* **2016**, *94*, 041408. [CrossRef]
2. Xiao, H.; Cao, W.; Ouyang, T.; Guo, S.; He, C.; Zhong, J. Lattice thermal conductivity of borophene from first principle calculation. *Sci. Rep.* **2017**, *7*, 45986. [CrossRef] [PubMed]
3. Yang, X.; Ding, Y.; Ni, J. Ab initio prediction of stable boron sheets and boron nanotubes: Structure, stability, and electronic properties. *Phys. Rev. B* **2008**, *77*, 041402. [CrossRef]
4. Feng, B.; Zhang, J.; Zhong, Q.; Li, W.; Li, S.; Li, H.; Cheng, P.; Meng, S.; Chen, L.; Wu, K. Experimental realization of two-dimensional boron sheets. *Nat. Chem.* **2016**, *8*, 563–568. [CrossRef] [PubMed]

5. Penev, E.S.; Kutana, A.; Yakobson, B.I. Can two-dimensional boron superconduct? *Nano Lett.* **2016**, *16*, 2522–2526. [CrossRef]
6. Zhong, Q.; Zhang, J.; Cheng, P.; Feng, B.; Li, W.; Sheng, S.; Li, H.; Meng, S.; Chen, L.; Wu, K. Metastable phases of 2D boron sheets on Ag(111). *J. Phys. Condens. Matter* **2017**, *29*, 095002. [CrossRef]
7. Tang, H.; Ismail-Beigi, S. Novel precursors for boron nanotubes: The competition of two-center and three-center bonding in boron sheets. *Phys. Rev. Lett.* **2007**, *99*, 115501. [CrossRef]
8. Wang, Y.; Kong, L.; Chen, C.; Cheng, P.; Feng, B.; Wu, K.; Chen, L. Realization of regular-mixed quasi-1D borophene chains with long-range order. *Adv. Mater.* **2020**, *32*, e2005128. [CrossRef]
9. Wu, X.; Dai, J.; Zhao, Y.; Zhuo, Z.; Yang, J.; Zeng, X.C. Two-Dimensional boron monolayer sheets. *ACS Nano* **2012**, *6*, 7443–7453. [CrossRef]
10. Li, D.; Gao, J.; Cheng, P.; He, J.; Yin, Y.; Hu, Y.; Chen, L.; Cheng, Y.; Zhao, J. 2D Boron sheets: Structure, growth, and electronic and thermal transport properties. *Adv. Funct. Mater.* **2019**, *30*, 1904349. [CrossRef]
11. Chen, C.; Lv, H.; Zhang, P.; Zhuo, Z.; Wang, Y.; Ma, C.; Li, W.; Wang, X.; Feng, B.; Cheng, P.; et al. Synthesis of bilayer borophene. *Nat. Chem.* **2022**, *14*, 25–31. [CrossRef]
12. Wu, R.; Drozdov, I.K.; Eltinge, S.; Zahl, P.; Ismail-Beigi, S.; Bozovic, I.; Gozar, A. Large-Area single-crystal sheets of borophene on Cu(111) surfaces. *Nat. Nanotechnol.* **2019**, *14*, 44–49. [CrossRef] [PubMed]
13. Zhong, Q.; Kong, L.; Gou, J.; Li, W.; Sheng, S.; Yang, S.; Cheng, P.; Li, H.; Wu, K.; Chen, L. Synthesis of borophene nanoribbons on Ag(110) surface. *Phys. Rev. Mater.* **2017**, *1*, 021001. [CrossRef]
14. Kiraly, B.; Liu, X.; Wang, L.; Zhang, Z.; Mannix, A.J.; Fisher, B.L.; Yakobson, B.I.; Hersam, M.C.; Guisinger, N.P. Borophene synthesis on Au(111). *ACS Nano* **2019**, *13*, 3816–3822. [CrossRef]
15. Li, W.; Kong, L.; Chen, C.; Gou, J.; Sheng, S.; Zhang, W.; Li, H.; Chen, L.; Cheng, P.; Wu, K. Experimental realization of honeycomb borophene. *Sci. Bull.* **2018**, *63*, 282–286. [CrossRef]
16. Vinogradov, N.A.; Lyalin, A.; Taketsugu, T.; Vinogradov, A.S.; Preobrajenski, A. Single-Phase Borophene on Ir(111): Formation, Structure, and Decoupling from the Support. *ACS Nano* **2019**, *13*, 14511–14518. [CrossRef]
17. Omambac, K.M.; Petrovic, M.; Bampoulis, P.; Brand, C.; Kriegel, M.A.; Dreher, P.; Janoschka, D.; Hagemann, U.; Hartmann, N.; Valerius, P.; et al. Segregation-Enhanced epitaxy of borophene on Ir(111) by thermal decomposition of borazine. *ACS Nano* **2021**, *15*, 7421–7429. [CrossRef]
18. Sheng, S.; Wu, J.B.; Cong, X.; Zhong, Q.; Li, W.; Hu, W.; Gou, J.; Cheng, P.; Tan, P.H.; Chen, L.; et al. Raman spectroscopy of two-dimensional borophene sheets. *ACS Nano* **2019**, *13*, 4133–4139. [CrossRef]
19. Feng, B.; Sugino, O.; Liu, R.Y.; Zhang, J.; Yukawa, R.; Kawamura, M.; Iimori, T.; Kim, H.; Hasegawa, Y.; Li, H.; et al. Dirac fermions in borophene. *Phys. Rev. Lett.* **2017**, *118*, 096401. [CrossRef]
20. Feng, B.; Zhang, J.; Ito, S.; Arita, M.; Cheng, C.; Chen, L.; Wu, K.; Komori, F.; Sugino, O.; Miyamoto, K.; et al. Discovery of 2D anisotropic dirac cones. *Adv. Mater.* **2018**, *30*, 1704025. [CrossRef]
21. Aizawa, T.; Suehara, S.; Otani, S. Phonon dispersion of a two-dimensional boron sheet on Ag(111). *Phys. Rev. Mater.* **2021**, *5*, 064004. [CrossRef]
22. Liu, X.; Li, Q.; Ruan, Q.; Rahn, M.S.; Yakobson, B.I.; Hersam, M.C. Borophene synthesis beyond the single-atomic-layer limit. *Nat. Mater.* **2022**, *21*, 35–40. [CrossRef]
23. Sheng, S.; Wu, J.B.; Cong, X.; Li, W.; Gou, J.; Zhong, Q.; Cheng, P.; Tan, P.H.; Chen, L.; Wu, K. Vibrational properties of a monolayer silicene sheet studied by tip-enhanced raman spectroscopy. *Phys. Rev. Lett.* **2017**, *119*, 196803. [CrossRef]
24. Sheng, S.; Li, W.; Gou, J.; Cheng, P.; Chen, L.; Wu, K. Low-Temperature, ultrahigh-vacuum tip-enhanced Raman spectroscopy combined with molecular beam epitaxy for in situ two-dimensional materials' studies. *Rev. Sci. Instrum.* **2018**, *89*, 053107. [CrossRef]
25. Blochl, P.E. Projector augmented-wave method. *Phys. Rev. B* **1994**, *50*, 17953–17979. [CrossRef] [PubMed]
26. Kresse, G.; Hafner, J. Ab initio molecular dynamics for open-shell transition metals. *Phys. Rev. B* **1993**, *48*, 13115–13118. [CrossRef] [PubMed]
27. Kresse, G.; Furthmüller, J. Efficiency of ab-initio total energy calculations for metals and semiconductors using a plane-wave basis set. *Comput. Mater. Sci.* **1996**, *6*, 15–50. [CrossRef]
28. Perdew, J.P.; Burke, K.; Ernzerhof, M. Generalized gradient approximation made simple. *Phys. Rev. Lett.* **1996**, *77*, 3865–3868. [CrossRef] [PubMed]
29. Grimme, S.; Antony, J.; Ehrlich, S.; Krieg, H. A consistent and accurate ab initio parametrization of density functional dispersion correction (DFT-D) for the 94 elements H-Pu. *J. Chem. Phys.* **2010**, *132*, 154104. [CrossRef]
30. Togo, A.; Tanaka, I. First principles phonon calculations in materials science. *Scr. Mater.* **2015**, *108*, 1–5. [CrossRef]
31. Momma, K.; Izumi, F. VESTA: A three-dimensional visualization system for electronic and structural analysis. *J. Appl. Crystallogr.* **2008**, *41*, 653–658. [CrossRef]

molecules MDPI

Article

Fabrication of Hydrogen Boride Thin Film by Ion Exchange in MgB_2

T. Hirabayashi [1,2,†], S. Yasuhara [1,†], S. Shoji [1,3], A. Yamaguchi [1], H. Abe [4], S. Ueda [5,6,7], H. Zhu [2], T. Kondo [8,9,*] and M. Miyauchi [1,*]

1 Department of Materials Science and Engineering, School of Materials and Chemical Technology, Tokyo Institute of Technology, Tokyo 152-8552, Japan; hirabayashi.t.ac@m.titech.ac.jp (T.H.); yasuhara.s.aa@m.titech.ac.jp (S.Y.); shoji.s.chem@gmail.com (S.S.); ayamaguchi@ceram.titech.ac.jp (A.Y.)
2 Department of Mechanical Engineering, School of Materials Science and Engineering, Tsinghua University, Beijing 100084, China; hongweizhu@tsinghua.edu.cn
3 Department of Materials Science & Engineering, Cornell University, Ithaca, NY 14853, USA
4 Center for Green Research on Energy and Environmental Materials, National Institute for Materials Science, Tsukuba 305-0044, Japan; ABE.Hideki@nims.go.jp
5 Synchrotron X-ray Station at SPring-8, National Institute for Materials Science, Hyogo 679-5148, Japan; UEDA.Shigenori@nims.go.jp
6 Research Center for Advanced Measurement and Characterization, National Institute for Materials Science, Tsukuba 305-0047, Japan
7 Research Center for Functional Materials, National Institute for Materials Science, Tsukuba 305-0044, Japan
8 Department of Materials Science and Tsukuba Research Center for Energy Materials Science, Faculty of Pure and Applied Sciences, University of Tsukuba, Tsukuba 305-8573, Japan
9 Materials Research Center for Element Strategy, Tokyo Institute of Technology, Yokohama 226-8503, Japan
* Correspondence: takahiro@ims.tsukuba.ac.jp (T.K.); mmiyauchi@ceram.titech.ac.jp (M.M.)
† The 1st and 2nd authors equally contributed to this work.

Citation: Hirabayashi, T.; Yasuhara, S.; Shoji, S.; Yamaguchi, A.; Abe, H.; Ueda, S.; Zhu, H.; Kondo, T.; Miyauchi, M. Fabrication of Hydrogen Boride Thin Film by Ion Exchange in MgB_2. *Molecules* **2021**, *26*, 6212. https://doi.org/10.3390/molecules26206212

Academic Editor: Mohammed Baalousha

Received: 11 September 2021
Accepted: 9 October 2021
Published: 14 October 2021

Abstract: In this study, hydrogen boride films are fabricated by ion-exchange treatment on magnesium diboride (MgB_2) films under ambient temperature and pressure. We prepared oriented MgB_2 films on strontium titanate ($SrTiO_3$) substrates using pulsed laser deposition (PLD). Subsequently, these films were treated with ion exchangers in acetonitrile solution. TOF-SIMS analysis evidenced that hydrogen species were introduced into the MgB_2 films by using two types of ion exchangers: proton exchange resin and formic acid. According to the HAXPES analysis, negatively charged boron species were preserved in the films after the ion-exchange treatment. In addition, the FT-IR analysis suggested that B-H bonds were formed in the MgB_2 films following the ion-exchange treatment. The ion-exchange treatment using formic acid was more efficient compared to the resin treatment; with respect to the amount of hydrogen species introduced into the MgB_2 films. These ion-exchanged films exhibited photoinduced hydrogen release as observed in a powder sample. Based on the present study, we expect to be able to control the morphology and hydrogen content of hydrogen boride thin films by optimising the ion-exchange treatment process, which will be useful for further studies and device applications.

Keywords: hydrogen boride; thin film; ion-exchange; hydrogen release; pulsed laser deposition

1. Introduction

In contrast to their bulk counterparts, two-dimensional materials, such as graphene, exhibit unique electronic properties and ultra-high specific surface areas. These unusual characteristics result in energy band structures sensitive to external perturbations and attract significant research interest [1]. Recently, Kondo et al. successfully synthesised hydrogen boride (HB) nanosheets [2]. The synthesis method involves the 3-day reaction of layered bulk magnesium diboride (MgB_2) powder with an ion-exchange resin in an organic solvent to promote the exfoliation of MgB_2. This is the first experimentally synthesised borophane (hydrogenated borophene), which is composed of sp^2-bonded boron, forming hexagonal boron networks with bridge hydrogens. The chemical composition of the material is HB, where the hydrogen/boron atomic ratio is 1/1. Previous experiments have shown that hydrogen boride sheets have a high H_2 content (8.5 wt %) [2], unique electronic properties [3,4], solid acid catalytic properties [5,6], photoinduced hydrogen release capability [7], and a unique ability to reduce metal ions to form nanocomposites [8–10]. In addition, theoretical models and computer simulations demonstrate that borophane is a promising candidate for applications such as anode material, alcohol gas sensors, current limiters, and photodetectors [11–14].

At present, studies on the synthesis of hydrogen boride have been carried out only for a powder sample obtained from finely dispersed MgB_2 [2–10]. However, hydrogen boride thin films deposited on substrates would offer better opportunities for the in-depth analysis of the physical properties and investigation of possible applications of this material. Thus far, there has been no research on well-defined thin films of hydrogen boride. The epitaxial growth of hydrides is quite a challenging objective, owing to their thermal and chemical instability in air and multiple charged states [15–17].

Herein, we have developed a facile method for the fabrication of thin films of hydrogen boride using an ion-exchange treatment of deposited MgB_2 films under ambient temperature and pressure. Firstly, oriented MgB_2 films on $SrTiO_3$ (STO) substrates were prepared by pulsed laser deposition (PLD). Afterwards, the deposited MgB_2 films were treated with an ion-exchanger in an organic solvent to fabricate hydrogen boride films. Thin films wherein B-H bonds were introduced were successfully synthesised by using two types of ion-exchangers: proton exchange resin and formic acid. We also observed significant hydrogen release from these films under ultraviolet (UV) light irradiation.

2. Results and Discussions

To fabricate the single-phase MgB_2 film, a two-step PLD method was used as described in the experimental section. Figure 1 shows the out-of-plane XRD patterns of the MgB_2 film deposited on the STO (100) substrate. As shown in the inset image, a black film was formed on the STO substrate. Only the *10-11* diffraction peak of the MgB_2 film was detected, except for the peaks associated with the substrate, indicating its strong orientation. Such oriented growth of MgB_2 is in agreement with previous reports [18–20]. Although a strong *10-11* peak appeared, our in-plane XRD analysis could not detect the diffraction peaks because the film thickness was insufficient for in-plane diffraction. Consecutively, we were unable to investigate the lattice matching relationship in the plane, and the film was described as a single-oriented thin film. The thickness of the MgB_2 film varied between 50 nm and 270 nm (see Supplementary Information, Figure S1). When the film was prepared by a one-step PLD method (MgB_2 deposition without Mg deposition), a peak of boron-rich magnesium boride (MgB_x) was observed in addition to the MgB_2 peak (Figure S2). The most likely reason for this is the lack of Mg in the one-step PLD method due to the high volatility of Mg. The expected candidates of the produced MgB_x compound are listed in the Supplementary Information (Table S1). When a different crystalline substrate, such as sapphire, was used in a two-step PLD process, no diffraction peaks of MgB_2 were observed (Figure S3). These results indicate that a dense and oriented MgB_2 film can be deposited on the STO (100) substrate by a two-step PLD method.

Next, the oriented MgB_2 film was treated with ion-exchange resin in acetonitrile. MgB_2 is a layered compound in which positive magnesium ions with a valence number close to divalent are intercalated between negatively charged hexagonal boron layers, forming a honeycomb structure [21–23]. In the present study, an ion-exchange resin was ground in an agate mortar to increase the reaction area for efficient ion-exchange in a reaction time equivalent to that of powder sample synthesis. During the ion-exchange treatment, Mg^{2+} ions in MgB_2 are changed with protons. Therefore, we expect that negatively charged boron states (B^-) can remain even after the process, similar to previous powder sample experiments [2]. HAXPES analysis was performed to characterise the valence states of the boron and magnesium species in the film. Figure 2a shows the HAXPES spectra of the

B 1s core-level before and after the ion-exchange treatment. As a result, the majority of the boron sites retained their negatively charged B^- state (peak at ~188.9 eV), even after the treatment. This indicates that oxidation does not occur during the ion-exchange, and impurities, such as boric acid, are not formed (B 1s of boric acid is known to appear at 193.2 eV [2]). As shown in Figure 2b, after the ion-exchange the Mg 2p core-level peak broadens, indicating that the surroundings of Mg in the thin film change significantly. X-ray diffraction data revealed a gradual decrease of the MgB_2 *10-11* peaks after the treatment, suggesting that its crystallinity becomes worse after ion-exchange (see Supplementary Information, Figures S4 and S5).

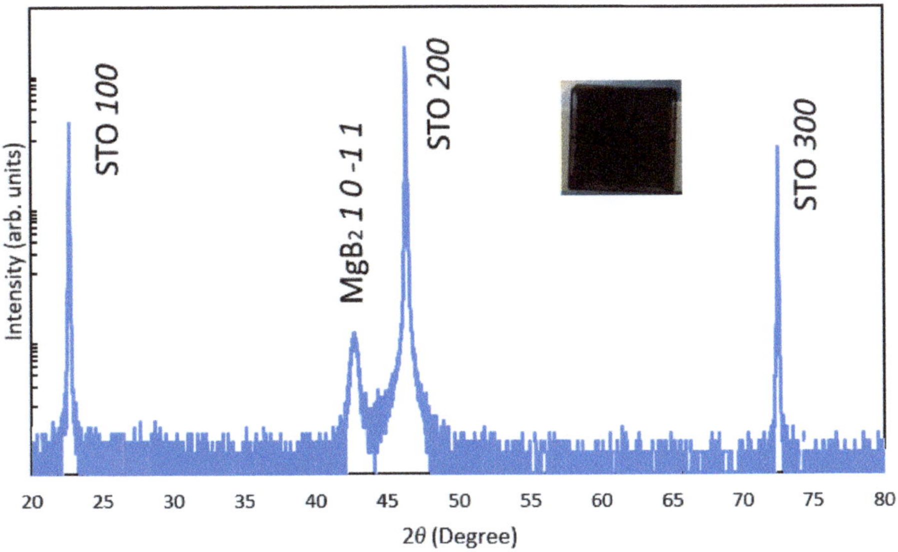

Figure 1. Out-of-plane XRD pattern of prepared MgB_2 film. The inset image shows the appearance of the as-deposited MgB_2 film on the STO substrate.

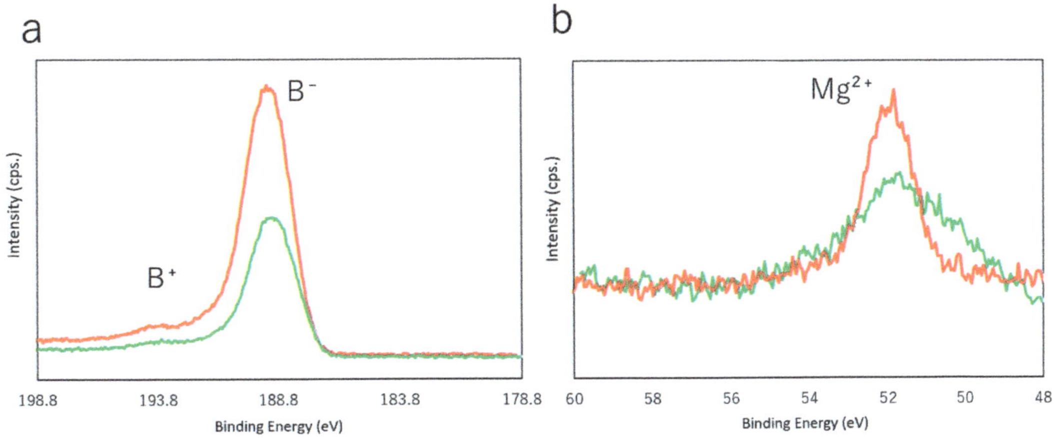

Figure 2. HAXPES spectra of B 1s (**a**) and Mg 2p (**b**) core-levels for the MgB_2 film before (red) and after (green) ion-exchange treatment using resin.

Since photo-emission spectroscopy analysis is known to have low sensitivity to hydrogen species, TOF-SIMS analysis of the thin MgB_2 films was performed. Figure 3 shows the TOF-SIMS depth profile of the ionised hydrogen species (H^+) in the films. The red line demonstrates the H^+ species profile of the MgB_2 film. In this case, the observed H^+ species on the surface originate from adsorbed water and/or hydrocarbon impurities. The ion-exchanged films exhibit a greater amount of hydrogen species than that of the untreated film. It can be assumed that the resin would not be an efficient reagent because of the limited exchange process at the solid-to-solid interface between MgB_2 and the resin. Therefore, as an alternative ion-exchange method, liquid formic acid was used as a proton donor for more efficient ion-exchange. It is worth noting that the film treated with formic acid displayed a higher hydrogen content than the film treated with ion-exchange resin. The etching rate of the TOF-SIMS analysis was approximately 0.35–0.4 nm/s and was determined by the depth of the rastered area and sputtering time (Figure S6). Considering the etching rate, the 200 s sputtering time corresponds to 70–80 nm etching depth. The depth distribution of the hydrogen species varied depending on the method. In the formic acid-treated thin films, hydrogen species were introduced at a depth of less than 100 nm from the surface.

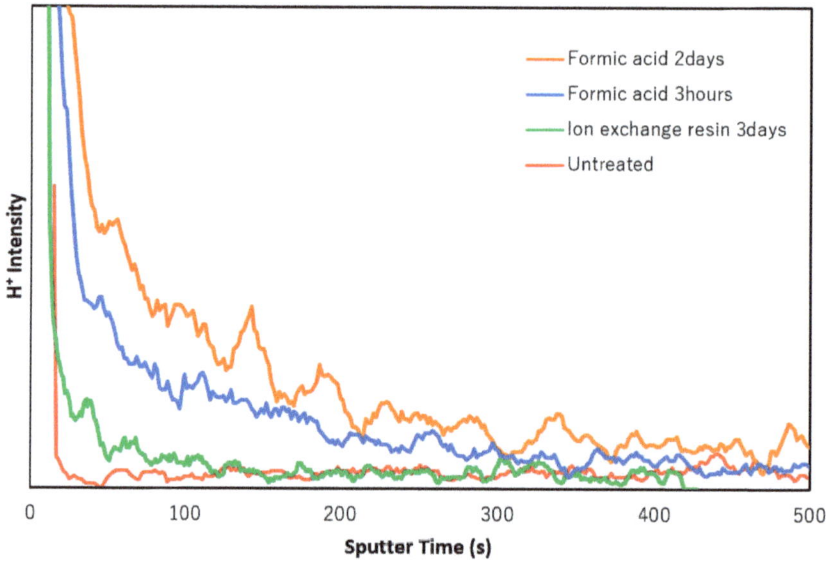

Figure 3. TOF-SIMS analysis of H^+ species distribution in thin films: untreated thin film (red line), thin film treated by ion-exchange resin for 3 days (green line), thin film treated by formic acid for 3 h (blue line), and thin film treated by formic acid for 2 days (orange line).

We also investigated the TOF-SIMS depth profiles of the boron and magnesium species (Figure S7). In the MgB_2 film before the ion-exchange treatment, enrichment in boron species was observed near the surface (Figure S7a, red line) due to the volatilization of Mg in the outermost surface of the MgB_2 thin film during the PLD process, as discussed above. After the ion-exchange process with the resin, both the boron and magnesium species' contents decreased. The decrease in the abundance of the magnesium species is due to leaching via the ion-exchange with protons inhomogeneously, while the lower boron species content can be explained by the exfoliation of boron-rich layers from the thin film. The decrease in the boron species signal was more significant compared to that of the magnesium species. The integrated values of the TOF-SIMS signal intensities from 0 to 20 s (corresponding to the 7–8 nm layer on the top surface) before and after the ion-exchange treatments were calculated and compared. Then, the relative values after the resin ion-exchange with respect to the values before the ion-exchange were 0.51 and 0.71 for the boron species and magnesium species, respectively.

TOF-SIMS analysis of the boron and magnesium species were also performed for thin films treated with formic acid (Figure S8). Similar to the resin treatment, the content of both boron and magnesium species was reduced after the formic acid treatment. In this case, the decrease in the magnesium signal was more significant than that of the boron species. These results indicate that a long-term treatment with formic acid can induce a proton ion-exchange process rather than an exfoliation of the boron layers in the MgB_2 film. We also evaluated the TOF-SIMS signal of the carbon species after the sample was treated with formic acid (Figure S9). The signal intensities of the carbon species were very low, even after the formic acid treatment. As shown in Figure 3, the hydrogen species signal was stronger and detectable at a greater depth below the surface compared to the carbon species signal (Figure S9). These results indicate that the dissociated protons from formic acid were successfully introduced into the MgB_2 film during the ion-exchange process.

Next, FT-IR measurements were conducted in an attempt to confirm the formation of B-H and/or B-H-B bonds in the ion-exchanged film as an additional spectroscopic evidence for the ion exchange of Mg^{2+} by protons. In spite of using the highly sensitive ATR method, no signals characteristic of these bonds were detected, presumably because the film was too thin. Instead, the FT-IR spectrum of MgB_2 powder was investigated (Figure S10), commercially available from Sigma-Aldrich and handled in the same manner as the thin films. Notably, after the ion-exchange treatment in formic acid, a peak appeared around 2500 cm^{-1}. According to a previous report, this signal corresponds to the terminal B-H stretching mode in hydrogen boride sheets [4]. Since we followed an identical ion-exchange process for both the powder sample and the thin film, we assume that a similar B-H bond was formed in the film sample, as the existence of hydrogen species was also confirmed in the film by TOF-SIMS measurements.

We also compared the surface morphologies of the as-deposited and ion-exchanged films by SEM and AFM, as shown in Figure 4. The SEM (Figure 4a) and AFM (Figure 4b) images of the as-deposited MgB_2 thin film (before the ion-exchange) reveal cross-shaped rectangular patterns. This implies that the MgB_2 film has an epitaxial relationship with the STO substrate, although we could not confirm this by in-plane XRD. The peak to valley (PTV) height difference in the AFM image was 50–100 nm (Figure 4b, Figure S11a). In contrast, the film treated with the ion-exchange resin exhibited a smoother surface compared to the as-deposited MgB_2 film, with a PTV height difference of less than 50 nm (Figures 4c,d and S11b). This is likely because of the physical contact (mechanical abrasion) between the MgB_2 thin film and the resin during the ion-exchange. The thickness of the film after the ion-exchange treatment was less than 200 nm (Figure S12). This might be one of the reasons for the lower H content and its narrower depth distribution observed in the TOF-SIMS profile. On the other hand, the film treated with formic acid for 2 days exhibited more pronounced crisscross patterns on the surface and a larger PTV height difference (50–200 nm), as shown in Figures 4e,f and S11c. This is because the ion-exchange and exfoliation were more efficient when using formic acid, as compared to the use of ion-exchange resin. These results reveal that both the hydrogen content and the surface morphology of the thin films can vary significantly depending on the type of ion-exchange treatment. This also suggests that, by choosing an appropriate ion-exchange method and conditions, the morphology and hydrogen content of the films can be controlled for various device applications.

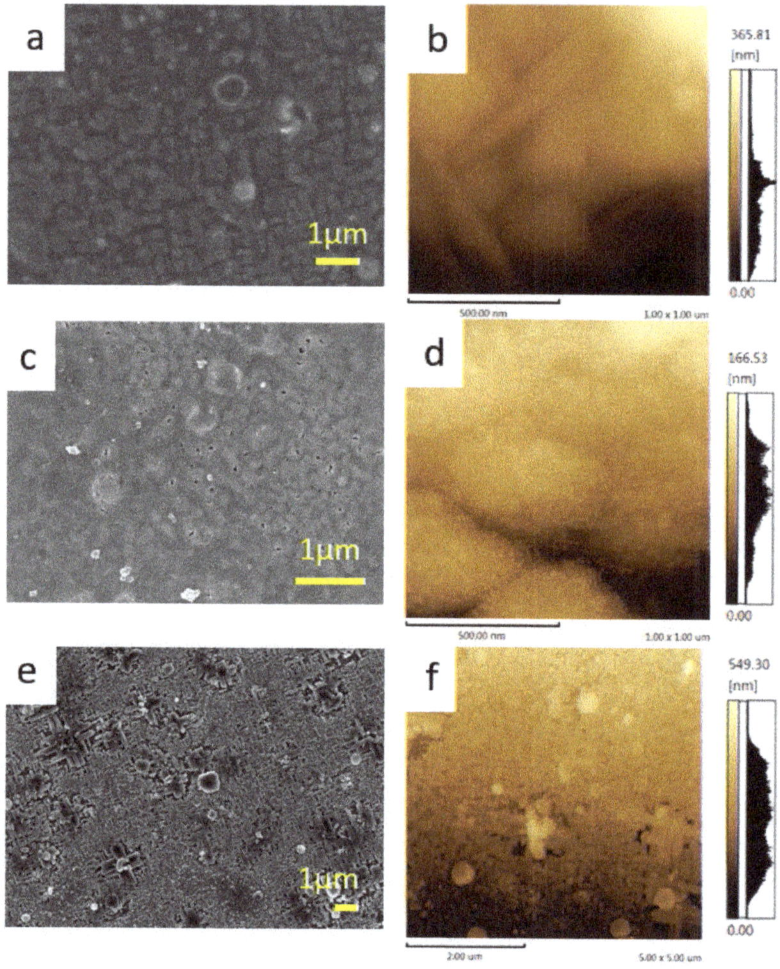

Figure 4. SEM (**a**,**c**,**e**) and AFM (**b**,**d**,**f**) images of thin films. As-deposited MgB_2 film without ion-exchange treatment (**a**,**b**), ion-exchanged film treated with resin for 3 days (**c**,**d**), and ion-exchanged film treated with formic acid for 2 days (**e**,**f**).

Previous experiments on powder systems have demonstrated that UV irradiation of the B-H bonds in hydrogen boride sheets can lead to hydrogen gas release [7]. In this study, the hydrogen release was induced by photoirradiation, which causes the electron transition from the σ-bonding state (valence band) to the anti-bonding state of boron and hydrogen orbitals (conduction band) with a gap of 3.8 eV and causes the self-reduction of protons together with oxidation of boron in HB, even under mild ambient conditions. We anticipated that a similar mechanism might lead to hydrogen emission from the ion-exchanged thin films upon UV irradiation. As shown in Figure 5, hydrogen molecules were detected from both types of ion-exchanged films upon UV light irradiation, while the as-deposited MgB_2 did not release hydrogen gas. The amount of hydrogen released from the formic acid treated film was larger than that released from the resin-treated film. These results agree well with the TOF-SIMS data, which showed a higher hydrogen species content in the film treated with formic acid compared to the other one. That strongly implies that B-H bonds were formed in the ion-exchanged film, as suggested by the FT-

IR analysis, and hydrogen molecules can be liberated by an electron transition from the bonding state to the anti-bonding state of hydrogen and boron orbitals under UV light irradiation [7].

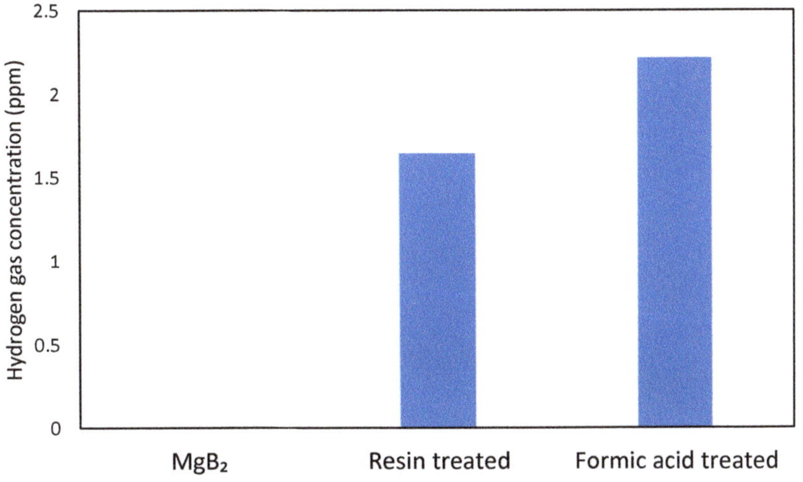

Figure 5. Hydrogen release from as-deposited MgB_2 and ion-exchanged thin films under UV irradiation.

Based on the experimental data, we can propose a reaction scheme for the ion-exchange process in MgB_2 films, as shown in Figure 6. The oriented MgB_2 thin film reacts with ion-exchange resin or formic acid (left panel) causing H^+ to intercalate inside the film and Mg^{2+} ions to be released from the film in the acetonitrile solution (middle panel), similar to a layered metal oxide case [24]. In addition to the ion-exchange, a partial exfoliation of boron layers occurred at the outermost surface, leaving H^+ inside the thin film to form B-H bonds (right panel). In terms of the chemical composition, the ion-exchanged film is depicted as an inhomogeneous hybrid material with a top layer mainly consisting of HB and a bottom layer mainly consisting of MgB_2.

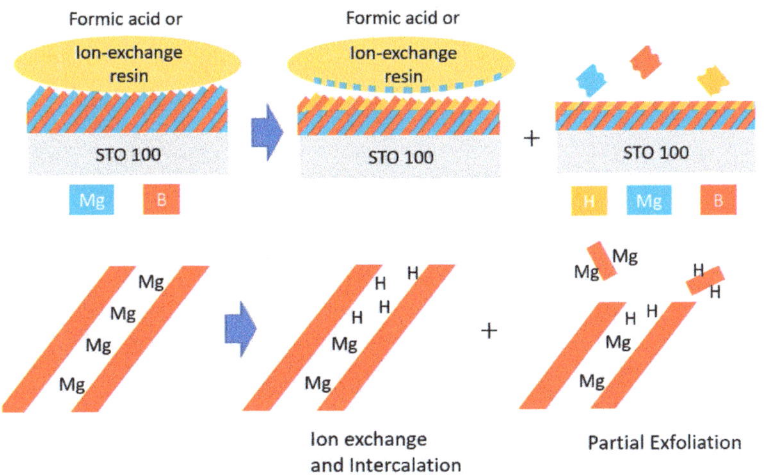

Figure 6. Proposed reaction scheme of the ion-exchange mechanism.

3. Materials and Methods

3.1. Materials and Equipment

$SrTiO_3$ (100) single-crystal substrates were purchased from SHINKOSHA CO., LTD. Mg powder (212–600 μm, 99.9%, Wako, Osaka, Japan) and MgB_2 powder (99%; Sigma-Aldrich, St. Louis, MO, USA) were purchased to use MgB_2 thin film deposition. Acetonitrile (99.5% (JIS Special grade), FUJIFILM Wako Pure Chemical Industries Ltd., Osaka, Japan), a cation-exchange resin (15JS-HG·DRY, Organo Corp., Tokyo, Japan) and formic acid (Kanto Chemical Co., Inc., Tokyo, Japan) were purchased for ion-exchange treatment.

A pulsed laser ablation device (PLFD-221-1R, Freedom Ltd., Kawasaki, Japan) was used for PLD. X-ray diffractometer (Smartlab, Rigaku Corporation, Tokyo, Japan), laser microscope (LEXT OLS5100, Olympus Corporation, Tokyo, Japan), TOF-SIMS 5-100-AD (ION-TOF GmbH, Germany), FT/IR-6100 (JASCO, Co., Ltd., Tokyo, Japan), atomic force microscope (SPM-9700, Shimadzu Corp., Kyoto, Japan) and scanning electron microscope (JEM-2010F, JEOL, Ltd., Tokyo, Japan) were used for characterization of prepared films.

3.2. Preparation of MgB_2 Thin Films

The MgB_2 thin films were deposited on $SrTiO_3$ (100) single-crystal substrates by PLD using the fourth harmonic wavelength (266 nm) of a neodymium: yttrium aluminium garnet (Nd: YAG) laser. The substrate size was 1.0×1.0 cm^2. The thin film deposition was performed under argon partial pressure of 1.0 mTorr and substrate temperature of 700 °C. A high-purity single-phase MgB_2 film was deposited via a two-step deposition using subsequently two targets: a metallic magnesium (Mg) target pellet for 10 min and an MgB_2 pellet for 60 min for the same substrate. The former process was necessary to compensate for the Mg supply in the MgB_2 film because of the high volatility of Mg. The Mg pellet was prepared by uniaxial pressing using Mg powder, while the MgB_2 pellet was prepared using commercial powder.

3.3. Ion-Exchange Treatment of MgB_2 Thin Films

Two different types of ion-exchange method were employed to treat the deposited MgB_2 thin films: the first method involved a 3-day reaction with a cation-exchange resin (400 mg) that was ground in an agate mortar in acetonitrile while referencing the previous report [2]. As an alternative ion-exchange method, a solution of 3.6 mL of acetonitrile and 0.4 mL of formic acid as a proton donor was prepared according to the report by Kawamura et al. [25], and MgB_2 thin films were immersed into the solution for either 3 h or 2 d. Both treatments were conducted at room temperature and atmospheric pressure in a glovebox (<1 ppm O_2) to prevent air exposure. The ion-exchanged film samples were washed with acetonitrile and dried under a nitrogen atmosphere.

3.4. Characterization of Prepared Films

X-ray diffraction (XRD) measurements were performed using a Rigaku Smartlab X-ray diffractometer (45 kV, 200 mA, radiation source: Cu($K\alpha_1$)). The film thickness was measured using a laser microscope. We evaluated the height difference between the MgB_2 thin film and the bare STO substrate near the edge of the thin film.

Hard X-ray photoemission spectroscopy (HAXPES) measurements were performed at BL15XU of SPring-8 (Super Photon Ring 8 GeV, Hyōgo Prefecture, Japan). The excitation photon energy and total energy resolution were set to 5.95 keV and 240 meV, respectively. The measurements were taken at room temperature and the pressure of the analysis chamber of HAXPES was 1.1×10^{-7} Pa. Details of HAXPES experiments at BL15XU were described elsewhere [26].

Time-of-flight secondary ion mass spectroscopy (TOF-SIMS) depth profile analyses were performed using a TOF-SIMS 5-100-AD. The analysis parameters were as follows: primary ion source, Bi_3^{++}; energy, 60 keV; area, 100×100 μm^2. The sputtering parameters were as follows: sputtering ion source, O_2; energy, 2 keV; area, 300×300 μm^2. To investigate the relationship between the actual depth and the sputtering time of the TOF-SIMS, the depth of the rastered area by ion beam was measured using a laser microscope and the sputtering rate was estimated.

Fourier-transform infrared spectroscopy (FT-IR) spectra were recorded using an FT/IR-6100 (MCT detector). The MgB_2 film was too thin for the detection of the FT-IR signals of B-H and/or B-H-B bonds. Thus, an additional FT-IR analysis was conducted using ion-exchanged MgB_2 powder by an attenuated total reflection (ATR) method using a diamond holder. For FT-IR analysis, MgB_2 powder was treated in the same manner as the MgB_2 film. Atomic force microscopy (AFM) images were obtained in dynamic imaging mode using an SPM-9700 with a silicon cantilever. Scanning electron microscope (SEM) images were obtained using a JEM-2010F (6.0 kV).

3.5. *Photoinduced Hydrogen Gas Release*

A film sample was placed in a quartz cell (3.5 mL with septum screw cap) under a nitrogen atmosphere without any solvent. Hydrogen production in the quartz cell under near-UV light irradiation was evaluated by gas chromatography (Tracera-GC-2010 Plus with a BID detector, Shimadzu, Co., Ltd., Japan). A film sample was placed under dark conditions for 1 h, followed by UV irradiation for 2 h. Hydrogen generation was determined by the difference in hydrogen amounts before and after the UV irradiation. UV irradiation was performed using a mercury-xenon (Hg-Xe) lamp with a 340 nm band-pass filter, similar to previous reports [7]. As a control experiment, the MgB_2 film without an ion-exchange treatment was also evaluated in the same manner as the ion-exchanged films.

4. Conclusions

We established a fabrication process for protonated boride films by using a facile ion-exchange under ambient temperature and atmospheric pressure. This is the first report of HB thin film fabrication on a substrate. The oriented MgB_2 thin films were prepared by a two-step PLD method and protons were successfully introduced into the films by treatment with either ion-exchange resin or formic acid in an organic solvent. Depending on the treatment conditions, the films with different characteristics such as proton content and surface structure have been obtained. The irradiation treatment on thin film with near-UV light produces hydrogen gas, similar to the powder HB samples. The present study is the first to demonstrate light-induced hydrogen release from a protonated boride thin film. This study will provide a method to fabricate thin films with B-H bonds, which will be useful for further studies or device applications.

Supplementary Materials: The followings are available online: Table S1: XRD peaks data, Figures S1, S6 and S12: microscope images, Figures S2–S5: XRD patterns, Figures S7–S9: SIMS depth profiles, Figure S10: FT-IR spectra, Figure S11: AFM images.

Author Contributions: Experiments, T.H. and S.Y.; conceptualization, S.S.; synchrotron analysis, H.A. and S.U.; writing—original draft preparation, T.H.; writing—review and editing, S.Y., S.S., A.Y., H.A., S.U., H.Z., T.K. and M.M.; supervision, H.Z., T.K. and M.M.; funding acquisition, T.K. and M.M. All authors have read and agreed to the published version of the manuscript.

Funding: This research was funded by JSPS Kakenhi (Grant No. JP18H02055, JP19H02551, JP19H 05046:A01, JP21H00015:B01). This research was also financially funded by the MEXT Element Strategy Initiative to Form Core Research Center (JPMXP0112101001), MHI Innovation Accelerator LLC, Iketani Science and Technology Foundation, the UBE Industries Foundation, and Sanoh Industrial Co., Ltd.

Acknowledgments: The HAXPES measurements were performed with an approval of NIMS Synchrotron X-ray Station (Proposal Nos. 2020A4600 and 2020A4650). We thank Genseki at the Open Facility Center, Tokyo Institute of Technology, for the TOF-SIMS analysis. This work was supported by JSPS Kakenhi (Grant No. JP18H02055, JP19H02551, JP19H05046:A01, JP21H00015:B01). This work was also financially supported by the MEXT Element Strategy Initiative to Form Core Research Center (JPMXP0112101001), MHI Innovation Accelerator LLC, Iketani Science and Technology Foundation, the UBE Industries Foundation, and Sanoh Industrial Co., Ltd.

Conflicts of Interest: The authors declare no conflict of interest.

Sample Availability: Samples of the compounds are not available from the authors.

References

1. Li, X.; Tao, L.; Chen, Z.; Fang, H.; Li, X.; Wang, X.; Xu, J.; Zhu, H. Graphene and related two-dimensional materials: Structure-property relationships for electronics and optoelectronics. *Appl. Phys. Rev.* **2017**, *4*, 021306. [CrossRef]
2. Nishino, H.; Fujita, T.; Cuong, N.T.; Tominaka, S.; Miyauchi, M.; Iimura, S.; Hirata, A.; Umezawa, N.; Okada, S.; Nishibori, E.; et al. Formation and characterization of hydrogen boride sheets derived from MgB_2 by cation exchange. *J. Am. Chem. Soc.* **2017**, *139*, 13761–13769. [CrossRef]
3. Tateishi, I.; Cuong, N.T.; Moura, C.A.S.; Cameau, M.; Ishibiki, R.; Fujino, A.; Okada, S.; Yamamoto, A.; Araki, M.; Ito, S.; et al. Semimetallicity of free-standing hydrogenated monolayer boron from MgB_2. *Phys. Rev. Mater.* **2019**, *3*, 024004. [CrossRef]
4. Tominaka, S.; Ishibiki, R.; Fujino, A.; Kawakami, K.; Ohara, K.; Masuda, T.; Matsuda, I.; Hosono, H.; Kondo, T. Geometrical Frustration of B-H Bonds in Layered Hydrogen Borides Accessible by Soft Chemistry. *Chem* **2020**, *6*, 406–418. [CrossRef]
5. Fujino, A.; Ito, S.; Goto, T.; Ishibiki, R.; Kondo, J.N.; Fujitani, T.; Nakamura, J.; Hosono, H.; Kondo, T. Hydrogenated Borophene Shows Catalytic Activity as Solid Acid. *ACS Omega* **2019**, *4*, 14100–14104. [CrossRef]
6. Fujino, A.; Ito, S.; Goto, T.; Ishibiki, R.; Osuga, R.; Kondo, J.N.; Fujitani, T.; Nakamura, J.; Hosono, H.; Kondo, T. Ethanol-ethylene conversion mechanism on hydrogen boride sheets probed byin situinfrared absorption spectroscopy. *Phys. Chem. Chem. Phys.* **2021**, *23*, 7724–7734. [CrossRef] [PubMed]
7. Kawamura, R.; Cuong, N.T.; Fujita, T.; Ishibiki, R.; Hirabayashi, T.; Yamaguchi, A.; Matsuda, I.; Okada, S.; Kondo, T.; Miyauch, M. Photoinduced hydrogen release from hydrogen boride sheets. *Nat. Commun.* **2019**, *10*, 4880. [CrossRef] [PubMed]
8. Ito, S.; Hirabayashi, T.; Ishibiki, R.; Kawamura, R.; Goto, T.; Fujita, T.; Yamaguchi, A.; Hosono, H.; Miyauchi, M.; Kondo, T. Hydrogen boride sheets as reductants and the formation of nanocomposites with metal nanoparticles. *Chem. Lett.* **2020**, *49*, 789–793. [CrossRef]
9. Saad, A.; Liu, D.; Wu, Y.; Song, Z.; Li, Y.; Najam, T.; Zong, K.; Tsiakaras, P.; Cai, X. Ag nanoparticles modified crumpled borophene supported Co3O4 catalyst showing superior oxygen evolution reaction (OER) performance. *Appl. Catal. B* **2021**, *298*, 120529. [CrossRef]
10. Gao, S.; Zhang, Y.; Bi, J.; Wang, B.; Li, C.; Liu, J.; Kong, C.; Yang, S.; Yang, S. 2D hydrogenated boride as a reductant and stabilizer forin situsynthesis of ultrafine and surfactant-free carbon supported noble metal electrocatalysts with enhanced activity and stability. *J. Mater. Chem. A* **2020**, *8*, 18856–18862. [CrossRef]
11. Xiang, P.; Chen, X.; Xiao, B.; Wang, Z.M. Highly Flexible Hydrogen Boride Monolayers as Potassium-Ion Battery Anodes for Wearable Electronics. *ACS Appl. Mater. Interfaces* **2019**, *11*, 8115–8125. [CrossRef] [PubMed]
12. Li, H.; Zhang, B.; Wu, Y.; Hou, J.; Jiang, D.; Duan, Q. Hydrogenated borophene/blue phosphorene: A novel two-dimensional donor-acceptor heterostructure with shrunken interlayer distance as a potential anode material for Li/Na ion batteries. *J. Phys. Chem. Solids* **2021**, *155*, 110108. [CrossRef]
13. Nagarajan, V.; Chandiramouli, R. Sensing properties of monolayer borophane nanosheet towards alcohol vapors: A first-principles study. *J. Mol. Graph. Model.* **2017**, *73*, 208–216. [CrossRef] [PubMed]
14. An, Y.; Hou, Y.; Wang, H.; Li, J.; Wu, R.; Wang, T.; Da, H.; Jiao, J. Unveiling the Electric-Current-Limiting and Photodetection Effect in Two-Dimensional Hydrogenated Borophene. *Phys. Rev. Appl.* **2019**, *11*, 064031. [CrossRef]
15. Oguchi, H.; Isobe, S.; Kuwano, H.; Shiraki, S.; Orimo, S.I.; Hitosugi, T. Pulsed laser deposition of air-sensitive hydride epitaxial thin films: LiH. *APL Mater.* **2015**, *3*, 096106. [CrossRef]
16. Shimizu, R.; Oguchi, H.; Hitosugi, T. Metal hydrides: Epitaxial growth and electronic properties. *J. Phys. Soc. Jpn.* **2020**, *89*, 051012. [CrossRef]
17. Gonzalez-Silveira, M.; Gremaud, R.; Schreuders, H.; van Setten, M.J.; Batyrev, E.; Rougier, A.; Dupont, L.; Bardají, E.G.; Lohstroh, W.; Dam, B. In-situ deposition of alkali and alkaline earth hydride thin films to investigate the formation of reactive hydride composites. *J. Phys. Chem. C* **2010**, *114*, 13895–13901. [CrossRef]
18. Kang, W.N.; Kim, H.J.; Choi, E.M.; Jung, C.U.; Lee, S.I. MgB_2 superconducting thin films with a transition temperature of 39 kelvin. *Science* **2001**, *292*, 1521–1523. [CrossRef] [PubMed]
19. Shinde, S.R.; Ogale, S.B.; Greene, R.L.; Venkatesan, T. Superconducting MgB_2 thin films by pulsed laser deposition. *Appl. Phys. Lett.* **2001**, *79*, 227–229. [CrossRef]
20. Chen, K.; Ma, P.; Nie, R.J.; Yang, T.; Xie, F.X.; Liu, L.Y.; Wang, S.Z.; Dai, Y.D.; Wang, F. Growth and superconductivity characteristics of MgB_2 thin films. *Supercond Sci. Technol.* **2002**, *15*, 1721–1724. [CrossRef]
21. Garg, K.B.; Chatterji, T.; Dalela, S.; Heinonnen, M.; Leiro, J.; Dalela, B.; Singhal, R.K. Core level photoemission study of polycrystalline MgB_2. *Solid State Commun.* **2004**, *131*, 343–347. [CrossRef]
22. Talapatra, A.; Bandyopadhyay, S.K.; Sen, P.; Barat, P.; Mukherjee, S.; Mukherjee, M. X-ray photoelectron spectroscopy studies of MgB_2 for valence state of Mg. *Phys. C Supercond* **2005**, *419*, 141–147. [CrossRef]
23. Kurmaev, E.Z.; Lyakhovskaya, I.I.; Kortus, J.; Moewes, A.; Miyata, N.; Demeter, M.; Neumann, M.; Yanagihara, M.; Watanabe, M.; Muranaka, T.; et al. Electronic structure of (formula presented)X-ray emission and absorption studies. *Phys. Rev. B Condens. Matter Mater. Phys.* **2002**, 1–4. [CrossRef]

24. Takayoshi, S.; Mamoru, W.; Yu, K.; Yoshinori, F. Na^+/H^+ Ion-exchange Process on Layered Hydrous Titanium Dioxide. *Bull. Chem. Soc. Jpn.* **1985**, *58*, 3500–3505. [CrossRef]
25. Kawamura, R.; Yamaguchi, A.; Shimada, C.; Ishibiki, R.; Fujita, T.; Kondo, T.; Miyauchi, M. Acid assisted synthesis of HB sheets through exfoliation of MgB_2 bulk in organic media. *Chem. Lett.* **2020**, *49*. [CrossRef]
26. Ueda, S. Application of hard X-ray photoelectron spectroscopy to electronic structure measurements for various functional materials. *J. Electron Spectrosc. Relat. Phenom.* **2013**, *190*, 235–241. [CrossRef]

molecules

Article

Highly Dispersed Ni Nanoclusters Spontaneously Formed on Hydrogen Boride Sheets

Natsumi Noguchi [1], Shin-ichi Ito [2], Miwa Hikichi [2], Yohei Cho [3], Kazuho Goto [1], Atsushi Kubo [2], Iwao Matsuda [4], Takeshi Fujita [5], Masahiro Miyauchi [3,*] and Takahiro Kondo [2,6,7,*]

1 Graduate School of Pure and Applied Sciences, University of Tsukuba, Tsukuba 305-8574, Japan
2 Faculty of Pure and Applied Sciences, University of Tsukuba, Tsukuba 305-8573, Japan
3 School of Materials and Chemical Technology, Tokyo Institute of Technology, Meguro, Tokyo 152-8552, Japan
4 Institute for Solid State Physics (ISSP), The University of Tokyo, Kashiwa 277-8581, Japan
5 School of Environmental Science and Engineering, Kochi University of Technology, 185 Miyanokuchi, Tosayamada, Kami 782-8502, Japan
6 R&D Center for Zero CO_2 Emission Functional Materials and Tsukuba Research Center for Energy Materials Science (TREMS), University of Tsukuba, Tsukuba 305-8573, Japan
7 The Advanced Institute for Materials Research, Tohoku University, Sendai 980-8577, Japan
* Correspondence: mmiyauchi@ceram.titech.ac.jp (M.M.); takahiro@ims.tsukuba.ac.jp (T.K.)

Abstract: Hydrogen boride (HB) sheets are two-dimensional materials comprising a negatively charged hexagonal boron network and positively charged hydrogen atoms with a stoichiometric ratio of 1:1. Herein, we report the spontaneous formation of highly dispersed Ni nanoclusters on HB sheets. The spontaneous reduction reaction of Ni ions by the HB sheets was monitored by in-situ measurements with an ultraviolet-visible spectrometer. Acetonitrile solutions of Ni complexes and acetonitrile dispersions of the HB sheets were mixed in several molar ratios (the HB:Ni molar ratio was varied from 100:0.5 to 100:20), and the changes in the absorbance were measured over time. In all cases, the results suggest that Ni metal clusters grow on the HB sheets, considering the increase in absorbance with time. The absorbance peak position shifts to the higher wavelength as the Ni ion concentration increases. Transmission electron microscopy images of the post-reaction products indicate the formation of Ni nanoclusters, with sizes of a few nanometers, on the HB sheets, regardless of the preparation conditions. These highly dispersed Ni nanoclusters supported on HB sheets will be used for catalytic and plasmonic applications and as hydrogen storage materials.

Keywords: hydrogen boride (HB) sheets; Ni nanoclusters; transmission electron microscopy

Citation: Noguchi, N.; Ito, S.-i.; Hikichi, M.; Cho, Y.; Goto, K.; Kubo, A.; Matsuda, I.; Fujita, T.; Miyauchi, M.; Kondo, T. Highly Dispersed Ni Nanoclusters Spontaneously Formed on Hydrogen Boride Sheets. *Molecules* **2022**, *27*, 8261. https://doi.org/10.3390/molecules27238261

Academic Editors: Josep M. Oliva-Enrich and Franck Rabilloud

Received: 20 October 2022
Accepted: 23 November 2022
Published: 26 November 2022

1. Introduction

Hydrogen boride (HB) sheets are two-dimensional (2D) nanosheets comprising a positively charged hydrogen and negatively charged boron network with an atomic ratio of $B/H = 1.0$. They were among the first synthesized hydrogenated borophenes (borophanes) [1]. Boron atoms form a hexagonal 2D network in HB sheets, wherein hydrogen atoms are bound to boron atoms by three-center two-electron (B–H–B) and two-center two-electron (B–H) bonds [2], with chemical stability against water [3]. HB sheets have been experimentally verified to exhibit excellent solid acid catalytic activity [4,5] and highly sensitive gas-sensing capability [2], as well as semimetal electronic [6], light-responsive hydrogen release [7], and carbon dioxide adsorption and conversion properties [8]. Furthermore, theoretical studies have revealed their intriguing electronic [9], optical, and thermal properties [10,11], as well as their possible applications in hydrogen release devices [12,13], reversible hydrogen storage [14], current limiters [15], photodetectors, individual amino acid sensors [16], rechargeable Li/Na ion battery electrodes [17,18], and anodes for rechargeable potassium-ion batteries, with high capacities, low voltages, and high rate performance [19]. Furthermore, the formation of HB sheets has paved the way for the conceptual development of new types of HB materials [20–28].

HB sheets also function as reductants for specific ions: the HB sheets had a redox potential between −0.277 and −0.257 V versus that of a standard hydrogen electrode (SHE) [29]. Therefore, metal nanoclusters can be supported by a single process, i.e., by mixing HB sheets with specific metal ions in a liquid [29], where metal ions with redox potentials lower than HB can be reduced, whereas those with redox potentials higher than HB cannot be reduced. This implies that only specific metals can be selectively reduced by HB sheets. By applying this property, HB sheets have been used to prepare superior composite electrode catalysts for oxygen evolution [30] and oxygen reduction reactions [31]. Therefore, a detailed understanding of the reduction properties of HB and the formation process of nanoclusters and/or nanocomposites is essential to promote the practical usage of HB sheets.

Herein, we report the spontaneous formation of highly dispersed Ni nanoclusters on HB sheets. Ni nanoclusters were formed on HB sheets by mixing Ni ions and HB sheets in an acetonitrile solution at different Ni ion concentrations. Although the sizes of the Ni nanoclusters were not significantly affected by the concentration of Ni ions, their densities were controlled by the concentration. The resulting highly dispersed Ni nanoclusters supported on the HB sheets may be useful for catalytic and plasmonic applications, as well as hydrogen storage materials [32,33].

2. Results and Discussion

Figure 1 shows the temporal change in the ultraviolet-visible (UV-vis) spectra of the HB dispersion at a 0.02 mol/L concentration in acetonitrile after mixing with Ni ions at various Ni ion concentrations (the molar ratio of the HB:Ni ions was adjusted from 100:0.5 to 100:20). The spectra shorter than 250 nm are not shown, since their photon numbers in a UV-Vis detector were too low for quantitative comparison (Figure S1). Moreover, the spectrum shorter than 330 nm for the high concentrated Ni ions conditions are not shown because the original absorbance intensities are too high to be saturated in this range. Across all concentrations, after mixing, the absorbance increases with time, at specific wavenumbers. This change can be attributed to the increase in the plasmon absorbance of the Ni nanoclusters on the HB sheets because of the spontaneous reduction reaction of the Ni ions on the HB sheets. In this reaction, the Ni ions may be exchanged with the protons of the HB sheets, and the electrons in the boron species would reduce the Ni ions, resulting in the formation of metallic Ni nanoclusters onto the HB sheets, while positively charged B would locally appear, similar to the results for Cu nanoclusters formed on HB sheets [29]. To evaluate the chemical states of Ni and B in the products, X-ray photoelectron spectroscopy (XPS) was performed for the samples of HB:Ni = 100:0.5 and 100:10, as shown in Figure 2. In both cases, the Ni peak and $B^{\delta+}$ component are detected, along with the $B^{\delta-}$ component, which originated from the pure HB sheets (for comparison, the XPS result for the pristine HB sheets is shown in Figure S2). The Ni intensity and $B^{\delta+}$ population are larger for HB:Ni = 100:10 than those for HB:Ni = 100:0.5, and a consistent correlation between the B:Ni ratio and $B^{\delta+}/B^{\delta-}$ is obtained (Table 1), indicating that Ni deposition occurs because of the spontaneous reduction reaction of the following process:

$$\begin{aligned} \mathrm{nH^+B^-} + \mathrm{Ni^{2+}} &\rightarrow \mathrm{2H^+} + \mathrm{Ni^{2+}}\cdot(\mathrm{n}-2)\mathrm{H^+nB^-} \\ &\rightarrow \mathrm{2H^+} + \mathrm{Ni}\cdot(\mathrm{n}-2)\mathrm{H^+}(\mathrm{n}-1)\mathrm{B^-}\cdot\mathrm{B^+} \end{aligned} \quad (1)$$

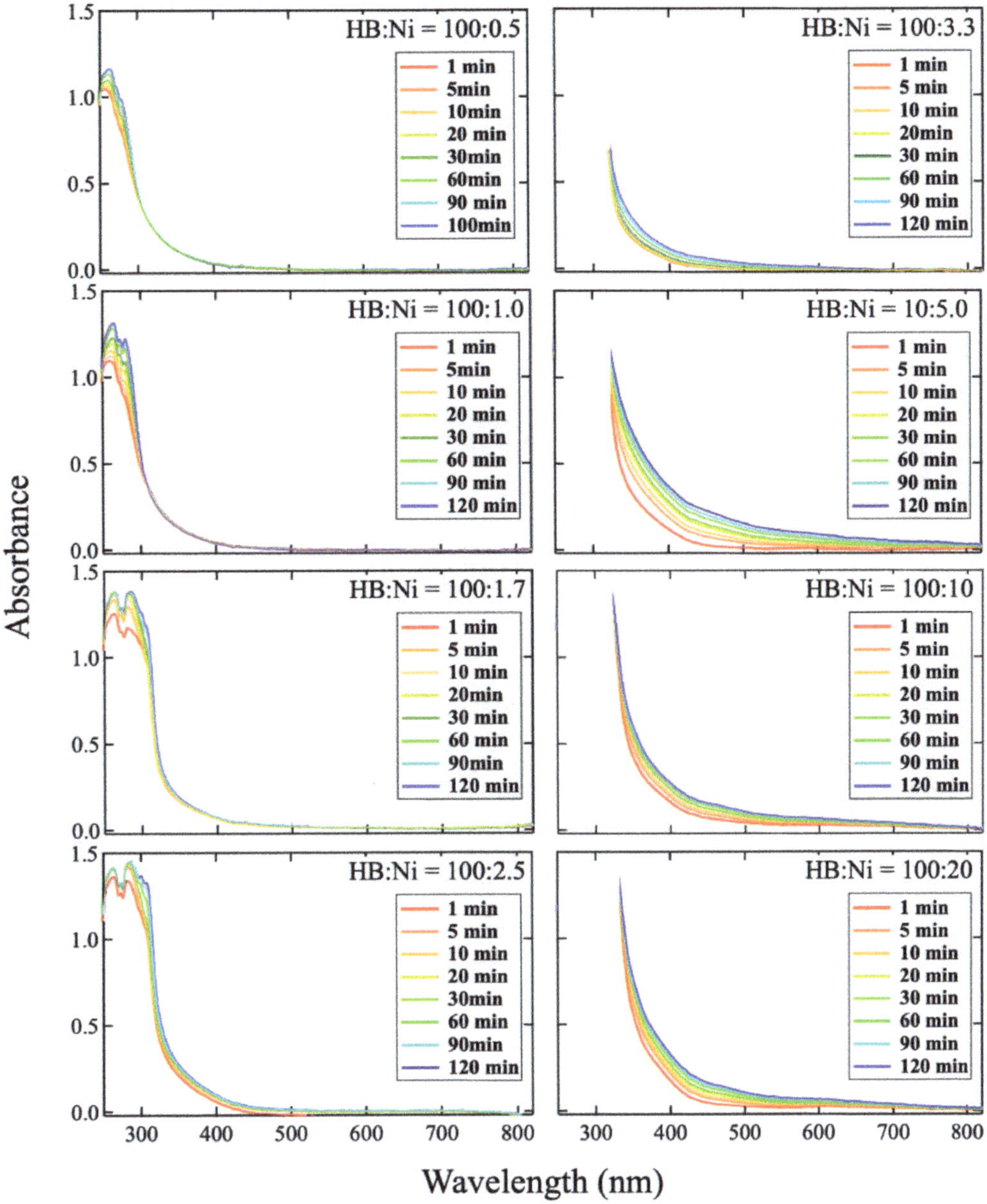

Figure 1. Temporal change in UV-vis spectra of HB dispersion at 0.02 mol/L in acetonitrile after mixing with Ni ions at various Ni ion concentrations. Molar ratio of the HB:Ni ions was adjusted from 100:0.5 to 100:20. Spectrum range under 330 nm for HB:Ni = 100:3.3, 100:5.0, 100:10, and 100:20 represents the range of intensity saturation, and is not shown here. Legend indicates time (1, 5, 10, 20, 30, 60, 90, and 120 min) elapsed from the start of mixing the two dispersions.

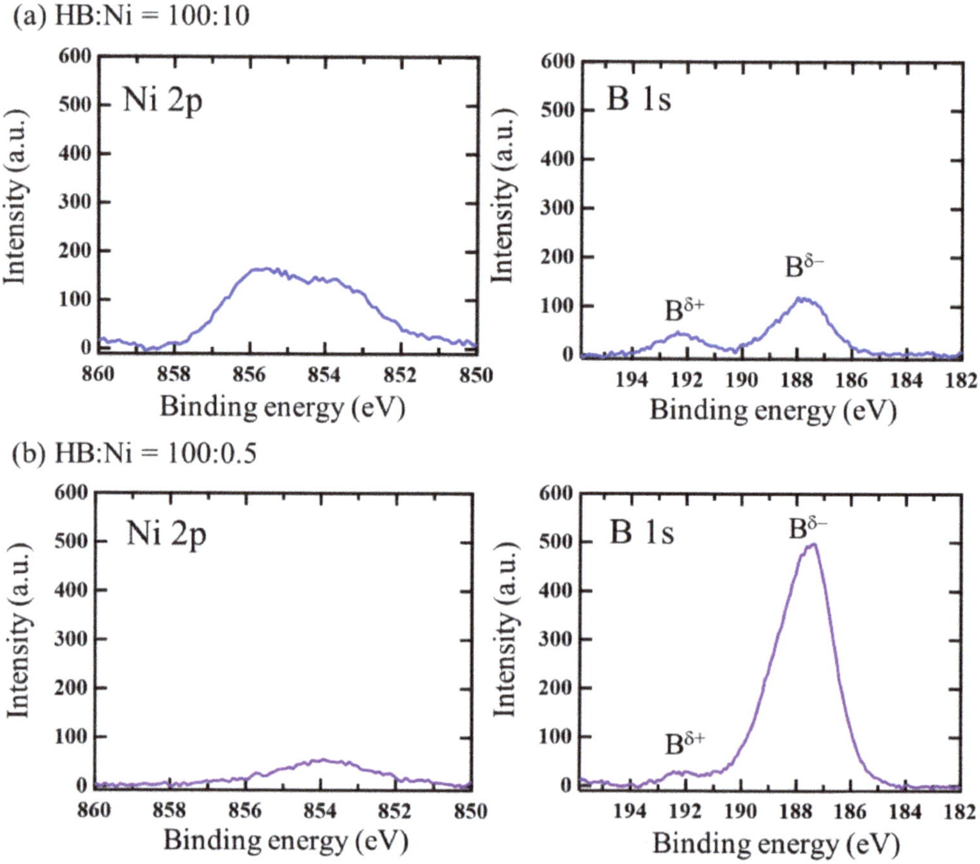

Figure 2. XPS results of the sample after drying the following mixtures in acetonitrile: (**a**) HB:Ni = 100:10, and (**b**) HB:Ni = 100:0.5.

Table 1. Analyzed results of XPS spectra.

Sample	Atomic Ratio Estimated from Peak Area (B:Ni)	B 1s Peak $B^{\delta-}$	$B^{\delta+}$
HB:Ni = 100:0.5	100:0.5	96%	4%
HB:Ni = 100:10	100:5.4	67%	33%

It should be noted that the appearance of oxide Ni peak for the sample with HB:Ni = 100:10, shown in Figure 2a, may arise from the surface oxidation of Ni nanoclusters because of the need to expose the sample to air before setting it in the XPS load-lock chamber. A similar spontaneous reduction reaction has also been reported in the formation of nanocomposites using graphene [34], layered $CaSi_2$ [35], polysilane [36], and Mg-deficient hydroxyl-functionalized boron nanosheets [37].

In Figure 1, the wavelength of the absorbance changes according to the concentration of Ni ions. To study the change, each spectrum was subtracted by the spectrum captured at 1 min after mixing, as shown in Figure 3. In the case of a lower Ni concentration (HB:Ni = 100:0.5, HB:Ni = 100:1.0, and HB:Ni = 100:1.7), the absorbance in the range of 250–300 nm increases with time after mixing. For a higher Ni ion concentration (HB:Ni = 100:2.5, HB:Ni = 100:3.3,

HB:Ni = 100:5.0, HB:Ni = 100:10, and HB:Ni = 100:20), the absorbance ranging from 320 to 800 nm increases with time after mixing. The change in the absorption feature as a function of the Ni ion density is highlighted in Figure 4, where a representative subtracted spectrum when the spectrum change is saturated is shown. The absorbance peak position shifts to the longer wavelength as the Ni ion concentration increases.

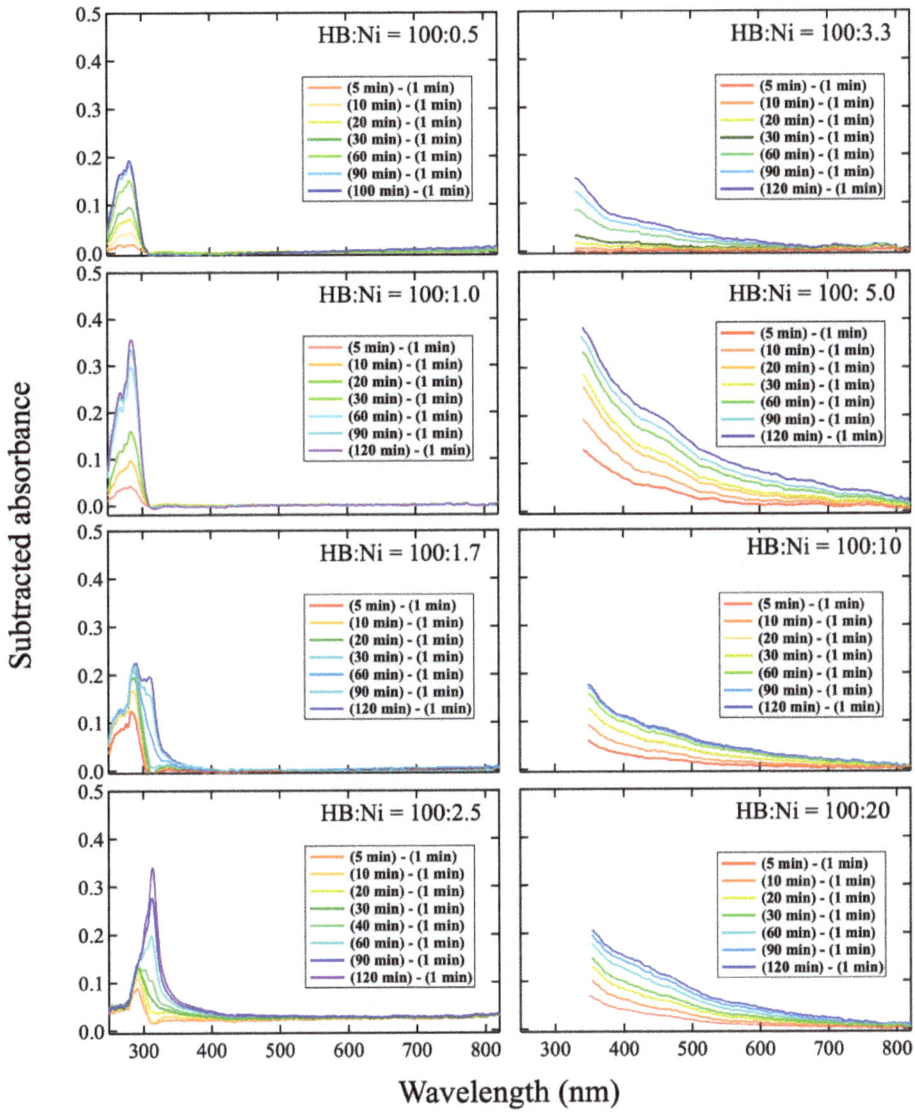

Figure 3. Subtracted spectra from UV-vis measurement; each spectrum shows a difference immediately after mixing (1 min). For the sample of HB:Ni = 100:2.5, the subtracted spectrum of "(40 min)-(1 min)" is also presented because it shows multiple components.

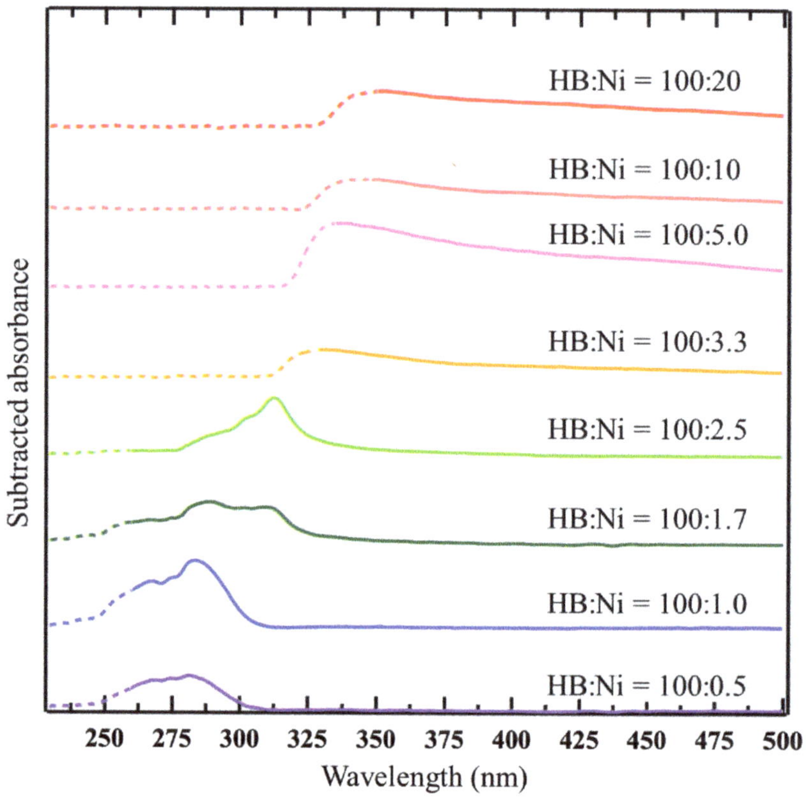

Figure 4. Respective subtracted spectra of HB:Ni = 100:20 (90 min), 100:10 (120 min), 100:5.0 (120 min), 100:3.3 (120 min), 100:2.5 (120 min), 100:1.7 (120 min), 100:1.0 (120 min), and 100:0.5 (120 min). The wavelength ranges for the intensity saturation in the original spectra and/or low detection sensitivity are shown by dots.

Two possible reasons account for the shift in the absorbance wavelength: (1) the sizes of the Ni nanoclusters formed on the HB sheets increase in proportion to the Ni ion concentration, and (2) the effect of localized surface plasmon resonance due to the change in the distance between the Ni nanoclusters on the HB sheets. Regarding (1), it has been reported that the absorption peak due to surface plasmon resonance shifts to longer wavelengths as the particle size of the metal nanoclusters increases [38–40]. For (2), the absorption peak shifts to longer wavelengths because of surface plasmon coupling when the distance between the nanoclusters decreases [41].

To clarify which of the two factors caused the shift in the absorption peaks in this study, samples were prepared with two distinctly different Ni ion concentrations (HB:Ni = 100:10 and 100:0.5, 42 d and 48 d after mixing) and observed via transmission electron microscopy (TEM); the results are shown in Figure 5. In both samples, several clusters are observed as dark spheres, suggesting that Ni atoms form clusters after the reduction reaction of Ni ions on the HB sheets, as reported previously [29]. Indeed, the presence of Ni with B is clearly seen in the obtained energy dispersive X-ray spectroscopy (EDS) conducted during TEM (Figure S3), which is consistent with the XPS results (Figure 2). The selected area electron diffraction (SAED), which was conducted during TEM, shows a halo pattern, as seen in Figure 5c. This indicates that the size of the Ni nanoclusters is considerably small.

Consistently, X-ray diffraction (XRD) patterns show no diffraction peaks, indicating that there are no larger sized Ni particles in the sample.

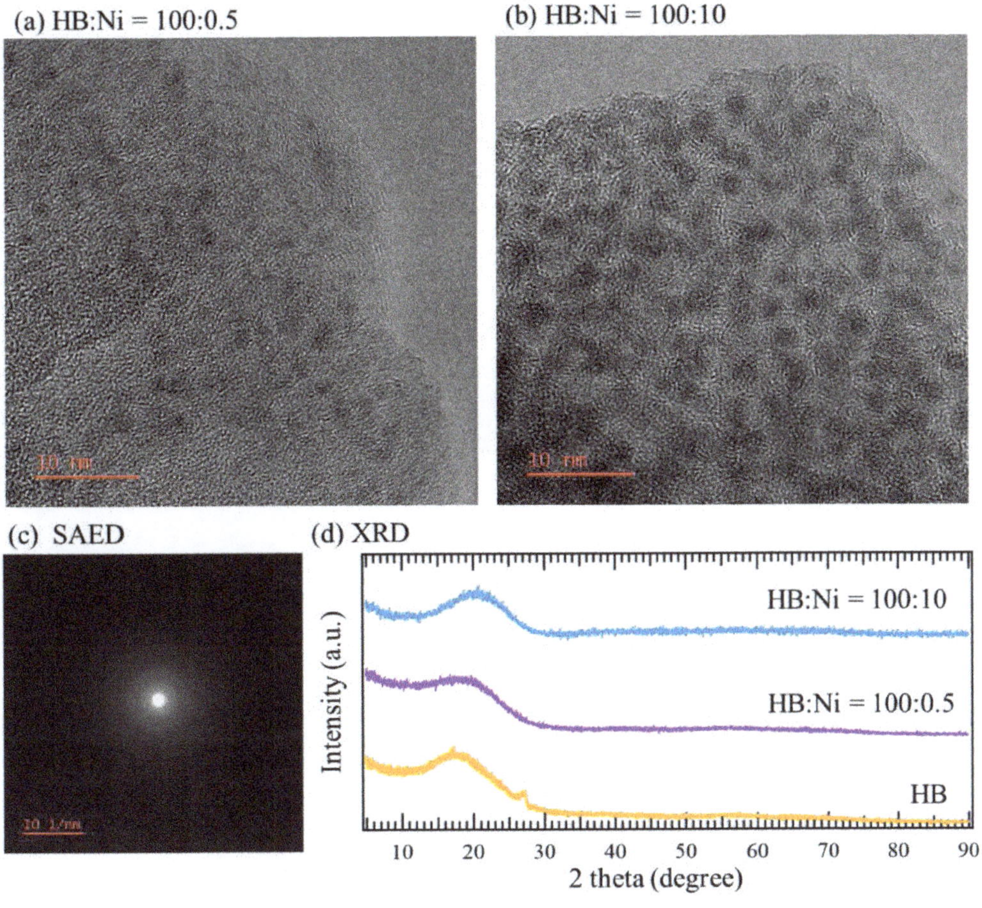

Figure 5. TEM images SAED pattern and XRD patterns. (**a**) TEM images of HB:Ni = 100:0.5, and (**b**) HB:Ni = 100:10. (**c**) SAED pattern simultaneously obtained with the TEM image shown in panel b. (**d**) XRD patterns of HB, HB:Ni = 100:0.5, and HB:Ni = 100:10 (a Kapton capsule was used as a cover of the sample to prevent the exposure of the sample to the atmosphere). The XRD pattern of HB shows a small peak at 27°, which is considered to be originated from the stacking of some of the sheets.

To quantitatively evaluate the size and dispersion of the Ni nanoclusters on the HB sheets, the diameter (D) and inter-center distance (L) of the dark spheres in 17 TEM images were statistically analyzed, as shown by the histogram in Figure 6. The mean D values of the Ni nanoclusters for HB:Ni = 100:0.5 and 100:10 were estimated as 1.59 ± 0.02 and 2.04 ± 0.03 nm, respectively, and the mean values of L were approximately 3.08 ± 0.06 and 3.08 ± 0.05 nm, respectively. This result indicates that the sizes of the Ni nanoclusters formed on the HB sheets do not vary significantly with the Ni ion concentration. This contrasts with the case of Pt clusters formed on graphene using a similar method, where the sizes of the Pt clusters were reported to vary proportionally to the amount of prepared Pt precursors [42]. In other words, HB can easily support highly-dispersed small metal nanoclusters with high density, in contrast to graphene. In this study, the Ni ions presumably interact strongly with the HB sheets when they are reduced. The Ni clusters are then considered to be anchored at the reacted sites on the HB sheets, preventing significant aggregation, which differs from the metal ions reduction process using a reducing agent under homogeneous media.

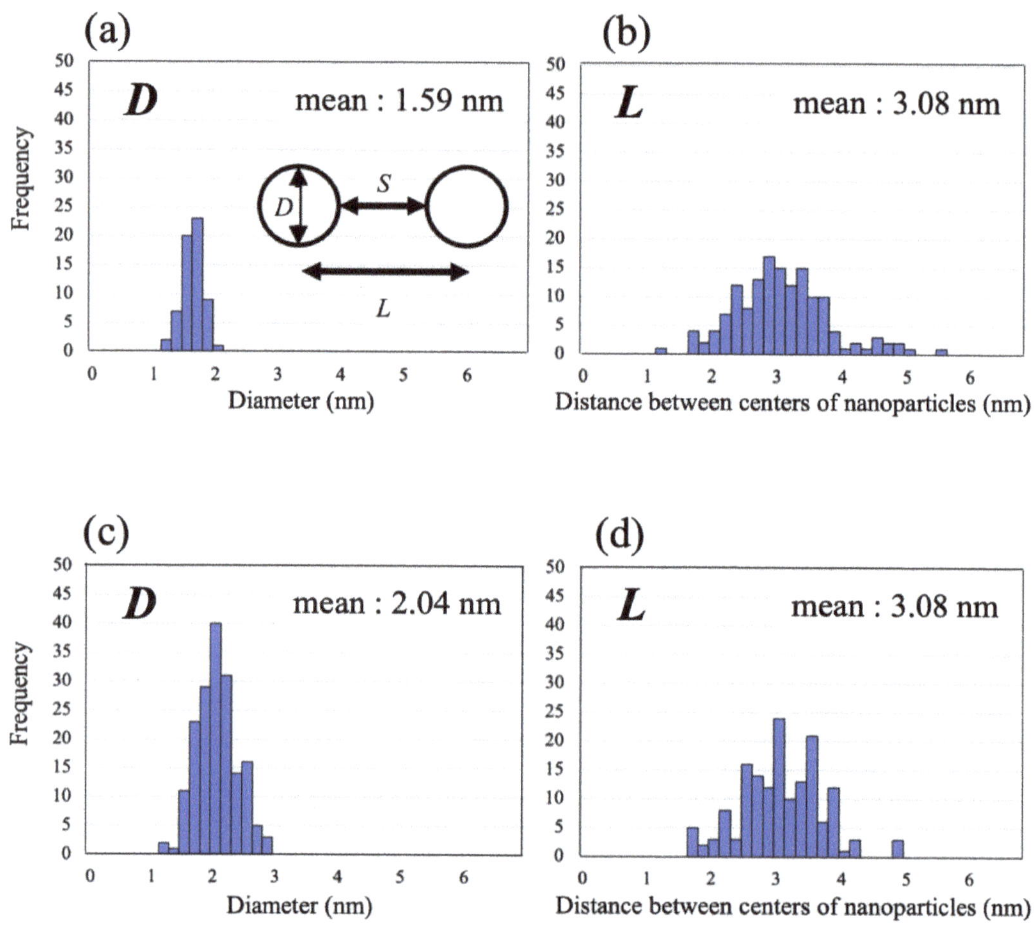

Figure 6. Histograms of (**a**,**c**) diameter, (**b**,**d**) distance between centers of nanoparticles. (**a**,**b**) are the results at HB:Ni = 100:0.5, (**c**,**d**) are the results at HB:Ni = 100:10.

From the average values of *L* and *D*, the inter-surface distance (*S*, the length between the nearest cluster surfaces) values can be derived as 1.04 ± 0.06 nm for HB:Ni = 100:10, and 1.49 ± 0.07 nm for HB:Ni = 100:0.5. Notably, as the particle size increases, due to plasmon resonance, the peak position of the spectrum shifts toward longer wavelengths [38–40,43]. For example, in the case of Au clusters, a change in the particle size from 9 to 99 nm results in a peak shift of 25 nm [43]. Conversely, the spectral peak position shifts by approximately 40 nm when the interparticle distance of the Au clusters changes from 12 to 6 nm [41]. Previous studies suggest that the plasmon resonance is more sensitive to the interparticle distance than to the particle size. As shown in Figure 4, the optical absorption peak shifts 56 nm from HB:Ni = 100:0.5 to HB:Ni = 100:10, where the difference in *D* of these samples is 0.45 ± 0.03 nm and that of *S* is 0.45 ± 0.09 nm. Based on the previous plasmonic studies, the present optical absorption shift is ascribed to the change in the interparticle distance

S, rather than particle size D. This peak shift is also observed during the nanoparticle formation for HB:Ni = 100:2.5, as shown in Figure 3 (the main peak position shifts from 290 to 320 nm).

The resultant highly dispersed Ni nanoclusters supported on the HB sheets may be useful for catalytic and plasmonic applications, and also as hydrogen storage materials, because highly dispersed Ni-clusters have been reported to contribute to the superior hydrogen desorption kinetics of MgH_2 [32,33].

3. Materials and Methods

3.1. *Materials*

MgB_2 powder (99%, RareMetallic Co., Ltd., Tokyo, Japan), acetonitrile (99.5%,Wako Pure ChemicalIndustries Ltd., Osaka, Japan), and an ion-exchange resin (AmberliteIR120B hydrogen, Organo Corp., Tokyo, Japan) were used to synthesize the HB sheets and the HB dispersion. $Ni(C_5H_7O_2)_2$ (98%, Merck) was used to prepare the $Ni(C_5H_7O_2)_2$ dispersion.

3.2. *Synthesis of HB Sheets*

The HB sheets were prepared using a previously reported ion-exchange method [1]. MgB_2 (1.0 g) powder in acetonitrile (200 mL) was added to the ion-exchange resin (60 mL) in acetonitrile (200 mL) in a Schlenk flask under a nitrogen atmosphere. The mixture was stirred with a magnetic stirrer at 310 rpm for 3 d at room temperature (~300 K). The process was sensitive to water because of the hydrolysis reaction of MgB_2 [44]. Thus, water was carefully removed beforehand. In this study, the recently reported acid-assisted reaction was not applied [45]. The supernatant was kept for 1 d at 255 K to solidify and separate the byproduct $B(OH)_3$. Dried HB sheets were prepared by heating the resulting liquid at 343 K while pumping with a cooling trap. We characterized the product rigorously using X-ray photoelectron spectroscopy (XPS) to confirm the absence of Mg, the presence of negatively charged B, and the absence of oxidized B, as shown in Figure S2 [1–8,29].

3.3. *XPS Measurements*

To characterize the sample, XPS measurement was performed using a JPS 9010 TR spectrometer (XPS; JPS 9010 TR, JEOL, Ltd., Tokyo, Japan) equipped with an ultrahigh vacuum chamber and an Mg Kα X-ray source (1253.6 eV). The pass energy was 10 eV, the energy resolution (estimated from the Ag 3d5/2 peak width of a clean Ag sample) was 0.635 eV, and the binding energy uncertainty was $\pm$0.05 eV. The sample was placed on a graphite tape. The Shirley background was subtracted from the spectrum using SpecSurf version 1.8.3.7 (JEOL, Ltd., Japan). The same software was used for estimating the quantitative atomic ratio of Ni and B of the reaction product, based on the peak area and sensitivity factors. The charge build-up in the sample (because of the incomplete contact of the graphite tape with the sample holder) resulted in a slight shift to higher binding energies for those spectra. Therefore, we calibrated the charge build-up based on the C1s peak of the graphite tape as 284.6 eV. The charge states of boron were judged based on the B 1s core level peak position by comparing the results with those in the literature (listed in the Supplementary Information of Ref. [46]).

3.4. *UV-vis Measurements*

A UV-vis spectroscope (DH-2000-BAL, Ocean Optics, Inc., Dunedin, FL, USA) was used to measure the absorbance of the $Ni(C_5H_7O_2)_2$ and HB dispersions. For this, $Ni(C_5H_7O_2)_2$ powder was dispersed in acetonitrile at concentrations of 0.04, 0.002, 0.001, 7×10^{-4}, 5×10^{-4}, 3×10^{-4}, 2×10^{-4}, and 1×10^{-4} mol/L. The HB sheets (powder, 0.0059 g) were dispersed into acetonitrile (25 mL) at a fixed concentration of 0.02 mol/L. The absorbance of the mixture of the $Ni(C_5H_7O_2)_2$ dispersion and HB dispersion was then measured using the UV-vis spectrometer. The intensity of the light source of the UV-vis device was measured beforehand (Figure S1) to obtain a reliable wavelength range for the UV-vis measurements. To study the change in the UV-vis spectra, the subtracted spectra were used, as reported previously [29], where each spectrum was subtracted by the spectrum that was captured at 1 min after mixing.

3.5. TEM, SAED, and EDS Measurements

Measurements were performed at room temperature using a JEM-2100F TEM/STEM apparatus (JEOL, Ltd., Japan) with double spherical aberration (Cs) correctors (CEOS GmbH, Heidelberg, Germany) to obtain high-contrast images with a point-to-point resolution of 1.4 Å. The lens aberrations were optimized by evaluating the Zemlin tableau of amorphous carbon. The residual spherical aberration was almost zero (Cs = -0.8 ± 1.2 μm with 95% certainty). The acceleration voltage was set to 120 kV, which is the lowest voltage that is effective with the Cs correctors in this system.

3.6. XRD Measurements

The XRD patterns were recorded at room temperature (~300 K) using a benchtop X-ray diffractometer (Rigaku MiniFlex, Tokyo, Japan), which employed Cu Kα radiation. The X-rays were generated using the line focus principle. A reflection-free Si plate was used as the sample stage. The Kapton capsule was used as a cover for the sample to prevent the exposure of the sample to the atmosphere. The diffraction patterns were recorded using a D/teX Ultra silicon strip detector (Rigaku) at 0.01° s^{-1} up to a 2θ value of 90°.

4. Conclusions

When a solution of Ni ions was mixed with an HB dispersion, it was confirmed that the Ni ions were reduced to form Ni nanoclusters on the HB sheets, without the use of reduction reagents. The TEM images showed that the Ni nanoclusters were small (1–2 nm) and highly dispersed, without aggregation. This was ascribed to the anchoring effect due to the strong interaction between the reduced Ni and HB sheets. By changing the concentration of Ni ions, the density of the particles could be changed without significantly changing the size of the particles. These highly dispersed Ni nanoclusters on HB sheets are expected to be used as catalysts and hydrogen storage materials, as well as in plasmonic applications.

Supplementary Materials: The following supporting information can be downloaded at: https://www.mdpi.com/article/10.3390/molecules27238261/s1, Figure S1: Intensity of the light source of the UV-vis device; Figure S2: XPS of HB; Figure S3: Energy-dispersive X-ray spectroscopy (EDS).

Author Contributions: Experiments, N.N., S.-i.I., M.H., K.G. and T.F.; Conceptualization, M.M. and T.K.; writing—original draft preparation, N.N.; discussion, N.N., S.-i.I., M.H., Y.C., K.G., A.K., I.M., T.F., M.M. and T.K.; writing—review and editing, N.N., S.-i.I., M.H., Y.C., K.G., A.K., I.M., T.F., M.M. and T.K.; supervision, M.M. and T.K. All authors have read and agreed to the published version of the manuscript.

Funding: This research was funded by the MEXT Element Strategy Initiative: To Form Core Research Centers (JPMXP0112101001), JSPS KAKENHI (Grant Nos. JP19H02551, JP21H05012, JP22K18964, and Hydrogenomics [JP19H05046:A01 and JP21H00015:B01]), JST CREST (Grant No. JPMJCR21O4), Honda Motor CO., LTD, and MHI Innovation Accelerator LLC.

Data Availability Statement: Data is available on request from the corresponding authors.

Conflicts of Interest: The authors declare no conflict of interest.

References

1. Nishino, H.; Fujita, T.; Cuong, N.T.; Tominaka, S.; Miyauchi, M.; Iimura, S.; Hirata, A.; Umezawa, N.; Okada, S.; Nishibori, E.; et al. Formation and Characterization of Hydrogen Boride Sheets Derived from MgB_2 by Cation Exchange. *J. Am. Chem. Soc.* **2017**, *139*, 13761–13769. [CrossRef] [PubMed]
2. Tominaka, S.; Ishibiki, R.; Fujino, A.; Kawakami, K.; Ohara, K.; Masuda, T.; Matsuda, I.; Hosono, H.; Kondo, T. Geometrical Frustration of B-H Bonds in Layered Hydrogen Borides Accessible by Soft Chemistry. *Chem* **2020**, *6*, 406–418. [CrossRef]
3. Rojas, K.I.M.; Cuong, N.T.; Nishino, H.; Ishibiki, R.; Ito, S.I.; Miyauchi, M.; Fujimoto, Y.; Tominaka, S.; Okada, S.; Hosono, H.; et al. Chemical Stability of Hydrogen Boride Nanosheets in Water. *Commun. Mater.* **2021**, *2*, 81. [CrossRef]

4. Fujino, A.; Ito, S.I.; Goto, T.; Ishibiki, R.; Kondo, J.N.; Fujitani, T.; Nakamura, J.; Hosono, H.; Kondo, T. Hydrogenated Borophene Shows Catalytic Activity as Solid Acid. *ACS Omega* **2019**, *4*, 14100–14104. [CrossRef] [PubMed]
5. Fujino, A.; Ito, S.I.; Goto, T.; Ishibiki, R.; Osuga, R.; Kondo, J.N.; Fujitani, T.; Nakamura, J.; Hosono, H.; Kondo, T. Ethanol-Ethylene Conversion Mechanism on Hydrogen Boride Sheets Probed By in Situ infrared Absorption Spectroscopy. *Phys. Chem. Chem. Phys.* **2021**, *23*, 7724–7734. [CrossRef]
6. Tateishi, I.; Cuong, N.T.; Moura, C.A.S.; Cameau, M.; Ishibiki, R.; Fujino, A.; Okada, S.; Yamamoto, A.; Araki, M.; Ito, S.; et al. Semimetallicity of Free-Standing Hydrogenated Monolayer Boron from MgB_2. *Phys. Rev. Mater.* **2019**, *3*, 024004. [CrossRef]
7. Kawamura, R.; Cuong, N.T.; Fujita, T.; Ishibiki, R.; Hirabayashi, T.; Yamaguchi, A.; Matsuda, I.; Okada, S.; Kondo, T.; Miyauchi, M. Photoinduced Hydrogen Release from Hydrogen Boride Sheets. *Nat. Commun.* **2019**, *10*, 4880. [CrossRef]
8. Goto, T.; Ito, S.; Laxman Shinde, S.; Ishibiki, R.; Hikita, Y.; Matsuda, I.; Hamada, I.; Hosono, H.; Kondo, T. Carbon Dioxide Adsorption and Conversion to Methane and Ethane on Hydrogen Boride Sheets. *Commun. Chem.* **2022**, *5*, 118. [CrossRef]
9. Jiao, Y.; Ma, F.; Bell, J.; Bilic, A.; Du, A. Two-Dimensional Boron Hydride Sheets: High Stability, Massless Dirac Fermions, and Excellent Mechanical Properties. *Angew. Chem. Int. Ed.* **2016**, *55*, 10292–10295. [CrossRef]
10. Mortazavi, B.; Makaremi, M.; Shahrokhi, M.; Raeisi, M.; Singh, C.V.; Rabczuk, T.; Pereira, L.F.C. Borophene Hydride: A Stiff 2D Material with High Thermal Conductivity and Attractive Optical and Electronic Properties. *Nanoscale* **2018**, *10*, 3759–3768. [CrossRef]
11. He, J.; Li, D.; Ying, Y.; Feng, C.; He, J.; Zhong, C.; Zhou, H.; Zhou, P.; Zhang, G. Orbitally Driven Giant Thermal Conductance Associated with Abnormal Strain Dependence in Hydrogenated Graphene-like Borophene. *NPJ Comput. Mater.* **2019**, *5*, 47. [CrossRef]
12. Abtew, T.A.; Shih, B.; Dev, P.; Crespi, V.H.; Zhang, P. Prediction of a Multicenter-Bonded Solid Boron Hydride for Hydrogen Storage. *Phys. Rev. B* **2011**, *83*, 094108. [CrossRef]
13. Abtew, T.A.; Zhang, P. Charging-Assisted Hydrogen Release Mechanism in Layered Boron Hydride. *Phys. Rev. B* **2011**, *84*, 094303. [CrossRef]
14. Chen, L.; Chen, X.; Duan, C.; Huang, Y.; Zhang, Q.; Xiao, B. Reversible Hydrogen Storage in Pristine and Li Decorated 2D Boron Hydride. *Phys. Chem. Chem. Phys.* **2018**, *20*, 30304–30311. [CrossRef]
15. An, Y.; Hou, Y.; Wang, H.; Li, J.; Wu, R.; Wang, T.; Da, H.; Jiao, J. Unveiling the Electric-Current-Limiting and Photodetection Effect in Two-Dimensional Hydrogenated Borophene. *Phys. Rev. Appl.* **2019**, *11*, 064031. [CrossRef]
16. Kumawat, R.L.; Jena, M.K.; Pathak, B. Individual Identification of Amino Acids on an Atomically Thin Hydrogen Boride System Using Electronic Transport Calculations. *J. Phys. Chem. C* **2020**, *124*, 27194–27202. [CrossRef]
17. Shukla, V.; Araujo, R.B.; Jena, N.K.; Ahuja, R. Borophene's Tryst with Stability: Exploring 2D Hydrogen Boride as an Electrode for Rechargeable Batteries. *Phys. Chem. Chem. Phys.* **2018**, *20*, 22008–22016. [CrossRef]
18. Makaremi, M.; Mortazavi, B.; Singh, C.V. 2D Hydrogenated Graphene-like Borophene as a High Capacity Anode Material for Improved Li/Na Ion Batteries: A First Principles Study. *Mater. Today Energy* **2018**, *8*, 22–28. [CrossRef]
19. Xiang, P.; Chen, X.; Xiao, B.; Wang, Z.M. Highly Flexible Hydrogen Boride Monolayers as Potassium-Ion Battery Anodes for Wearable Electronics. *ACS Appl. Mater. Interfaces* **2019**, *11*, 8115–8125. [CrossRef]
20. Hirabayashi, T.; Yasuhara, S.; Shoji, S.; Yamaguchi, A.; Abe, H.; Ueda, S.; Zhu, H.; Kondo, T.; Miyauchi, M. Fabrication of Hydrogen Boride Thin Film by Ion Exchange in MgB_2. *Molecules* **2021**, *26*, 6212. [CrossRef]
21. Cuong, N.T.; Tateishi, I.; Cameau, M.; Niibe, M.; Umezawa, N.; Slater, B.; Yubuta, K.; Kondo, T.; Ogata, M.; Okada, S.; et al. Topological Dirac Nodal Loops in Nonsymmorphic Hydrogenated Monolayer Boron. *Phys. Rev. B* **2020**, *101*, 195412. [CrossRef]
22. Zhang, X.; Tsujikawa, Y.; Tateishi, I.; Niibe, M.; Wada, T.; Horio, M.; Hikichi, M.; Ando, Y.; Yubuta, K.; Kondo, T.; et al. Electronic Topological Transition of 2D Boron by the Ion Exchange Reaction. *J. Phys. Chem. C* **2022**, *126*, 12802–12808. [CrossRef]
23. Oliva-Enrich, J.M.; Kondo, T.; Alkorta, I.; Elguero, J.; Klein, D.J. Diborane Concatenation Leads to New Planar Boron Chemistry. *ChemPhysChem* **2020**, *21*, 2460–2467. [CrossRef] [PubMed]
24. Ploysongsri, N.; Vchirawongkwin, V.; Ruangpornvisuti, V. Hydrogen Boride Nanotubes and Their C, N, O Decoration and Doping Derivatives as Materials for Hydrogen-Containing Gases Storage and Sensing: A SCC–DFTB Study. *Vacuum* **2021**, *187*, 110140. [CrossRef]
25. Lei, B.; Zhang, Y.-Y.; Du, S.-X. Band Engineering of B_2H_2 Nanoribbons. *Chin. Phys. B* **2019**, *28*, 046803. [CrossRef]
26. Kondo, T. Recent progress in boron nanomaterials. *Sci. Technol. Adv. Mater.* **2017**, *18*, 780–804. [CrossRef] [PubMed]
27. Ozin, G.; Siler, T. Catalyst: New Materials Discovery: Machine-Enhanced Human Creativity. *Chem* **2018**, *4*, 1183–1189. [CrossRef]
28. Niibe, M.; Cameau, M.; Cuong, N.T.; Sunday, O.I.; Zhang, X.; Tsujikawa, Y.; Okada, S.; Yubuta, K.; Kondo, T.; Matsuda, I. Electronic Structure of a Borophene Layer in Rare-Earth Aluminum/Chromium Boride and Its Hydrogenated Derivative Borophane. *Phys. Rev. Mater.* **2021**, *5*, 084007. [CrossRef]
29. Ito, S.I.; Hirabayashi, T.; Ishibiki, R.; Kawamura, R.; Goto, T.; Fujita, T.; Yamaguchi, A.; Hosono, H.; Miyauchi, M.; Kondo, T. Hydrogen Boride Sheets as Reductants and the Formation of Nanocomposites with Metal Nanoparticles. *Chem. Lett.* **2020**, *49*, 789–793. [CrossRef]
30. Saad, A.; Liu, D.; Wu, Y.; Song, Z.; Li, Y.; Najam, T.; Zong, K.; Tsiakaras, P.; Cai, X. Ag Nanoparticles Modified Crumpled Borophene Supported Co_3O_4 Catalyst Showing Superior Oxygen Evolution Reaction (OER) Performance. *Appl. Catal. B* **2021**, *298*, 120529. [CrossRef]

31. Gao, S.; Zhang, Y.; Bi, J.; Wang, B.; Li, C.; Liu, J.; Kong, C.; Yang, S.; Yang, S. 2D Hydrogenated Boride as a Reductant and Stabilizer for *in Situ* Synthesis of Ultrafine and Surfactant-Free Carbon Supported Noble Metal Electrocatalysts with Enhanced Activity and Stability. *J. Mater. Chem. A Mater* **2020**, *8*, 18856–18862. [CrossRef]
32. Peng, C.; Li, Y.; Zhang, Q. Enhanced Hydrogen Desorption Properties of MgH_2 by Highly Dispersed Ni: The Role of in-Situ Hydrogenolysis of Nickelocene in Ball Milling Process. *J. Alloys Compd.* **2022**, *900*, 163547. [CrossRef]
33. Zhu, W.; Panda, S.; Lu, C.; Ma, Z.W.; Khan, D.; Dong, J.J.; Sun, F.Z.; Xu, H.; Zhang, Q.Y.; Zou, J.X. Using a Self-Assembled Two-Dimensional MXene-Based Catalyst (2D-Ni@Ti_3C_2) to Enhance Hydrogen Storage Properties of MgH_2. *ACS Appl. Mater. Interfaces* **2020**, *12*, 50333–50343. [CrossRef] [PubMed]
34. Zhang, X.; Ooki, W.; Kosaka, Y.R.; Okonogi, A.; Marzun, G.; Wagener, P.; Barcikowski, S.; Kondo, T.; Nakamura, J. Effect of pH on the spontaneous synthesis of palladium nanoparticles on reduced graphene oxide. *Appl. Surf. Sci.* **2016**, *389*, 911. [CrossRef]
35. Itahara, H.; Nakano, H. Formation mechanism of Cu_2O particles using layered $CaSi_2$ as a reduction/oxidation mediator. *J. Am. CerAm. Soc.* **2019**, *102*, 5738. [CrossRef]
36. Ohashi, M.; Yaokawa, R.; Takatani, Y.; Nakano, H. Versatile Reducing Reaction Field within Layered Polysilane for Efficient One-Pot Synthesis of Metal Nanoparticles. *ChemNanoMat* **2017**, *3*, 534. [CrossRef]
37. James, A.L.; Khandelwal, S.; Dutta, A.; Jasuja, K. Boron based nanosheets as reducing templates in aqueous solutions: Towards novel nanohybrids with gold nanoparticles and graphene. *Nanoscale* **2018**, *10*, 20514. [CrossRef] [PubMed]
38. Kreibig, U.; Vollmer, M. *Optical Properties of Metal Clusters*; Theoretical Considerations; Springer: Berlin, Germany, 1995; pp. 13–201. ISBN 978-3-642-08191-0.
39. Heath, J.R. Size-Dependent Surface-Plasmon Resonances of Bare Silver Particles. *Phys. Rev. B* **1989**, *40*, 9982–9985. [CrossRef]
40. Vaishali, S.S.; Doss, K.; Banu, A.S.; Kurian, G.A. Desmodium Gangeticum Root Aqueous Extract Mediated Synthesis of Ni Nanoparticle and Its Biological Evaluation. *Int. J. Pharm. Sci.* **2015**, *7*, 141–146.
41. Romero, I.; Aizpurua, J.; Bryant, G.W.; de Abajo, F.J.G. Plasmons in Nearly Touching Metallic Nanoparticles: Singular Response in the Limit of Touching Dimers. *Opt. Express* **2006**, *14*, 9988–9999. [CrossRef]
42. Siburian, R.; Kondo, T.; Nakamura, J. Size Control to a Sub-Nanometer Scale in Platinum Catalysts on Graphene. *J. Phys. Chem. C* **2013**, *117*, 3635–3645. [CrossRef]
43. Link, S.; El-Sayed, M.A. Size and Temperature Dependence of the Plasmon Absorption of Colloidal Gold Nanoparticles. *J. Phys. Chem. B* **1999**, *103*, 4212–4217. [CrossRef]
44. Nishino, H.; Fujita, T.; Yamamoto, A.; Fujimori, T.; Fujino, A.; Ito, S.; Nakamura, J.; Hosono, H.; Kondo, T. Formation Mechanism of Boron-Based Nanosheet through the Reaction of MgB_2 with Water. *J. Phys. Chem. C* **2017**, *121*, 10587–10593. [CrossRef]
45. Kawamura, R.; Yamaguchi, A.; Shimada, C.; Ishibiki, R.; Fujita, T.; Kondo, T.; Miyauchi, M. Acid Assisted Synthesis of HB Sheets through Exfoliation of MgB_2 Bulk in Organic Media. *Chem. Lett.* **2020**, *49*, 1194–1196. [CrossRef]
46. Kusaka, H.; Ishibiki, R.; Toyoda, M.; Fujita, T.; Tokunaga, T.; Yamamoto, A.; Miyakawa, M.; Matsushita, K.; Miyazaki, K.; Li, L.; et al. Crystalline Boron Monosulfide Nanosheets with Tunable Bandgaps. *J. Mater. Chem. A* **2021**, *9*, 24631–24640. [CrossRef]

molecules

MDPI

Article

Accelerated Synthesis of Borophane (HB) Sheets through HCl-Assisted Ion-Exchange Reaction with $YCrB_4$

Xiaoni Zhang [1], Miwa Hikichi [2], Takushi Iimori [1], Yuki Tsujikawa [1], Mei Yuan [2], Masafumi Horio [1], Kunio Yubuta [3], Fumio Komori [1], Masahiro Miyauchi [4], Takahiro Kondo [2] and Iwao Matsuda [1,*]

1 Institute for Solid State Physics (ISSP), The University of Tokyo, Kashiwa 277-8581, Japan
2 Faculty of Pure and Applied Sciences, University of Tsukuba, Tsukuba 305-8571, Japan
3 Faculty of Engineering, Kyushu University, Fukuoka 819-0395, Japan
4 Department of Materials Science and Engineering, School of Materials and Chemical Technology, Tokyo Institute of Technology, 2-12-1 Ookayama, Meguro-ku, Tokyo 152-8552, Japan
* Correspondence: imatsuda@issp.u-tokyo.ac.jp

Abstract: We present an enhanced method for synthesizing sheets of borophane. Despite the challenges associated with low efficiency, we discovered that incorporating hydrochloric acid into the ion-exchange reaction significantly improved the production yield from 20% to over 50%. After a thorough examination of the reaction, we gained insight into the underlying mechanisms and found that the use of hydrochloric acid provides two key benefits: accelerated production of borophene and isolation of high-purity products. This method has the potential to pave the way for the production of novel topological 2D materials with potential industrial applications.

Keywords: two-dimensional material; hydrogen boride (HB); borophane; ion exchange; infrared spectroscopy

Citation: Zhang, X.; Hikichi, M.; Iimori, T.; Tsujikawa, Y.; Yuan, M.; Horio, M.; Yubuta, K.; Komori, F.; Miyauchi, M.; Kondo, T.; et al. Accelerated Synthesis of Borophane (HB) Sheets through HCl-Assisted Ion-Exchange Reaction with $YCrB_4$. *Molecules* **2023**, *28*, 2985. https://doi.org/10.3390/molecules28072985

Academic Editor: Federico Cesano

Received: 27 February 2023
Revised: 16 March 2023
Accepted: 21 March 2023
Published: 27 March 2023

1. Introduction

Explorations of two-dimensional (2D) materials beyond graphene have pushed the cutting edge of materials science, leading to the development of next-generation devices [1–6]. Recently, atomic sheets of boron, known as borophene, were synthesized as a promising counterpart to graphene [7–9]. In contrast to carbon, boron atoms can combine under diverse bonding schemes, resulting in rich allotropes that display a variety of electronic structures [10,11]. The unique characteristics of 2D boron offer exotic quantum states and intriguing features, such as anisotropic metallicity and phonon-mediated superconductivity [12–14]. However, the layers of borophene require epitaxial growth on a crystal surface under a ultrahigh vacuum, a process that is prone to oxidization under ambient conditions [11–15]. This disadvantage makes novel materials made from borophene difficult for industrial applications.

To passivate the boron atoms chemically and to extract the free-standing layer, the synthesis of hydrogen boride (HB, borophane, or hydrogenated borophene) has recently come under scrutiny through a variety of hydrogenation techniques. The most simple and energy-saving approach is the liquid exfoliation method at room temperature, which is associated with an ion-exchange reaction of materials that incorporate a 2D framework of boron [16–20]. In metal boride crystals, such as MgB_2 and $YCrB_4$, borophene layers are sandwiched by metal atoms. The ion-exchange reaction proceeds through the deintercalation of the metal cations and hydrogenation of the boron layers, resulting in the formation and subsequent extraction of HB layers. The hydrogen terminations are held on both sides of the free-standing layer and make the material robust within ambient environments. This extreme surface treatment of an atomic layer was achieved with MgB_2 crystals in forming the honeycomb borophane composed of six-membered (hexagon) rings of boron [16,17]. Recently, a different borophane was synthesized by the same method from the $YCrB_4$ crystals [18,20], which is a Dirac nodal-loop semimetal characterized by a $\mathbb{Z}_2$ topology [20,21]. The "topological borophane" is made of five-membered (pentagon) and seven-membered (heptagon) rings, arranged to exhibit non-symmorphic symmetry. For simplicity, we abbreviate the name of the material as "5,7-HB". The atomic structure of 5,7-HB, as verified by calculation, indicates that the material has a low area mass density of only 21 % compared to graphene, highlighting its potential as a versatile material for various applications [21]. With the existence of various 2D polymorphs of boron known to exist in crystals of metallic borides, numerous borophene structures are possible [10].

The basic procedure of the ion-exchange reaction is to mix the metal boride crystals and ion resins in an organic solvent such as acetonitrile [16–20]. The reaction typically results in a production yield of 42.3% and takes 2–4 days at room temperature under the atmospheric N_2 pressure, to obtain macroscopic samples of, for example, honeycomb HB developed from MgB_2 [16]. This method was also found to be applicable to $YCrB_4$ crystals [18–20]; the synthesis generally required over three weeks of reaction time to achieve a 20% production yield. The simple mixing process has the potential to shorten the reaction time and increase efficiency. In the research on honeycomb HB synthesis, a 50% production yield was achieved within two hours by adding formic acid to the solution [22]. This improvement is believed to occur because the acid plays the role of a mediator for efficient proton exchange between boride and resin.

In the present research, we examined the acidic effects from ion-exchange reactions during the synthesis of 5,7-HB from a $YCrB_4$ crystal, as schematically drawn in Figure 1. The reaction was accelerated by adding hydrochloric acid to obtain macroscopic samples and a yield of >50% after several hours, whereas, by comparison, the original ion-exchange reaction provided a production yield of 20% after three weeks. Quantitative characterizations were made through X-ray photoelectron spectroscopy (XPS) and infrared-ray absorption spectroscopy (IRAS). Microscopic observations were made to evaluate the size of the product sheet. Details of liquid exfoliation were uncovered at separate steps in the synthesis. When subjected to the ion-exchange reaction, the metal atoms Y and Cr from the mother material $YCrB_4$ become metal chlorides and precipitate as residues with byproducts, namely boric acid and remnants of the mother material, $YCrB_4$. The high-speed reaction and ease of isolating the pure product offer a new strategy for the efficient synthesis and mass production of functional borophene sheets.

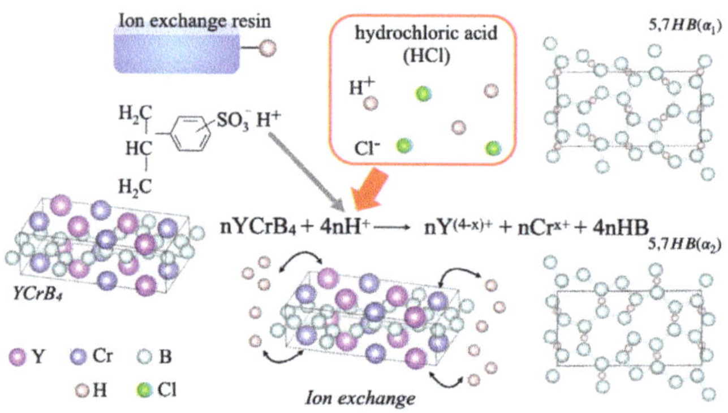

Figure 1. Concept of the synthesis of HB sheets with the 5- and 7-membered rings (5,7-HB) from a $YCrB_4$ crystal through the addition of hydrochloric acid (HCl).

2. Results and Discussion

2.1. Synthesis Process

A sheet of HB with a boron network of 5- and 7-membered rings can be prepared from crystals of $YCrB_4$ through the liquid exfoliation method with an ion-exchange reaction [18–20]. The reaction is described as

$$n\mathrm{YCrB_4} + 4n\mathrm{H^+} \rightarrow n\mathrm{Y^{(4-x)+}} + n\mathrm{Cr^{x+}} + 4n\mathrm{HB},$$

where HB represents a sheet of hydrogen boride or borophane. The reaction yield of the reaction can be evaluated by a ratio between n mol $YCrB_4$ and $4n$ mol HB.

Figure 2a illustrates the experimental procedures. The ion-exchange reaction was performed by stirring $YCrB_4$ crystals and ion-exchange resin in a solvent containing hydrochloric acid. The resulting reaction products were filtered and separated into solution and residues. The solution was then filtered again to remove any boric acid and other possible by-products. Fourier-transform (FT)-IRAS measurements on the solution, residues, and filtrate, while XPS measurements were conducted on the filtrate and residues. During the ion-exchange reaction, 0.2 mL samples of the solutions were taken using a volumetric pipette after reaction times lasting one hour, three hours, one day, and three days to conduct FT-IRAS measurements. We note that the solution contained layers of 5,7-HB that could be extracted through centrifugation (5000 rmp for 5 min) and filtration. The filtrate was also dropped and dried on a sample holder in Ar atmosphere for the XPS analysis. Figure 2b shows a series of photos taken after three days of reaction. Transparent and dark regions are observed in the sample depending on the amount of hydrochloric acid used. After removal of the ion-exchange resin and centrifugation, the sample was separated into a residue/precipitate [Figure 2c] and a solution that eventually provided the borophene sheet after filtration.

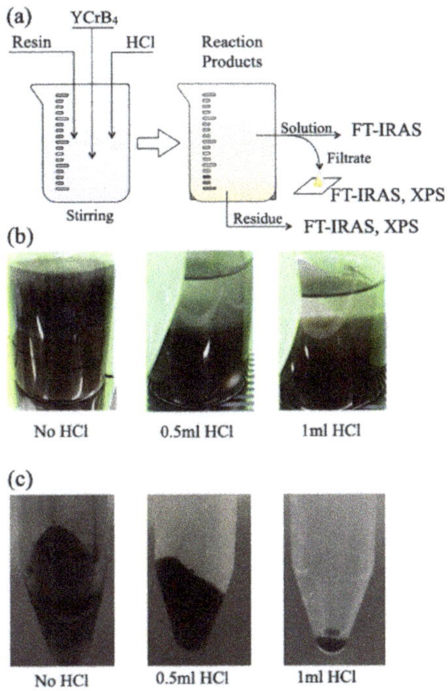

Figure 2. (**a**) Schematic drawing of the synthesis and characterization processes; (**b**,**c**) photos of (**b**) the reaction solutions and (**c**) the residue after three days of reaction with different amounts of hydrochloric acid.

Confirmations of the sheet morphology were made by two imaging approaches. For the first case, we isolated a single flake of the HB sheet from the sample filtrate (Figure 3a) by the spin-coating method, followed by observation with an optical microscope. As shown in Figure 3c, the sheet was clearly visible and had a size of 10 µm. For the second case, we dried the sample filtrate into powder (Figure 3b) and conducted the observation with a transmission electron microscope (TEM, operated at 200 kV). An edge of the HB sheet is presented in Figure 3d, providing concrete evidence of the formation of he 2D sheets of borophane. To further characterize the flakes of borophane, we used scanning electron microscopy (SEM) and an electron probe microanalysis (EPMA). Flakes with sizes of 10 µm are shown in the right panel of Figure 3e. Moreover, the elemental mapping in Figure 3e confirms the absence of possible impurities and byproducts, including boric acid. The HB flakes are nonuniform with various thicknesses.

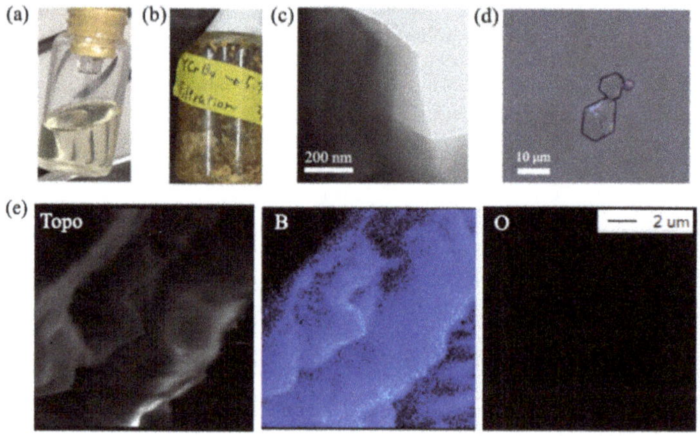

Figure 3. (**a**) A photo of the filtrate sample, synthesized by the addition of 1 mL HCl; (**b**) a photo of the powder sample after the drying treatment of (**a**); (**c**) an optical microscopy image of the single HB flake obtained after spin-coating of (**a**); (**d**) TEM images of the powder sample of (**b**). (**e**) SEM-EPMA images of the topographic (Topo) map and elemental maps of boron (B) and oxygen (O) of the HB flake.

Figure 4 shows a collection of X-ray diffraction (XRD) patterns of the HB and $YCrB_4$ samples. Diffraction peaks of the $YCrB_4$ crystals were found at angles, as expected. One can also capture the differences in the XRD signals of $YCrB_4$ before and after ball milling. After the process, the peak width became broader and new peaks appeared. These additional features indicate generations of the small crystals with more random orientations. This confirms that the ball milling procedure successfully grinds down crystals of the mother material and reduces their sizes for efficient ion exchange reaction. An XRD pattern of the HB sample, in contrast, shows broad features with a minor diffraction peak at $2\theta = 28.1°$ ($\pm 0.2°$) (an arrow in the figure). Based on Bragg's condition, the spacing corresponds to 3.17 Å ($\pm$0.03 Å). These results suggest that the HB sheets do not feature a long-range order but partially stack with each other with the interlayer spacing of 3.17 Å ($\pm$0.03 Å). This is consistent with the microscopic images of the powder samples that show the flexible nature of the sheet. It is, thus, reasonable to consider that the HB material essentially has an amorphous phase with partial stacking of the layers in contrast to a crystalline phase of $YCrB_4$. This is likely due to the peeling-off of boron layers by hydrogen adsorption during the reaction, which destroys the long-range order, rather than replacing metal and hydrogen atoms in the crystal framework.

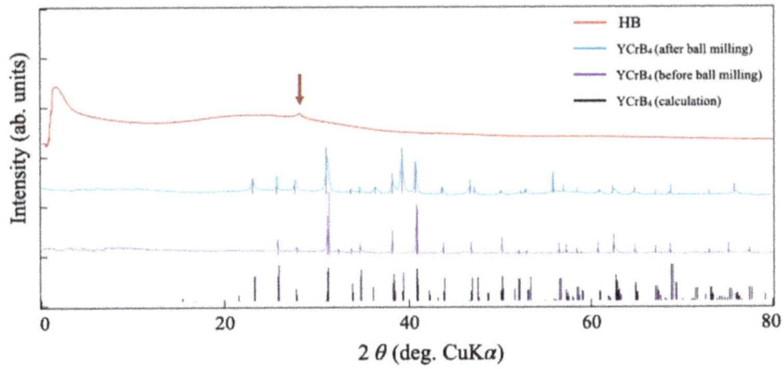

Figure 4. A collection of XRD patterns of the $YCrB_4$ crystals (before and after ball milling) and the HB powder. An arrow indicates a diffraction peak of the HB sample.

2.2. Enhancement of the HB Product by the Acid

Figure 5 compares the FT-IRAS spectra over a range of 450–4000 cm^{-1}, for the solvent (acetonitrile) and two types of reaction solutions, with and without hydrochloric acid, after one hour. The spectral peaks of the sample at 2250, 1450, 1050, 900, and 750 cm^{-1} match those of the solvent and are therefore assigned to the vibration modes of acetonitrile. The prominent peaks at around 640, 1640, and 3000–3500 cm^{-1} originate from the reaction products and were enhanced when the solution with hydrochloric acid was added. Two peaks at 640 and 1640 cm^{-1} are attributed to the HB samples [16,20,22–24]. The broad peak at 3000–3500 cm^{-1} is ascribed to the O-H stretching mode of the water molecule in the hydrochloric acid that remained in the sample. It is noteworthy that an IRAS peak around 1600 cm^{-1} may be attributed to vibrations of the H-O-H bending mode [25]. However, the peak intensity appears differently in previous reports, where the peak of the H-O-H bending mode appears much smaller in comparison to that of the O-H stretching mode [25]. Therefore, one can assign the peak at 1600 cm^{-1} as the B-H-B linkage mode [16,20,22–24]. In addition, the peak at 640 cm^{-1} is close to both the B-H stretching and B-B skeletal vibrational bonds, which are present in the borides [16,20,22–24,26,27]. In the previous reports, the peaks were assigned to the B-H stretching mode [16,20,22–24]. In the powder sample, the IRAS signal at 640 cm^{-1} is associated with the other peak at 2500 cm^{-1} [20]. When the HB powder is put back into the acetonitrile solution, the peak becomes undetectable. The FT-IRAS signal is ascribed to the stretching mode of the free terminal B-H bonds, located at the edges of the HB sheet. The absence of the signal at 2500 cm^{-1} in the liquid sample indicates that the free B-H bonds at the sheet edges are likely influenced by the solvent molecules (acetonitrile), making no signal detection at 2500 cm^{-1} in Figure 5. In this study, the peak observed at 640 cm^{-1} in the FT-IRAS spectrum is, thus, attributed to the B-B skeletal vibrational mode of the HB sheet, as previously reported in the literature, at 740 cm^{-1} in pure boron clusters [26]. The slight shift in the wave number can be attributed to the presence of hydrogen in the HB structure, which has an effect on the vibrational modes of the boron atoms. The IRAS signals provide evidence for the formation of the HB sheet, and the enhancement directly indicates an accelerated ion-exchange reaction by the acid.

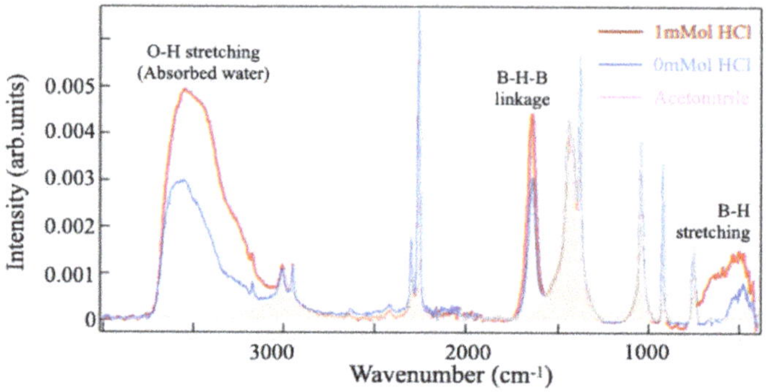

Figure 5. Comparison of the FT-IRAS spectra from the sample filtrate synthesized with (red curve) and without (blue curve) 1 mL of hydrochloric acid. For reference, the spectrum of the acetonitrile solvent (yellow shadow) is displayed.

To quantitatively reveal peak enhancements, differences between two FT-IRAS spectra were taken at different reaction times (Figure 6). Specifically, the spectrum of a sample solution without acid (0 mL HCl) was subtracted from that of a sample with 1 mL HCl at different reaction times [see Figure 6a–c], as well as from that of the sample filtrate obtained after three days of reaction [Figure 6d]. The labeled spectral peaks in the figure provide evidence of enhancements in the component due to the addition of HCl to the solution. The peaks labeled as A and B in Figure 6 at approximately 640 cm^{-1} and 1640 cm^{-1} are attributed to the vibrations of the B-B skeletal and B-H-B linkages modes in the HB sheet, respectively [16,22–24,26]. The peaks labeled C at ~3500 cm^{-1} are assigned to the O-H stretching mode of water molecules. The observed peaks confirm the accelerated reaction by hydrochloric acid. In the spectrum of the solution sample after three days of reaction (Figure 6c), one finds the broadening of peak A and the appearance of peaks D at 1050 cm^{-1} and E at 1540 cm^{-1}. After filtration, the spectrum shows an absence of both D and E peaks, in addition to the sharpening of peak C. Because the post-filtration process removes boric acid, $B(OH)_3$, from the sample, the spectral changes correspond to the vanishing of the FT-IRAS signal of this impurity. We also note that peaks A and B, which correspond to the HB sheet, become sharper and larger after filtration, suggesting that boric acid, a byproduct of the ion-exchange reaction, affects the neighboring HB sheets. Additionally, the intensity of the O-H stretching bond at 3500 cm^{-1} decreases after filtration, as shown in Figure 6c,d, while the intensity of the 1640 cm^{-1} signal increases, providing further evidence for the assignment of the signal at 1640 cm^{-1} to the B-H-B linkage mode rather than the H-O-H bending mode. The assignments are summarized in Table 1.

To examine the chemical composition of the filtrate sample shown in Figure 6d, XPS measurements were performed on the core-level binding energies of the key elements, B, Y, Cr, and Cl. Figure 7 displays a set of XPS spectra, including a reference spectrum of the filtrate sample prepared via the ion-exchange reaction without hydrochloric acid. After three days of reaction, no detectable amount of any compound was produced using the conventional method. In contrast, the new synthesis procedure with hydrochloric acid results in the appearance of a B 1s peak at a binding energy of 187.6 eV, providing evidence for the formation of sheets of hydrogen boride [16,18,20,22]. The XPS chemical shift of negatively charged boron, in combination with the detected HB bonds in FT-IRAS, clearly indicates the formation of the HB sheet. Additionally, the TEM image of the sample in Figure 3 further supports that it is not composed of elemental boron, as elemental boron sheets cannot exist in the air due to oxidation. Moreover, no signal of the Y 3d and Cr 2p core levels means that there is no remnant of the $YCrB_4$ crystal, Y cation, or Cr cation in the sample. In addition, the absence of the Cl 2p peak indicates that the counter ion in hydrochloric acid does not make any compound as a byproduct.

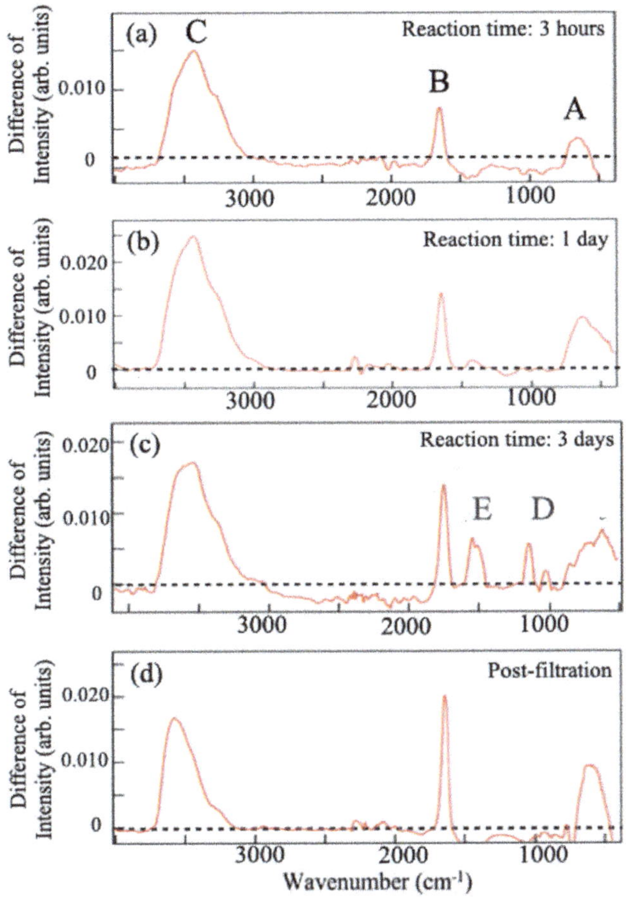

Figure 6. (**a**–**c**) FT-IRAS difference spectra of the solution samples with or without hydrochloric acid (HCl) taken at various reaction times: (**a**) 3 h, (**b**) 1 day, and (**c**) 3 days. (**d**) FT-IRAS difference spectrum of the filtrate after filtration of the sample (**c**). Peaks are labeled with capital letters.

Table 1. Summary of the assignment of the FT-IRAS peaks (Figure 6); the assignments are based on references [16,22–24,26,27] (unit: cm^{-1}).

Label	3 h	1 day	3 days	Post-Filtration	Reference (Vibration Mode)
A	634	631	635	632	B-B skeletal [26,27]
B	1652	1654	1645	1639	B-H-B linkage [16]
C	3232	3524	3014	3429	O-H stretching [28]
D	-	-	1050	-	H-O-B deforming [28]
E	-	-	1540	-	B-O stretching [29]

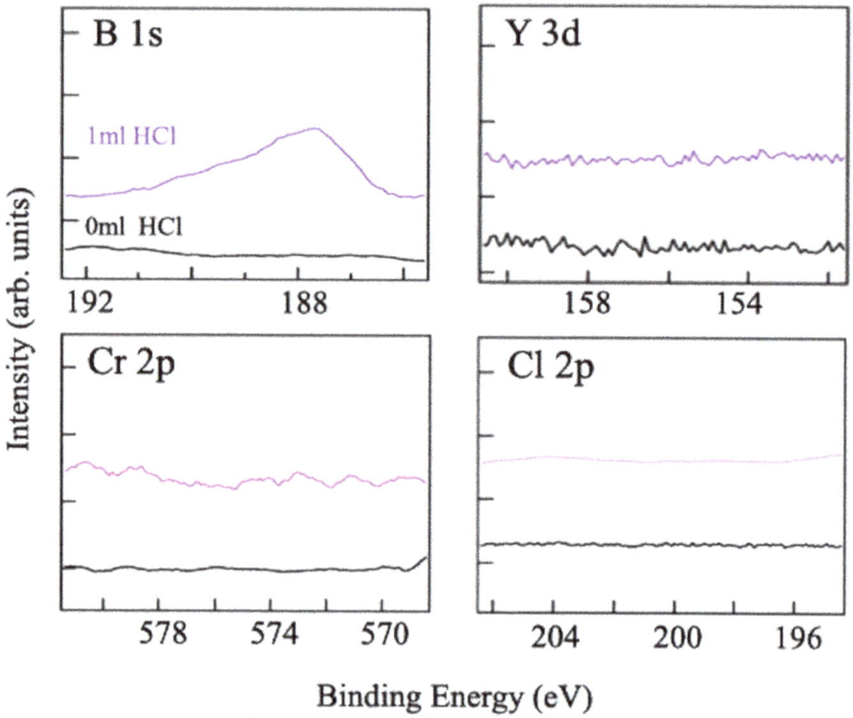

Figure 7. Collection of XPS spectra (purple curves) at the binding energy regions of B 1s, Y 3d, Cr 2p, and Cl 2p core levels for the filtrate sample after three days of reaction with hydrochloric acid (1 mL HCl). For comparison, the individual figures contain the XPS spectra (black curves) of the sample, prepared using the conventional method without hydrochloric acid (0 mL HCl).

Figure 8a,b show the variations in the intensity of the FT-IRAS peaks around 1640 cm^{-1} (B-H-B linkage mode) and 630 cm^{-1} (B-B skeletal mode) for the ion-exchange reaction processes with and without hydrochloric acid. The addition of acid clearly enhances the intensity of both peaks. Moreover, the modes exhibit contrasting behavior. The signal from the B-B skeletal mode increases consistently throughout the process, while the B-H-B linkage mode saturates at an early stage and decreases after several days of reaction. Given that boric acid was identified at this stage in synthesis [Figure 6c], the impurity likely intervenes in the formation of the HB sheet. The product has also shown its increase with the amount of hydrochloric acid (Figure 8c). The enhancement can be quantified by taking ratios of the FT-IRAS intensity at the vibration modes from the data obtained with and without HCl. Increments in the ratio [Figure 8d] are larger for the B-B skeletal mode than for the B-H-B linkage mode. The result indicates that during hydrogenation in the ion-exchange reaction, the formation of the B-H-B bond on the HB sheet begins with B-B bonding, and molecules of boric acid most likely intervene in the subsequent generation of the three-center two-electron (3c-2e) scheme in the B-H-B bond.

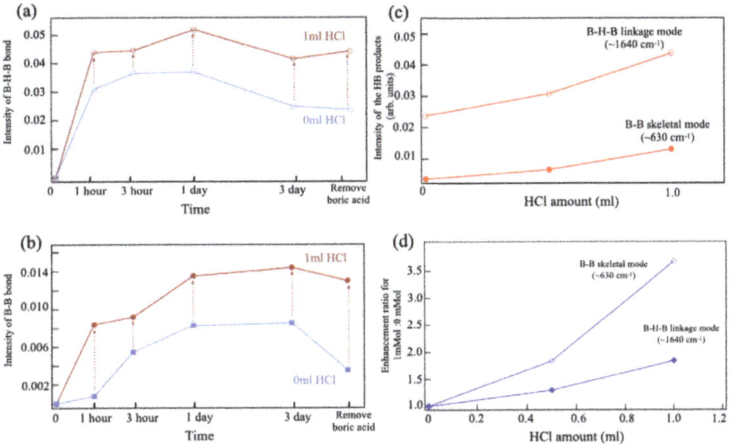

Figure 8. (**a**,**b**) Variations of the FT-IRAS intensity at around (**a**) 630 cm^{-1} (B-B skeletal mode) and (**b**) 1640 cm^{-1} (B-H-B linkage mode) at steps in the synthesis for different reaction times and post-treatments. (**c**,**d**) Changes in (**c**) intensity and (**d**) the ratio of the FT-IRAS signals around 630 and 1640 cm^{-1} with hydrochloric acid concentrations.

With the addition of HCl, the production yield of the reaction significantly improved from around 20% (0 mL HCl), 34% (0.5 mL HCl) to over 50% (1 mL HCl). These results provide evidence for the accelerated synthesis of borophane with 5- and 7-membered boron rings through the use of hydrochloric acid. The use of the acid in HB synthesis has been most notably conducted for the honeycomb borophane with 6-membered boron rings. The synthesis was made through the ion-exchange reaction of MgB_2 and is accelerated by formic acid [22]. In the present research, the synthesis of 5,7-HB was examined with formic acid but the reaction process and production yield remained unchanged. The difference arises because the mother materials of metal borides react differently with the acid type. The right combination is the key factor in synthesizing the variety of HB sheets through the acid-accelerated ion-exchange reaction.

2.3. Analysis of the Precipitate Residue after the Reaction

The addition of hydrochloric acid in the ion-exchange reaction results in an apparent separation of the solution and the precipitate [Figure 2b]. Drying the precipitate results in a solid residue, Figure 2c, which allows us to handle it for analysis or post-treatment. Figure 9 shows a collection of the FT-IRAS spectra of the various types of the residue obtained after three days of reaction. For comparison, the figure provides the spectra of the solvent (acetonitrile). All residues have similar spectral features, share the same origins, and contain no signal of the volatile solvent. Peaks are found at 630 and 1640 cm^{-1}, which can be assigned to the vibration modes of the HB sheet (Table 1, peaks A and D). The spectral appearance implies the existence of 5,7-HB in the residue. The other vibration peaks at 1000–1500 cm^{-1} (D) and the broadband at 3500 cm^{-1} (E) indicates that there are also compounds with O-B and O-H bonds, in comparison with the references. In Figure 9, a spectrum of a mixture of boric acid and $YCrB_4$ is given as a reference to make comparisons with the residue spectra. The main features in the spectra of the residue show similarities with those of the starting materials, $YCrB_4$, and the byproducts, such as boric acid, $B(OH)_3$, confirming their presence in the residue.

After confirming the absence of Y, Cr, and Cl atoms in the filtrate sample of the HB sheet, we conducted an XPS analysis of the residue (Figure 10) to trace the elements. The appearance of the Cl core-level peaks indicates the production of chlorides and, accordingly, peaks of YCl_3 and $CrCl_3$ are found together with the remnant $YCrB_4$ in the individual spectra [30,31]. Two peaks are observed in the B 1s core-level spectrum at 187 and 192 eV. The former corresponds to negatively charged boron atoms that can be assigned to HB and $YCrB_4$, whereas the latter is assigned to positively charged boron of boric acid. These results indicate that Y and Cr atoms are efficiently captured as chlorides or are kept in the mother material that remains. By gathering the residue after the reaction, the Y and Cr atoms can be collected easily in solids. Then they become raw materials to produce $YCrB_4$, which can be used in the next synthesis of the 5,7-HB. Thus, this HCl-assisted ion-exchange reaction has the advantage of recycling rare-earth metal.

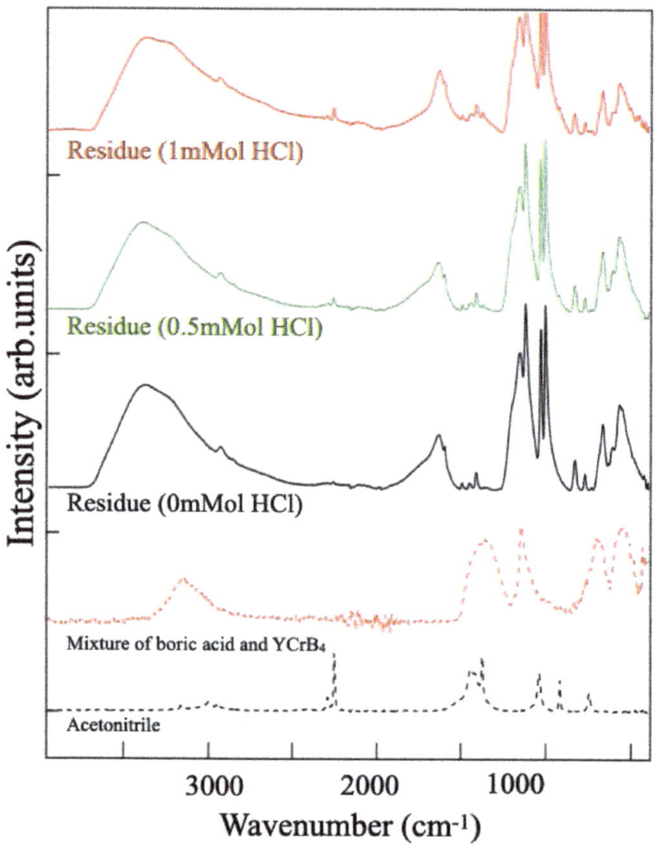

Figure 9. FT-IRAS spectra of the residue obtained from different amounts of HCl, taken after three days of reaction. Reference spectra of a mixture (boric acid and YCrB4) and the acetonitrile solvent are also shown.

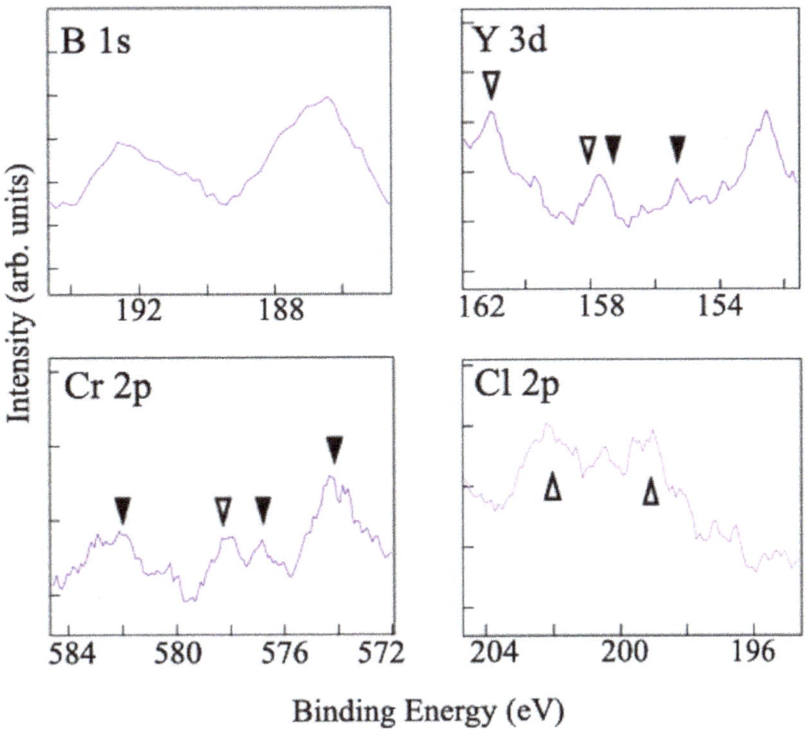

Figure 10. XPS spectra of the residue, taken after three days of reactions, for the core levels of B 1s, Y 3d, Cr 2p, and Cl 2p. Peaks of YCl_3 and $CrCl_3$ are labeled by open triangles [30,31], whereas remnants of $YCrB_4$ are labeled with black triangles [16,18,20].

To deepen the understanding of the ion-exchange reaction, the synthesis was made without the ball-milling process, leaving the mother crystals of $YCrB_4$ at mm size. Figure 11 shows several FT-IRAS difference spectra taken at separate reaction times and after the post-filtration of a sample after three days of reaction. The assignments of FT-IRAS peaks are summarized in Table 2. Peaks corresponding to the B-B skeletal (A'), as well as B-H-B linkage (B') vibrational modes of the HB·sheet were observed after a three-hour duration at 639 and 1627 cm^{-1}, respectively. Around 3500 cm^{-1}, a peak appears that is associated with the O-H stretching mode, C'. The spectral appearances are similar to those in Figure 6a,b. However, after one day, the spectrum is governed by peaks of boric acid, D' and E', at 1050 and 1512 cm^{-1}, respectively, with the weak B' peak. The FT-IRAS spectra show the presence of the feature at 2500 cm^{-1} that can be assigned to the free B-H bonding mode [20]. Since the spectral appearance is associated with the reduction of the A' and B' signals, it may be ascribed to broken fragments of the small HB molecules from the sheet. Eventually, after three days, no appearance of the FT-IRAS signal of the boron material was detected in the solution or the filtrate. The products precipitate at the bottom of the reaction beaker.

These results reveal significant details of the ion-exchange reaction underlying the synthesis of 5,7-HB. From Figures 6 and 11, one finds that the production of the HB sheet is followed by that of boric acid and the crossover time depends on the size of the mother material, $YCrB_4$. The reaction time required for HB synthesis is longer when the crystal size is reduced through ball milling. This is because as the crystal size decreases, the ratio between surface area and volume increases, and the ion-exchange reaction primarily occurs at the surface of the crystal. Exfoliation results in a larger amount of product when the crystal size is smaller. After the reaction, molecules of borophane and boric acid precipitate and mix to form the residue. Therefore, to obtain high-purity products of 5,7-HB, the reaction time needs to be optimized and depends on the size of the $YCrB_4$ crystal and the desired level of filtrate extraction.

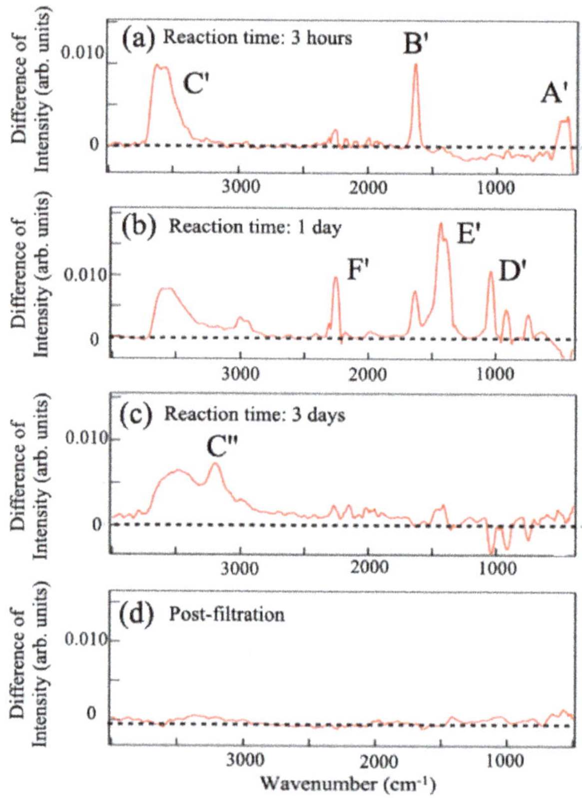

Figure 11. Difference spectrum of FT-IRAS, obtained by subtracting the spectrum of the normal ion-exchange sample from the spectrum of the acid-ion exchange sample, synthesized from bulk $YCrB_4$. The measurements were taken at (**a**) 3 h, (**b**) 1 day, (**c**) 3 days, and (**d**) post-filtration. Peaks of the vibrational modes are labeled.

Table 2. Summary of the assignments for the FT-IR peaks (Figure 11, sample synthesized from bulk $YCrB_4$); the assignments are based on references [16,22–24] (unit: cm^{-1}).

Label	3 h	1 day	3 days	Post-Filtration	Reference (Vibration Mode)
A′	639	-	-	-	B-B skeletal [26]
B′	1627	1642	-	-	B-H-B linkage [16]
C′	3540	3550	3500	-	O-H stretching [29]
D′	-	1050	-	-	B-H stretching [22,29]
E′	-	1512	-	-	B-O stretching [28]
F′	-	2364	-	-	B-H stretching [23]
C″	-	-	3225	-	O-H stretching [29]

3. Materials and Methods

3.1. Synthesis Methods

Polycrystalline $YCrB_4$ was fabricated by the arc melting method [20]. Rare earth metals, Y (890 mg), Cr (520 mg), and B (432 mg) were loaded in a Cu hearth with an atomic ratio of 1:1:4, and melted under the Ar atmosphere. In total, the resulting polycrystals were 1.632 g and were milled in an attritor by 5 mm diameter ZrO_2-based balls, reduced to μm-sized powders. The crystalline features were confirmed by X-ray diffraction (XRD).

The ion exchange reaction was prepared as follows. 100 mg powders of $YCrB_4$ were mixed with a 20 mL cation ion-exchange resin (0.5–1.0 mm, Amberlite IR120B hydrogen form, Organo Corp., Tokyo, Japan) into a 50 mL solvent of acetonitrile (99.5%, Wako Pure Chemical Industries, Ltd., Osaka, Japan). Then, 1 mol/L HCl (0 mL/0.5 mL/1 mL) of hydrochloric acid was added to the solvent. The reaction was continuously stirred (250 rpm) using a magnetic stirrer under inert gas (Ar) at room temperature. The filtration was conducted after ion exchange reaction (0.2 μm pore filter, Omnipore Membrane Filters, Merck Millipore, Billerica, MA, USA).

3.2. *Characterization Methods*

XRD (SmartLab 3 kW) measurements were conducted with CuKα operated at 40 kV and 30 mA at room temperature in an ambient atmosphere. The TEM (JEOL-2100) was operated with 200 keV. The SEM-EPMA (JEOL JXA-8530F) was conducted with 20 keV.

FT-IRAS measurements were performed to investigate the vibrational modes of the sample that evidence the formation of vibrational bonds in the HB sheet. The experiment was conducted using a commercial system (BRUKER ALPHA II FT-IR spectrometer). The FT-IRAS spectra were recorded by applying the attenuated total reflectance method and using a prism holder that was filled with a solution of either a sample of the filtrate or a dried sample of the residue. All measurements were made at room temperature.

Elemental compositions and chemical states of the samples were examined by XPS. Measurements were performed with the incident X-ray beam of energy hν = 1253.6 eV (Mg Kα) and a photoelectron analyzer (Scienta Omicron DA30-L-8000). Before their transfer into the UHV chamber, samples were annealed at 373 K in nitrogen gas to remove water molecules as they violate the UHV condition for XPS measurements. The XPS spectra were recorded at room temperature.

4. Conclusions

We discovered that adding hydrochloric acid to the ion-exchange reaction significantly improves the efficiency of producing topological borophane, a novel 2D material. The production yield increases to 50% compared to the reaction without the acid (20%). The high-purity borophane sheets are dispersed in solution and can be extracted through filtration. The metal atoms of the mother material, $YCrB_4$, Y and Cr, form metal chlorides and precipitate as a residue along with byproducts such as boric acid and remnants of $YCrB_4$. The use of hydrochloric acid provides advantages in both accelerating production yield and isolating high-purity products. The mass-production method developed in this research paves the way for expanding fundamental research on novel topological 2D materials that may have a wide range of industrial applications.

Author Contributions: Conceptualization, X.Z., T.K., M.M., and I.M.; methodology, X.Z., M.H. (Miwa Hikichi), T.I., Y.T., M.Y., M.H. (Masafumi Horio), F.K., K.Y., and M.M.; validation, X.Z., M.H. (Miwa Hikichi), Y.T., and T.I.; formal analysis, X.Z.; investigation, X.Z., M.H. (Miwa Hikichi), and T.I.; resources, K.Y.; data curation, X.Z.; writing—original draft preparation, X.Z.; writing—review and editing, X.Z. and I.M.; visualization, X.Z. and I.M.; supervision, T.K. and I.M.; project administration, I.M.; funding acquisition, I.M. All authors have read and agreed to the published version of the manuscript.

Funding: This work was supported by JSPS KAKENHI grants JP21H05012, JP19H04398, and JP18H03874, and by a JST, CREST grant JPMJCR21O4.

Data Availability Statement: Not applicable.

Acknowledgments: The TEM measurement was carried out at the Institute for Solid State Physics, the University of Tokyo. Daisuke Nishio-Hamane is acknowledged for his technical invaluable contributions in the operation of TEM and optical microscopy measurements. Kazuki Yamaguchi is acknowledged for his support of the spin-coating procedures and optical microscopy measurement. Rieko Ishii is acknowledged for her technical support of the arc-melting procedure to produce the $YCrB_4$ crystals. Hikari Yoshioka is acknowledged for her support during the synthesis and the measurement of infrared-ray absorption spectroscopy.

Conflicts of Interest: The authors declare no conflict of interest.

References

1. Novoselov, K.; Geim, A.; Morozov, S.; Jiang, D.; Katsnelson, M.; Grigorieva, I.; Dubonos, S.; Firsov, A. Two-dimensional gas of massless Dirac fermions in graphene. *Nature* **2005**, *438*, 197–200. [CrossRef] [PubMed]
2. Neto, A.; Guinea, F.; Peres, N.; Novoselov, K.; Geim, K. The electronic properties of graphene. *Rev. Mod. Phys.* **2009**, *81*, 109. [CrossRef]
3. Matsuda, I. *Monatomic Two-Dimensional Layers: Modern Experimental Approaches for Structure, Properties, and Industrial Use*; Elsevier: Amsterdam, The Netherlands, 2018.
4. Chhowalla, M.; Shin, H.; Eda, G.; Li, L.; Loh, L.; Zhang, H. The chemistry of two-dimensional layered transition metal dichalcogenide nanosheets. *Nat. Chem.* **2013**, *5*, 263. [CrossRef] [PubMed]
5. Shekhar, C.; Nayak, A.; Sun, Y.; Schmidt, M.; Nicklas, M.; Leermakers, I.; Zeitler, U.; Skourski, Y.; Wosnitza, J.; Liu, Z.; et al. Extremely large magnetoresistance and ultrahigh mobility in the topological Weyl semimetal candidate NbP. *Nat. Phys.* **2015**, *11*, 645. [CrossRef]
6. Nair, R.; Blake, P.; Grigorenko, A.; Novoselov, K.; Booth, T.; Stauber, T.; Peres, N.; Geim, A. Fine Structure Constant Defines Visual Transparency of Graphene. *Science* **2008**, *320*, 1308. [CrossRef] [PubMed]
7. Matsuda, I.; Wu, K. (Eds.) *2D Boron: Boraphene, Borophene, Boronene*, 1st ed.; Springer: Cham, Switzerland, 2021.
8. Feng, B.; Zhang, J.; Ito, S.; Arita, M.; Cheng, C.; Chen, L.; Wu, K.; Komori, F.; Sugino, O.; Miyamoto, K.; et al. Discovery of 2D Anisotropic Dirac Cones. *Adv. Mater.* **2018**, *30*, 1704025. [CrossRef] [PubMed]
9. Feng, B.; Sugino, O.; Liu, R.Y.; Zhang, J.; Yukawa, R.; Kawamura, M.; Iimori, T.; Kim, H.; Hasegawa, Y.; Li, H.; et al. Dirac Fermions in Borophene. *Phys. Rev. Lett.* **2017**, *118*, 096401. [CrossRef]
10. Ando, Y.; Zhang, X.; Tsujikawa, Y.; Sato, Y.; Horio, M.; Haruyama, J.; Sugino, O.; Kondo, T.; Matsuda, I. Homotopic Analysis of Quantum States in Two-Dimensional Polymorphs by a Herringbone Lattice Model. *Phys. Rev. B* **2022**, *106*, 195106. [CrossRef]
11. Tsujikawa, Y.; Horio, M.; Zhang, X.; Senoo, T.; Nakashima, T.; Ando, Y.; Ozaki, T.; Mochizuki, I.; Wada, K.; Hyodo, T.; et al. Structural and electronic evidence of boron atomic chains. *Phys. Rev. B* **2022**, *106*, 205406. [CrossRef]
12. Lopez-Bezanilla, A.; Littlewood, P.B. Electronic Properties of 8-Pmmn Borophene. *Phys. Rev. B* **2016**, *93*, 241405. [CrossRef]
13. Gao, M.; Li, Q.-Z.; Yan, X.-W.; Wang, J. Prediction of Phonon-Mediated Superconductivity in Borophene. *Phys. Rev. B* **2017**, *95*, 024505. [CrossRef]
14. Feng, B.; Zhang, J.; Liu, R.-Y.; Iimori, T.; Lian, C.; Li, H.; Chen, L.; Wu, K.; Meng, S.; Komori, F.; et al. Direct Evidence of Metallic Bands in a Monolayer Boron Sheet. *Phys. Rev. B* **2016**, *94*, 041408. [CrossRef]
15. Mannix, A.; Zhou, X.; Kiraly, B.; Wood, J.; Alducin, D.; Myers, B.D.; Liu, X.; Fisher, B.; Santiago, U.; Guest, J.; et al. Synthesis of Borophenes: Anisotropic, Two-dimensional Boron Polymorphs. *Science* **2015**, *350*, 15131516. [CrossRef] [PubMed]
16. Nishino, H.; Fujita, T.; Cuong, N.T.; Tominaka, S.; Miyauchi, M.; Iimura, S.; Hirata, A.; Umezawa, N.; Okada, S.; Nishibori, E.; et al. Formation and Characterization of Hydrogen Boride Sheets Derived from MgB_2 by Cation Exchange. *J. Am. Chem. Soc.* **2017**, *139*, 13761–13769. [CrossRef] [PubMed]
17. Tateishi, I.; Cuong, N.T.; Moura, C.A.S.; Cameau, M.; Ishibiki, R.; Fujino, A.; Okada, S.; Yamamoto, A.; Araki, M.; Ito, S.; et al. Semimetallicity of Free-standing Hydroenated Monolayer Boron from MgB_2. *Phys. Rev. Mat.* **2019**, *3*, 024004.
18. Niibe, M.; Cameau, M.; Cuong, N.T.; Sunday, O.I.; Zhang, X.; Tsujikawa, Y.; Okada, S.; Yubuta, K.; Kondo, T.; Matsuda, I. Electronic Structure of a Borophene Layer in Rare-earth Aluminum/Chromium Boride and its Hydrogenated Derivative Borophane. *Phys. Rev. Mat.* **2020**, *5*, 084007. [CrossRef]
19. Tateishi, I.; Zhang, X.; Matsuda, I. Electronic Structures of Polymorphic Layers of Borophane. *Molecules* **2022**, *27*, 1808. [CrossRef]
20. Zhang, X.; Tsujikawa, Y.; Tateishi, I.; Niibe, M.; Wada, T.; Horio, M.; Hikichi, M.; Ando, Y.; Yubuta, K.; Kondo, T.; et al. Electronic Topological Transition of 2D Boron by the Ion Exchange Reaction. *J. Phys. Chem. C.* **2022**, *126*, 12802–12808. [CrossRef]
21. Cuong, N.T.; Tateishi, I.; Cameau, M.; Niibe, M.; Umezawa, N.; Slater, B.; Yubuta, K.; Kondo, T.; Ogata, M.; Okada, S.; et al. Topological Dirac Nodal Loops in Nonsym-morphic Hydrogenated Monolayer Boron. *Phys. Rev. B* **2020**, *101*, 195412. [CrossRef]
22. Kawamura, R.; Yamaguchi, A.; Shimada, C.; Ishibiki, R.; Fujita, T.; Kondo, T.; Miyauchi, M. Acid Assisted Synthesis of HB Sheets Through Exfoliation of MgB_2 Bulk in Organic Media. *Chem. Lett.* **2020**, *49*, 1194–1196. [CrossRef]
23. Tominaka, S.; Ishibiki, R.; Fujino, A.; Kawakami, K.; Ohara, K.; Masuda, T.; Matsuda, I.; Hosono, H.; Kondo, T. Geometrical Frustration of BH Bonds in Layered Hydrogen Borides Accessible by Soft Chemistry. *Chem* **2020**, *6*, 406. [CrossRef]
24. Rojas, K.I.M.; Cuong, N.T.; Nishino, H.; Ishibiki, R.; Ito, S.; Miyauchi, Y.; Fujimoto, Y.; Tominaka, S.; Okada, S.; Hosono, H.; et al. Chemical Stability of Hydrogen Boride Nanosheets in Water. *Commun. Mater.* **2021**, *2*, 81. [CrossRef]
25. Seki, T.; Chiang, K.Y.; Yu, C.C.; Yu, X.; Okuno, M.; Hunger, J.; Nagata, Y.; Bonn, M. The Bending Mode of Water: A Powerful Probe for Hydrogen Bond Structure of Aqueous Systems. *J. Phys. Chem. Lett.* **2020**, *11*, 8459–8469. [CrossRef]

26. Hou, C.; Tai, G.; Hao, J.; Sheng, L.; Liu, B.; Wu, Z. Ultrastable Crystalline Semiconducting Hydrogenated Borophene. *Angew. Chem. Int. Ed.* **2020**, *59*, 10819–10825. [CrossRef] [PubMed]
27. Zhang, P.; Tian, X.; Sheng, S.; Ma, C.; Chen, L.; Feng, B.; Zhang, Y.; Chen, L.; Zhao, J.; Wu, K. Vibrational Property of α-Borophene Determined by Tip-Enhanced Raman Spectroscopy. *Molecules* **2022**, *27*, 834. [CrossRef] [PubMed]
28. Andrews, L.; Burkholder, T.R. Infrared Spectra of Molecular $B(OH)_3$ and HOBO in Solid Argon. *J. Chem. Phys.* **1992**, *97*, 7203. [CrossRef]
29. Kawamura, R.; Cuong, N.T.; Fujita, T.; Ishibiki, R.; Hirabayashi, T.; Yamaguchi, A.; Matsuda, I.; Okada, S.; Kondo, T.; Miyauchi, M. Photoinduced Hydrogen Release from Hydrogen Boride Sheets. *Nat. Commun.* **2019**, *10*, 4880. [CrossRef]
30. Uwamino, Y.; Tsuge, A.; Ishizuka, T.; Yamatera, H. X-ray Photoelectron Spectroscopy of Rare Earth Halides. *Bull. Chem. Soc. Jpn.* **1986**, *59*, 2263–2267. [CrossRef]
31. Sleigh, C.; Pijpers, A.P.; Jaspers, A.; Coussens, B.; Meier, R.J. On the Determination of Atomic Charge via ESCA Including Application to Organometallics. *J. Electron Spectrosc. Relat. Phenom.* **1996**, *77*, 41–57. [CrossRef]

molecules

Review

Electronic Structures of Polymorphic Layers of Borophane

Ikuma Tateishi [1,*], Xiaoni Zhang [2] and Iwao Matsuda [2]

1 RIKEN Center for Emergent Matter Science, Wako 351-0198, Saitama, Japan
2 Institute for Solid State Physics, The University of Tokyo, Kashiwa 277-8581, Chiba, Japan; xnzhang17@g.ecc.u-tokyo.ac.jp (X.Z.); imatsuda@issp.u-tokyo.ac.jp (I.M.)
* Correspondence:ikuma.tateishi@riken.jp

Abstract: The search for free-standing 2D materials has been one of the most important subjects in the field of studies on 2D materials and their applications. Recently, a free-standing monolayer of hydrogenated boron (HB) sheet has been synthesized by hydrogenation of borophene. The HB sheet is also called borophane, and its application is actively studied in many aspects. Here, we review recent studies on the electronic structures of polymorphic sheets of borophane. A hydrogenated boron sheet with a hexagonal boron frame was shown to have a semimetallic electronic structure by experimental and theoretical analyses. A tight-binding model that reproduces the electronic structure was given and it allows easy estimation of the properties of the material. Hydrogenated boron sheets with more complicated nonsymmorphic boron frames were also analyzed. Using the symmetry restrictions from the nonsymmorphic symmetry and the filling factor of hydrogenated boron sheets, the existence of a Dirac nodal line was suggested. These studies provide basic insights for research on and device applications of hydrogenated boron sheets.

Keywords: borophene; borophane; 2D material; topological material; crystalline symmetry; Dirac electron; Dirac nodal line material

Citation: Tateishi, I.; Zhang, X.; Matsuda, I. Electronic Structures of Polymorphic Layers of Borophane. *Molecules* **2022**, *27*, 1808. https://doi.org/10.3390/molecules27061808

Academic Editor: Athanassios C. Tsipis

Received: 31 December 2021
Accepted: 22 February 2022
Published: 10 March 2022

1. Introduction

In recent years, two-dimensional (2D) materials have attracted much interest due to their unique physical properties and potential applications in a variety of quantum devices [1–5]. One of the most famous 2D material is graphene. Graphene shows a wide variety of interesting properties, such as a high mobility, the anomalous integer quantum Hall effect, edge-dependent mesoscopic effect, and so on [6]. In particular, it is known that the linear dispersive electronic band of graphene, which is called Dirac electron, and the topological properties of the wave function around it play an important role in the unique physical properties of graphene. Recently, as an analog of graphene, honeycomb-lattice materials with other atoms (Xenes) such as silicene [7–11], phosphorene [12–16], germanene [11,17–19], arsenene [14,20,21], antimonene [14,20,22], and bismuthene [14,23,24], have also been actively studied. These materials are expected to be a new platform to explore unique physical properties qualitatively different from those of graphene, taking advantage of the difference in model parameters such as the strength of spin–orbit interaction. As a development in another direction, a material with a similar topological electronic band structure has been explored. The typical example is a Dirac nodal line material [25–27]. While graphene has a band dispersion with degenerate points with a linear dispersion around it, so-called Dirac points and Dirac cones, Dirac nodal line materials have a degenerate line in the momentum space in the band dispersion. Many nodal line materials have been proposed in 3D crystals [27–32], and Cu_2Si [5] and CuSe [33] are also known as 2D Dirac nodal line materials. While these 2D materials possess attractive properties, most of them are restricted to preparation on substrates. For convenience in experiments and applications, the search for free-standing 2D materials has been one of the important subjects in this field [34]. Very recently, a monolayer of boron sheet has been synthesized by hydrogenation of a borophene frame in a MgB_2 crystal [35,36], which is one of the

Xenes. The obtained monolayer hydrogenated boron (HB) sheet, or "borophane", has been attracting attention as a new 2D material [37–40]. Because the HB sheets are synthesized from boron frames in a 3D crystal material, another type of HB with a different boron frame has also been synthesized [41]. Not only a variety of borophene systems synthesized on metal substrates [42–47], but the free-standing HB sheets are also an important platform to study the 2D boron materials.

In this review, we first review what kind of boron frame we can find in 3D crystals that exist in nature. Next, we review recent works on the electronic structure of a HB sheet with a honeycomb boron frame, which is the most basic example of HB. Based on that, we present two subsequent studies. One of them is a study on the electronic structure of honeycomb HB nanoribbons. The other is a study on HB sheets with other types of boron frame, nonsymmorphic HB sheets. Finally, we review the results of these experimental studies on the HB sheets.

2. Borophene in Natural 3D Crystals

The synthesis of a HB sheet is realized by the proton ion-exchange reaction of magnesium diboride (MgB_2): $MgB_2 + 2H^+ \rightarrow Mg^{2+} + 2HB$ [35]. By this method, the honeycomb boron frame in MgB_2 is peeled and obtained as a free-standing monolayer with the hydrogen termination. This method can be used in other metal-boron-layered materials to obtain other types of HB sheets. Therefore, in this section, we first review what kind of borophene layer we can find in 3D crystals in nature.

The most standard one is the hexagonal boron frame in MgB_2 [48] (Figure 1a), as introduced above. In a MgB_2 crystal, there are two layers in the unit cell, a Mg layer and a hexagonal borophene layer. It is noteworthy that the borophene layer has the same symmetry as graphene. Because graphene exhibits a great variety of quantum phenomena, the hexagonal borophene and HB from MgB_2 is the most actively studied material in the field of borophene and HB.

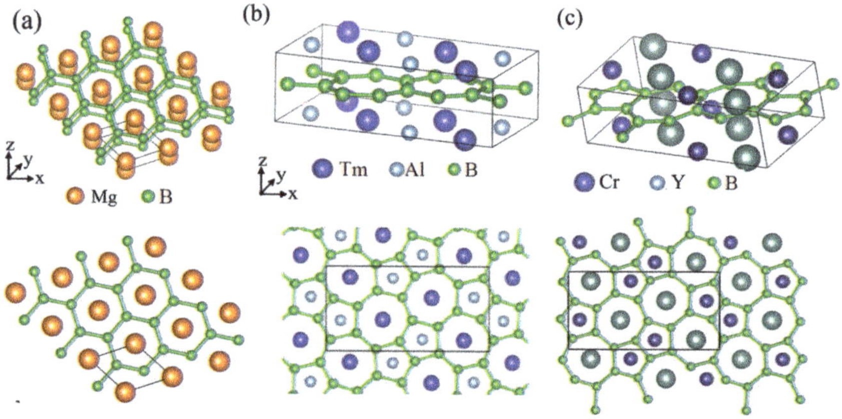

Figure 1. Crystal structures of metal-boron-layered materials. (**a**) Crystal structure of MgB_2. (**b**) Crystal structure of $TmAlB_4$. (**c**) Crystal structure of $YCrB_4$.

The next example is $TmAlB_4$ [49], which consists of a metal atoms (Tm,Al) layer and a borophene layer (Figure 1b). The characteristic feature of this material is that the crystal and borophene frame have a nonsymmorphic symmetry. A nonsymmorphic symmetry is a symmetry described by a screw or glide, which are a rotation or mirror followed by a fractional translation, respectively. For example, when basic translation vectors $\boldsymbol{a}_1$, $\boldsymbol{a}_2$, and $\boldsymbol{a}_3$ are, respectively, defined along the x, y, and z axes, the crystal structure of $TmAlB_4$ is invariant for a screw $\left\{C_{2x}\middle|\frac{1}{2}\frac{1}{2}0\right\}$, which is a two-fold rotation along the x axis (C_{2x})

followed by a fractional translation $\frac{1}{2}(a_1 + a_2)$. Because this nonsymmorphic symmetry gives a strong restriction on the electronic band structure of the material, the borophene and HB from $TmAlB_4$ is also an interesting target of research in this field. The same borophene frame is also found in $YCrB_4$ (Figure 1c) [50]. These materials can be used as parent materials of nonsymmorphic borophene and HB.

In the following sections, we see electronic structures of the borophene and HB obtained from these materials. First, we review a study on the most standard ones, hexagonal borophene and HB.

3. HB Sheet with Hexagonal Boron Frame

In this section, we review the hexagonal borophene and HB sheet [37]. As easily expected, a hydrogenation generally has a significant impact on the electronic structure [51,52]. The first question to be considered is how this hydrogenation affects the electronic structure in the hexagonal HB sheet. In particular, the presence or absence of topologically nontrivial band structures such as graphene is of interest. The Dirac electron at the K point of graphene is attributed to the symmetry of the honeycomb structure of graphene and its filling factor. Because the hexagonal borophene from MgB_2 has the same structure as graphene, the electronic band structure is really similar to that of graphene but the Fermi energy is not located on the Dirac point. Figure 2 is the electronic structure of borophene obtained by the first-principles calculation (GGA-PBE exchange correlation functional). We can see there is a Dirac electron around 3 eV at the K point as in graphene. Since boron has fewer electrons than carbon, the Dirac cone is located above the Fermi level.

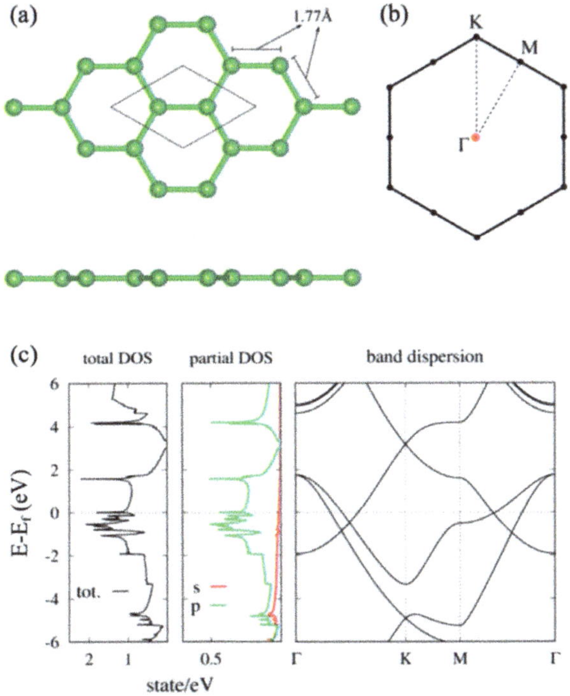

Figure 2. (**a**) Lattice structure and (**b**) BZ of the honeycomb boron sheet. (**c**) Density of states and electronic band dispersion of the honeycomb boron sheet (adapted from [37]).

When the borophene is modified into HB, the honeycomb symmetry is broken by hydrogenation. Therefore, the electronic structure of the hexagonal HB sheet should be analyzed in detail. Figure 3 is the electronic structure of the hexagonal HB sheet. The electronic band structure of the hexagonal HB sheet (Figure 3) does not resemble that of the honeycomb boron sheet and Dirac electrons are no longer found. We can see that the hexagonal HB sheet is a semimetal with a hole pocket at the Γ point and an electron pocket at the Y point in the Brillouin zone (BZ). The effective mass of the carrier is estimated as $(m^{\Gamma}_{xx}, m^{\Gamma}_{yy}) \simeq (-4.02m_e, -2.23m_e)$ for a hole and $(m^{Y}_{xx}, m^{Y}_{yy}) \simeq (1.65m_e, 11.4m_e)$ for an electron, respectively (m_e is the bare electron mass). In [37], it was shown that the electron and hole pockets originate from some electronic bands in the honeycomb boron sheet by varying the position of the hydrogen and checking the symmetry of the wave functions. As a result, it was found that the hole pocket originates from one of the doubly degenerate hole pockets at the Γ point in the honeycomb boron sheet, while the electron pocket originates from the band around 2 eV at the M point.

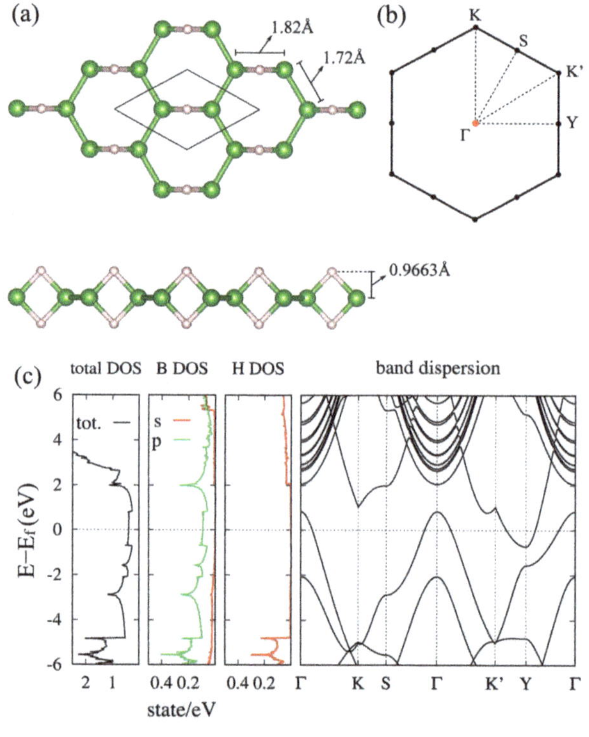

Figure 3. (**a**) Lattice structure and (**b**) BZ of the hexagonal HB. (**c**) Density of states and electronic band dispersion of the hexagonal HB (adapted from [37]).

From this result, the effects of hydrogenation and the B–H–B bond on the electronic structure can be generally discussed. Although the symmetry and electronic structure are significantly changed by hydrogenation, the bonding and antibonding states of the B–H–B bond appear in the energy region far from the Fermi level, and thus there is almost no density of state (DOS) of hydrogen near the Fermi level (Figure 3). As a result, the electronic states derived from the boron frame remain around the Fermi level even after hydrogenation. This allows partial inheritance of the properties of the boron frame before hydrogenation. For example, the possibility of two-dimensional superconductivity is pointed out in [37]. Although not all the properties are necessarily inherited from the boron frame because the electronic state is significantly changed by hydrogenation, it may provide potential applications for hydrogenated boron sheets.

A tight-binding model of the hexagonal HB sheet has also been proposed in [37]. The basis of the tight-binding model consist of p_z orbitals on the boron atoms and bonding states of the B–B bonds. In [37], the electronic structure has been qualitatively discussed only with the nearest-neighbor hoppings, and it is shown that hopping along the B–H–B bond is larger than others. This tight-binding model can be used as a convenient tool for the analysis of the HB sheet.

Based on the results of the analysis on the hexagonal borophene and HB, here we show two ways to extend the study on borophene and HB. The first one is aimed at nanodevice applications of the hexagonal HB. We show electronic structures of HB nanoribbons in Section 4 as an example. The other way is an investigation of topological bands that are given by the nonsymmorphic symmetry in other HB sheets, which is reviewed in Section 5.

4. HB Nanoribbon

In this section, we calculate the electronic structure of a hexagonal HB nanoribbon. To discuss the anisotropy and the spreading of the wave function due to the H–B–H coupling in detail, we used the same calculation method as in [37], but including the long-range hoppings. The tight-binding model was obtained by Wannier90 [53], and electronic band structures of nanoribbons are calculated using the WannierTools package [54]. We considere four edge truncation types, zigzag-1 (ZZ1), zigzag-2 (ZZ2), armchair-1 (AC1), and armchair-2 (AC2) shown in Figure 4. We calculate the electronic structures of the four types of nanoribbons by changing the ribbon width. The ribbon width is given as a number of unit cell (N), and the unit cell is shown as a red dashed cell in the left figure of each row in Figure 4. The 2D BZ (gray shaded rhombus) and corresponding 1D BZ of each nanoribbon are shown in the second column of each row. The red lines in the 2D BZ are Fermi surfaces of the electron and hole pockets. The right three panels are electronic band structures of the nanoribbon for $N = 3$, $N = 6$, and $N = 10$. In the case of ZZ1, the electron and hole pockets overlap at the G point, and the system becomes a 1D conductor when the ribbon width is large enough ($N = 10$). As the ribbon width gets smaller, the electronic state is modified by a finite size effect that comes from a spread of the Wannier functions. Around $N = 3 \sim 6$, the finite effect becomes significant and the electronic structure becomes gapped in $N = 3$. This result indicates that hoppings between the third nearest cells have a non-negligible effect on the low-energy electronic structure. In the case of ZZ2, the hole pocket is projected on the G point while the electron pocket is projected on the X point in the 1D BZ. Therefore, the electronic structure of the ZZ2 nanoribbon is semimetallic. Furthermore, in the ZZ2 case, the ribbon width dependence is almost the same as that of the ZZ1 case, and the electronic structure becomes gapped around $N = 3$. Next, we move to the AC1 and AC2 cases. In the AC1 case, the electronic structure in a large ribbon width case is metallic, as in the case of ZZ1. The AC2 case is in principle semimetallic, but the electron and hole pockets are so large that the system becomes metallic in a large ribbon width case as $N > 10$. In the cases of AC1 and AC2, the finite size effect is significant in $N < 6 \sim 10$. This is simply because the unit cells that make up the AC1 and AC2 nanoribbons are heavily distorted and their width is small compared to the case of ZZ1 and ZZ2. Note that the ribbon widths of the $N = 6$ ribbons in the AC1 and AC2 cases are comparable to those of the $N = 3$ ribbons in the ZZ1 and ZZ2 cases (roughly 8 Å). These results are consistent with the first-principles calculations in [55].

In the above, as an example of using the tight-binding model, we have calculated the electronic structure of HB nanoribbons and their ribbon width dependence. In the small ribbon width case, the finite size effect gets significant and the electronic structure becomes gapped. When the ribbon width is smaller than 7 Å, the HB nanoribbon is expected to be a 1D small-gap semiconductor. By changing the edge truncation, one can choose a direct or indirect gap.

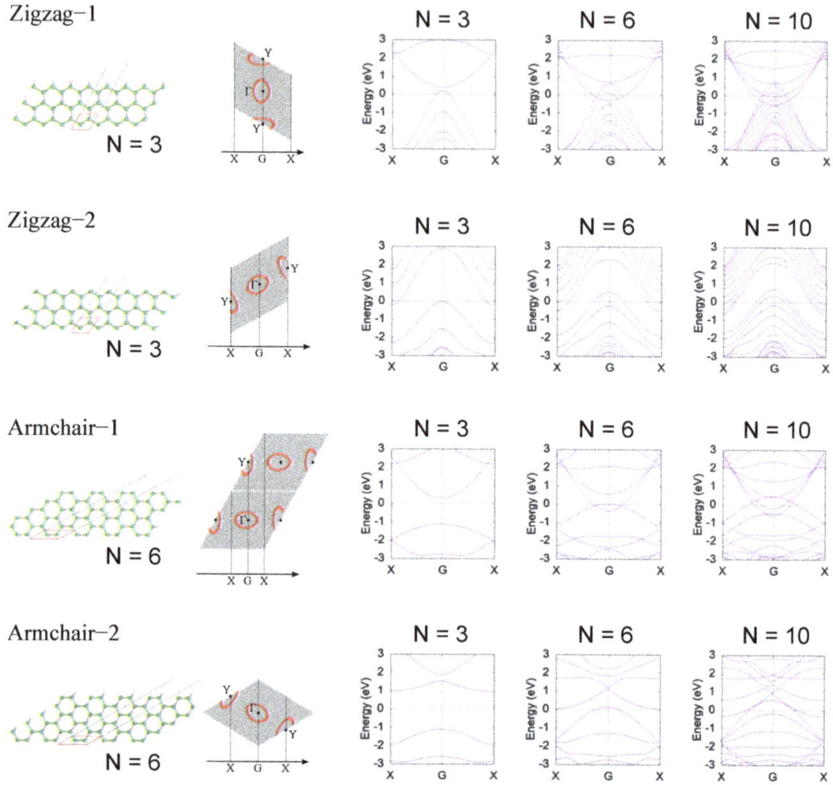

Figure 4. HB nanoribbon and its electron structure. Red cells in lattice structure figures are the unit cells that construct the nanoribbon. In each row (each truncation type), the definition of the 1D BZ is shown with the 2D BZ (gray shaded rhombus) in the second right figure. For four edge truncation types, electron band structures of nanoribbons are calculated for a three-unit cell case ($N = 3$), a six-unit cell case ($N = 6$), and a ten-unit cell case ($N = 10$).

5. HB Sheet with Nonsymmorphic Boron Frame

In this section, we review the electronic structures of nonsymmorphic HB sheets from $TmAlB_4$ or $YCrB_4$ [38]. In the nonsymmorphic HB sheets, the existence of a Dirac nodal line has been proposed. First, we explain the symmetry restrictions on the electronic band structure given by the nonsymmorphic symmetry. Because the nonsymmorphic operation is combined with a fractional translation, a quasi-band-folding is generally seen in the band structure. As a result, these nonsymmorphic operators generally guarantee double degeneracies of electronic bands on some parts of the BZ boundary. If the band folding is given by a pure fractional translation, all momenta $\boldsymbol{k}$ are invariant for the pure translation and thus an eigenvalue of the pure translation is defined for all Bloch wave functions $\psi_{\boldsymbol{k}}$. Even if a Dirac nodal line is obtained by the band folding, an optical excitation between bands with different translation eigenvalues is basically prohibited, and thus the obtained Dirac nodal line is not usually used as a linear-dispersive band. On the other hand, the nonsymmorphic operations are combined with rotations or mirrors, and thus a generic momentum $\boldsymbol{k}$ is not invariant for the operation. Therefore, the Dirac nodal line obtained by the quasi-band-folding of nonsymmorphic symmetry can be used as a linear-dispersive band.

The nonsymmorphic boron frame obtained from $TmAlB_4$ is named as (5–7)-α-borophene (Figure 5), which belongs to the layer group *Pbam* (layer group no. 44, or 3D space group no. 55). Here, (5–7)-α indicates that the α lattice is composed of pentagons and heptagons. Especially in this layer group, it is proved that the double degeneracy given by the nonsymmorphic symmetry occurs on all points on the BZ boundary. This general symmetry restriction contributes to the appearance of a Dirac nodal line when the filling factor of the HB sheet is additionally considered. Generally, the ratio of B and H atoms is always B:H = 1:1. Further, the number of B atoms in the unit cell in a *Pbam* symmetric system is an even number due to the multiplicity from the nonsymmorphic operation. As a result, the filling factor (number of electrons in the unit cell) is written as $12n = 4n'$, where n and n' are integers. Because each doubly degenerate band on the BZ boundary consists of four states, the lower n' degenerate bands are completely occupied and the others are completely empty. These restrictions on the band degeneracy and its filling play an important role to determine the presence or absence of the Dirac nodal line in the material. By using this restriction, in [38], a mirror-eigenvalue-based index has been proposed to determine the presence or absence of a Dirac nodal line. Note that the well-known inversion-based index (Fu–Kane index [26,56]) does not correctly work in 2D systems. It is noteworthy that one needs to check wave functions only at the Γ point to calculate the mirror-eigenvalue-based index. Generally, in a 2D system, a possible Dirac nodal line is always protected by a mirror symmetry on the xy plane and thus the presence or absence is determined by checking the parity of the number of occupied bands that have a mirror eigenvalue -1 in each momentum. In [38], it was proved that the method gets even easier especially in the layer group *Pbam* with a filling factor $4n'$. By the representation theory of the layer group *Pbam*, it is proved that the doubly degenerate bands on the BZ boundary always have the same mirror eigenvalue. As a result, the number of occupied bands with a mirror eigenvalue -1 on the BZ boundary must be an even number. Due to this symmetry restriction on the BZ boundary, one only needs to check the parity at the Γ point to determine the presence or absence of a Dirac nodal line. When the number of occupied bands with a mirror eigenvalue -1 at the Γ point is odd, there is a Dirac nodal line and thus the material is a Dirac nodal line semimetal. Among all 80 layer groups, the layer groups *Pbam* (no. 44) and $P4/mbm$ (no. 64) have these properties.

After this general discussion, the electronic band structures and appearance of a Dirac nodal line were specifically shown for two types of HB sheets with different hydrogenation patterns for the α-borophene frames. Both of α_1 and α_2 were predicted to be Dirac nodal line semimetals (Figures 6 and 7). Figure 6 shows (a) the lattice structure, (b) the BZ and high-symmetry points and lines, (c) the total and partial density of states and the band dispersion of the (5–7)-α_1-HB. Figure 7 shows those of the (5–7)-α_2-HB. Note that the electronic bands on the BZ boundary (the S–Y line and S–X line) are doubly degenerate due to the symmetry restriction explained above. The mirror eigenvalues of the wave functions are also shown in Figures 6c and 7c. Because a band inversion between two bands with the opposite mirror eigenvalues occurs at the Γ point, the number of occupied bands with a mirror eigenvalue -1 at the Γ point is odd in these cases. We can see there are Dirac points (DP) on the Γ–Y and Γ–X lines, which are parts of the Dirac nodal line. Not only do these materials have Dirac nodal lines, but there is also no band other than the Dirac nodal line around the Fermi level and thus they are ideal 2D Dirac nodal line semimetals. Although an easy model has not been given in [38], the bonding types are almost identical with the hexagonal HB case and thus the finite size effect in a nanoribbon is expected to appear in the same ribbon width.

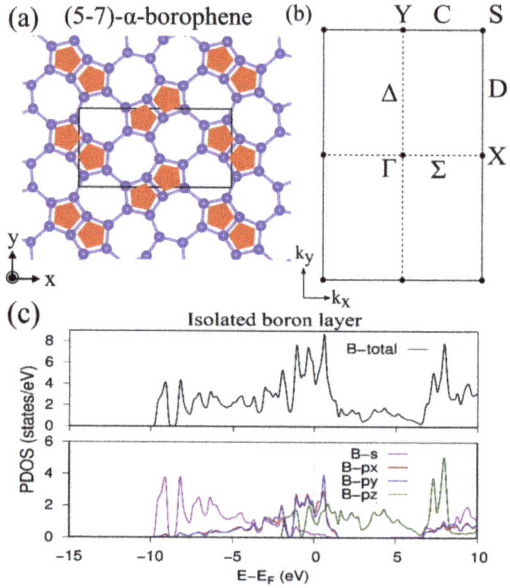

Figure 5. (5–7)-α-borophene. (**a**) Lattice structure of (5–7)-α-borophene. (**b**) BZ of the (5–7)-α-borophene. (**c**) Total and partial density of state of (5–7)-α-borophene obtained by the first-principles calculation (adapted from [38,41]).

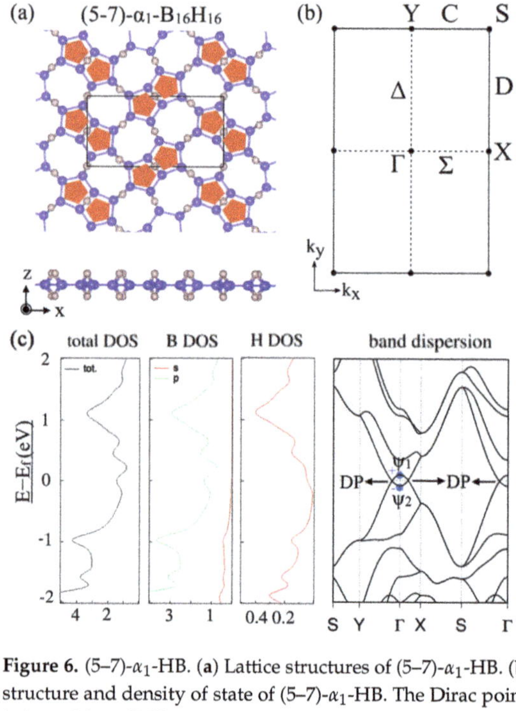

Figure 6. (5–7)-α_1-HB. (**a**) Lattice structures of (5–7)-α_1-HB. (**b**) BZ of (5–7)-α_1-HB. (**c**) Electronic band structure and density of state of (5–7)-α_1-HB. The Dirac points (DP) are parts of the Dirac nodal line (adapted from [38]).

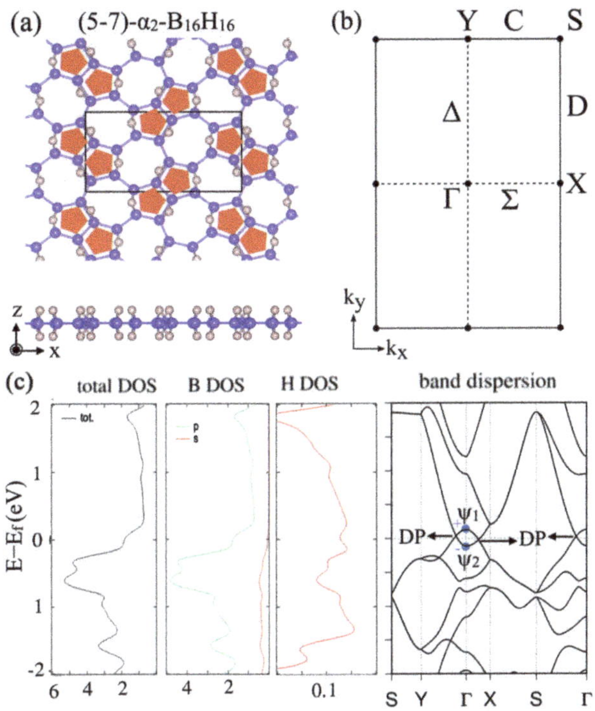

Figure 7. (5–7)-α_2-HB. (**a**) Lattice structures of (5–7)-α_2-HB. (**b**) BZ of (5–7)-α_2-HB. (**c**) Electronic band structure and density of state of (5–7)-α_2-HB. The Dirac points (DP) are parts of the Dirac nodal line (adapted from [38]).

In the above, we have reviewed the electronic band structures of (5–7)-α-HB sheets that are obtained from $TmAlB_4$. Due to the symmetry restriction in the band structure and filling factors of the systems, these materials were predicted to be good examples of 2D Dirac nodal semimetals. In [38], another type of nonsymmorphic HB sheet, (5,6,7)-γ-HB, and the existence of a Dirac nodal line were also discussed. Additionally, even if a boron frame is symmorphic, the hydrogenation pattern can break the symmetry of the boron frame and consequently a nonsymmorphic HB sheet can be obtained. As we have seen in this section, the nonsymmorphic HB sheets have a potential as a topological material while the hexagonal HB sheet does not have a topological band.

6. Experiments and Applications

Finally in this section, we review experimental studies of HB sheets and application studies.

The electronic structures of HB sheets have also been studied by X-ray spectroscopy experiments [37,41]. Both of the hexagonal and (5–7)-α HB sheets have been confirmed to be semimetallic at least with power samples. For example, the spectroscopy data of the (5–7)-α HB sheet is shown in Figure 8. The HB sheet was synthesized by the proton ion-exchange method, prepared by liquid exfoliation for measurement [35]. The spectra of the soft X-ray emission and absorption (SXE and SXA) at B K-edge were conducted in the BL-09A at the NewSUBARU Synchrotron Radiation Facility [57]. The spectroscopy measurements were carried out at room temperature. The calculated $\sigma/\sigma*$ and $\pi/\pi*$ bands for the (5–7)-α_1 and (5–7)-α_2 HB sheets are also shown for comparison. It is of note that the SXE and SXA spectra indicate the dipole transition between 2p and 1s states, reflecting the occupied valence band and unoccupied conduction band, respectively, [41,49], as shown in Figure 8.

One can find that the spectroscopy results agree well with calculated bands, especially with the $\alpha 1$ type. The consistent result provides evidence of semimetallicity and implies the Dirac nodal loop exists near the Fermi level.

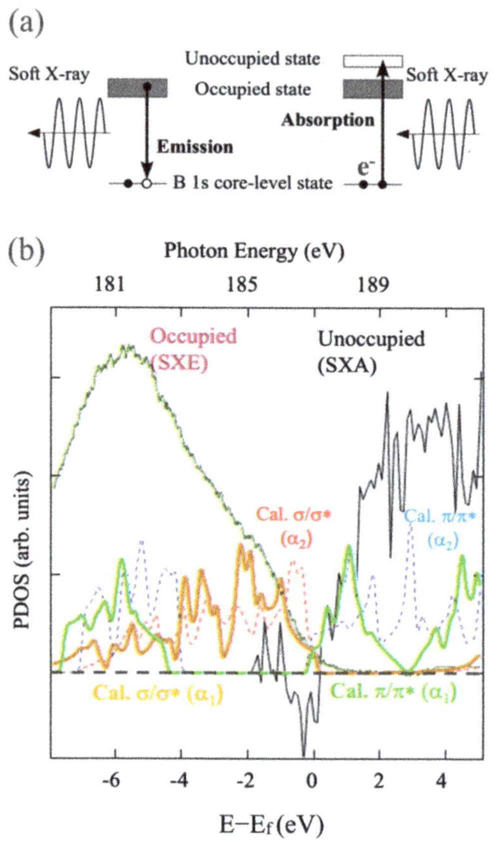

Figure 8. X-ray spectroscopy experiments on HB sheets. (**a**) Schematic picture of the soft X-ray emission and absorption (SXE and SXA) spectroscopy. (**b**) Spectroscopy data of the (5–7)-α HB sheet and calculated $\sigma/\sigma*$ and $\pi/\pi*$ bands for the (5–7)-α_1 and (5–7)-α_2 HB sheets (adapted from [41]).

Lastly, we mention the potential for other applications. The hexagonal HB has been studied for applications as catalysts [58–62] and hydrogen storage [63,64]. As a catalyst, an ethanol–ethylene conversion mechanism on the HB sheet has been studied in [59]. As hydrogen storage, decorated HB nanotubes have analytically been studied in [64], and a photoinduced hydrogen release has been experimentally reported in [63].

7. Summary and Outlook

We reviewed recent studies on electronic structures of free-standing monolayer hydrogenated boron (HB) sheets. HB sheets have recently been synthesized by the proton ion-exchange reaction and have been attracting attention as a new 2D material in both experimental and theoretical studies. A hexagonal hydrogenated boron sheet, which is synthesized from MgB_2, was revealed to have a semimetallic electronic structure. A tight-binding model that reproduces the electronic state was given and it allowed us to easily estimate the properties of the system, for example, the electronic band structures of nanoribbons. Nonsymmorphic hydrogenated boron sheets, which are synthesized from $TmAlB_4$, were also proposed as 2D materials that host Dirac nodal lines around the Fermi level. The appearance of the Dirac nodal line was strongly related to the symmetry restrictions on the band dispersion. The semimetallic electronic structure or the existence of the Dirac nodal line was also experimentally confirmed. In the future, it is expected that large sheet samples of hydrogenated boron sheets will be obtained that will allow a more detailed electronic structure analysis and application studies. In addition, if it becomes possible to synthesize more diverse HB sheets, it is expected that we will be able to select and use an ideal HB sheet as a 2D material for various applications.

Author Contributions: All the authors contributed to writing this review paper. All authors have read and agreed to the published version of the manuscript.

Funding: This work was supported by JST, CREST Grant Number JPMJCR21O4, Japan, and by JSPS KAKENHI Grant Numbers JP19H04398 and JP18H03874.

Data Availability Statement: No datasets were generated during the current study.

Acknowledgments: We acknowledge N. T. Cuong, M. Cameau, C. A. S. Moura, O. I. Sunday, Y. Tsujikawa, S. Ito, R. Ishibiki, A. Fujino, A. Yamamoto, M. Araki, S. Yamamoto, T. Tokushima, D. E. Weibel, M. Niibe, N. Umezawa, B. Slater, K. Yubuta, T. Kondo, S. Okada, and M. Ogata for their collaborations in the research. The experiments were supported at SPring-8 BL07LSU of the Synchrotron Radiation Research Organization and the Institute for Solid State Physics, the University of Tokyo.

Conflicts of Interest: The authors declare no conflict of interest.

References

1. Matsuda, I. *Monatomic Two-Dimensional Layers: Modern Experimental Approaches for Structure, Properties, and Industrial Use*; Elsevier: Amsterdam, The Netherlands, 2018.
2. Novoselov, K.; Mishchenko, A.; Carvalho, A.; Neto, A.C. 2D materials and van der Waals heterostructures. *Science* **2016**, *353*, aac9439. [CrossRef] [PubMed]
3. Franklin, A.D. Nanomaterials in transistors: From high-performance to thin-film applications. *Science* **2015**, *349*, aab2750. [CrossRef] [PubMed]
4. Yamazaki, S.; Hosomura, Y.; Matsuda, I.; Hobara, R.; Eguchi, T.; Hasegawa, Y.; Hasegawa, S. Metallic Transport in a Monatomic Layer of In on a Silicon Surface. *Phys. Rev. Lett.* **2011**, *106*, 116802. [CrossRef] [PubMed]
5. Feng, B.; Fu, B.; Kasamatsu, S.; Ito, S.; Cheng, P.; Liu, C.C.; Feng, Y.; Wu, S.; Mahatha, S.K.; Sheverdyaeva, P.; et al. Experimental realization of two-dimensional Dirac nodal line fermions in monolayer Cu 2 Si. *Nat. Commun.* **2017**, *8*, 1–6. [CrossRef] [PubMed]
6. Castro Neto, A.H.; Guinea, F.; Peres, N.M.R.; Novoselov, K.S.; Geim, A.K. The electronic properties of graphene. *Rev. Mod. Phys.* **2009**, *81*, 109–162. [CrossRef]
7. Vogt, P.; De Padova, P.; Quaresima, C.; Avila, J.; Frantzeskakis, E.; Asensio, M.C.; Resta, A.; Ealet, B.; Le Lay, G. Silicene: Compelling Experimental Evidence for Graphenelike Two-Dimensional Silicon. *Phys. Rev. Lett.* **2012**, *108*, 155501. [CrossRef] [PubMed]
8. Fleurence, A.; Friedlein, R.; Ozaki, T.; Kawai, H.; Wang, Y.; Yamada-Takamura, Y. Experimental Evidence for Epitaxial Silicene on Diboride Thin Films. *Phys. Rev. Lett.* **2012**, *108*, 245501. [CrossRef] [PubMed]
9. Lin, C.L.; Arafune, R.; Kawahara, K.; Tsukahara, N.; Minamitani, E.; Kim, Y.; Takagi, N.; Kawai, M. Structure of silicene grown on Ag (111). *Appl. Phys. Express* **2012**, *5*, 045802. [CrossRef]
10. Fukaya, Y.; Mochizuki, I.; Maekawa, M.; Wada, K.; Hyodo, T.; Matsuda, I.; Kawasuso, A. Structure of silicene on a Ag(111) surface studied by reflection high-energy positron diffraction. *Phys. Rev. B* **2013**, *88*, 205413. [CrossRef]
11. Xue, Y.; Huan, H.; Zhao, B.; Luo, Y.; Zhang, Z.; Yang, Z. Higher-order topological insulators in two-dimensional Dirac materials. *Phys. Rev. Res.* **2021**, *3*, L042044. [CrossRef]
12. Kou, L.; Chen, C.; Smith, S.C. Phosphorene: Fabrication, properties, and applications. *J. Phys. Chem. Lett.* **2015**, *6*, 2794–2805. [CrossRef] [PubMed]
13. Carvalho, A.; Wang, M.; Zhu, X.; Rodin, A.S.; Su, H.; Castro Neto, A.H. Phosphorene: From theory to applications. *Nat. Rev. Mater.* **2016**, *1*, 1–16. [CrossRef]
14. Zhang, S.; Xie, M.; Li, F.; Yan, Z.; Li, Y.; Kan, E.; Liu, W.; Chen, Z.; Zeng, H. Semiconducting group 15 monolayers: A broad range of band gaps and high carrier mobilities. *Angew. Chem.* **2016**, *128*, 1698–1701. [CrossRef]
15. Ezawa, M. Minimal models for Wannier-type higher-order topological insulators and phosphorene. *Phys. Rev. B* **2018**, *98*, 045125. [CrossRef]
16. Watts, M.C.; Picco, L.; Russell-Pavier, F.S.; Cullen, P.L.; Miller, T.S.; Bartuś, S.P.; Payton, O.D.; Skipper, N.T.; Tileli, V.; Howard, C.A. Production of phosphorene nanoribbons. *Nature* **2019**, *568*, 216–220. [CrossRef] [PubMed]

17. Dávila, M.; Xian, L.; Cahangirov, S.; Rubio, A.; Le Lay, G. Germanene: A novel two-dimensional germanium allotrope akin to graphene and silicene. *New J. Phys.* **2014**, *16*, 095002. [CrossRef]
18. Lin, C.H.; Huang, A.; Pai, W.W.; Chen, W.C.; Chen, T.Y.; Chang, T.R.; Yukawa, R.; Cheng, C.M.; Mou, C.Y.; Matsuda, I.; et al. Single-layer dual germanene phases on Ag(111). *Phys. Rev. Mater.* **2018**, *2*, 024003. [CrossRef]
19. Fukaya, Y.; Matsuda, I.; Feng, B.; Mochizuki, I.; Hyodo, T.; Shamoto, S. Asymmetric structure of germanene on an Al (111) surface studied by total-reflection high-energy positron diffraction. *2D Mater.* **2016**, *3*, 035019. [CrossRef]
20. Zhang, S.; Yan, Z.; Li, Y.; Chen, Z.; Zeng, H. Atomically thin arsenene and antimonene: Semimetal–semiconductor and indirect–direct band-gap transitions. *Angew. Chem.* **2015**, *127*, 3155–3158. [CrossRef]
21. Kamal, C.; Ezawa, M. Arsenene: Two-dimensional buckled and puckered honeycomb arsenic systems. *Phys. Rev. B* **2015**, *91*, 085423. [CrossRef]
22. Ares, P.; Palacios, J.J.; Abellán, G.; Gómez-Herrero, J.; Zamora, F. Recent progress on antimonene: A new bidimensional material. *Adv. Mater.* **2018**, *30*, 1703771. [CrossRef] [PubMed]
23. Reis, F.; Li, G.; Dudy, L.; Bauernfeind, M.; Glass, S.; Hanke, W.; Thomale, R.; Schäfer, J.; Claessen, R. Bismuthene on a SiC substrate: A candidate for a high-temperature quantum spin Hall material. *Science* **2017**, *357*, 287–290. [CrossRef] [PubMed]
24. Kowalczyk, P.J.; Brown, S.A.; Maerkl, T.; Lu, Q.; Chiu, C.K.; Liu, Y.; Yang, S.A.; Wang, X.; Zasada, I.; Genuzio, F.; et al. Realization of symmetry-enforced two-dimensional dirac fermions in nonsymmorphic α-bismuthene. *ACS Nano* **2020**, *14*, 1888–1894. [CrossRef] [PubMed]
25. Burkov, A.A.; Hook, M.D.; Balents, L. Topological nodal semimetals. *Phys. Rev. B* **2011**, *84*, 235126. [CrossRef]
26. Kim, Y.; Wieder, B.J.; Kane, C.L.; Rappe, A.M. Dirac Line Nodes in Inversion-Symmetric Crystals. *Phys. Rev. Lett.* **2015**, *115*, 036806. [CrossRef] [PubMed]
27. Yamakage, A.; Yamakawa, Y.; Tanaka, Y.; Okamoto, Y. Line-node Dirac semimetal and topological insulating phase in noncentrosymmetric pnictides CaAg X (X= P, As). *J. Phys. Soc. Jpn.* **2016**, *85*, 013708. [CrossRef]
28. Schoop, L.M.; Ali, M.N.; Straßer, C.; Topp, A.; Varykhalov, A.; Marchenko, D.; Duppel, V.; Parkin, S.S.; Lotsch, B.V.; Ast, C.R. Dirac cone protected by non-symmorphic symmetry and three-dimensional Dirac line node in ZrSiS. *Nat. Commun.* **2016**, *7*, 1–7. [CrossRef] [PubMed]
29. Wu, Y.; Wang, L.L.; Mun, E.; Johnson, D.D.; Mou, D.; Huang, L.; Lee, Y.; Bud'ko, S.L.; Canfield, P.C.; Kaminski, A. Dirac node arcs in PtSn 4. *Nat. Phys.* **2016**, *12*, 667–671. [CrossRef]
30. Bian, G.; Chang, T.R.; Sankar, R.; Xu, S.Y.; Zheng, H.; Neupert, T.; Chiu, C.K.; Huang, S.M.; Chang, G.; Belopolski, I.; et al. Topological nodal-line fermions in spin–orbit metal PbTaSe 2. *Nat. Commun.* **2016**, *7*, 1–8. [CrossRef] [PubMed]
31. Hirayama, M.; Okugawa, R.; Miyake, T.; Murakami, S. Topological Dirac nodal lines and surface charges in fcc alkaline earth metals. *Nat. Commun.* **2017**, *8*, 1–9. [CrossRef] [PubMed]
32. Tateishi, I.; Matsuura, H. Face Centered Cubic SnSe as a Trivial Dirac Nodal Line Material. *J. Phys. Soc. Jpn.* **2018**, *87*, 073702. [CrossRef]
33. Gao, L.; Sun, J.T.; Lu, J.C.; Li, H.; Qian, K.; Zhang, S.; Zhang, Y.Y.; Qian, T.; Ding, H.; Lin, X.; et al. Epitaxial growth of honeycomb monolayer CuSe with dirac nodal line fermions. *Adv. Mater.* **2018**, *30*, 1707055. [CrossRef] [PubMed]
34. Mannix, A.J.; Zhang, Z.; Guisinger, N.P.; Yakobson, B.I.; Hersam, M.C. Borophene as a prototype for synthetic 2D materials development. *Nat. Nanotechnol.* **2018**, *13*, 444–450. [CrossRef] [PubMed]
35. Nishino, H.; Fujita, T.; Cuong, N.T.; Tominaka, S.; Miyauchi, M.; Iimura, S.; Hirata, A.; Umezawa, N.; Okada, S.; Nishibori, E.; et al. Formation and characterization of hydrogen boride sheets derived from MgB2 by cation exchange. *J. Am. Chem. Soc.* **2017**, *139*, 13761–13769. [CrossRef] [PubMed]
36. Matsuda, I.; Wu, K. *2D Boron: Boraphene, Borophene, Boronene*; Springer: Berlin/Heidelberg, Germany, 2021.
37. Tateishi, I.; Cuong, N.; Moura, C.; Cameau, M.; Ishibiki, R.; Fujino, A.; Okada, S.; Yamamoto, A.; Araki, M.; Ito, S.; et al. Semimetallicity of free-standing hydrogenated monolayer boron from MgB 2. *Phys. Rev. Mater.* **2019**, *3*, 024004. [CrossRef]
38. Cuong, N.; Tateishi, I.; Cameau, M.; Niibe, M.; Umezawa, N.; Slater, B.; Yubuta, K.; Kondo, T.; Ogata, M.; Okada, S.; et al. Topological Dirac nodal loops in nonsymmorphic hydrogenated monolayer boron. *Phys. Rev. B* **2020**, *101*, 195412. [CrossRef]
39. Hou, C.; Tai, G.; Hao, J.; Sheng, L.; Liu, B.; Wu, Z. Ultrastable crystalline semiconducting hydrogenated borophene. *Angew. Chem. Int. Ed.* **2020**, *59*, 10819–10825. [CrossRef] [PubMed]
40. Li, Q.; Kolluru, V.S.C.; Rahn, M.S.; Schwenker, E.; Li, S.; Hennig, R.G.; Darancet, P.; Chan, M.K.; Hersam, M.C. Synthesis of borophane polymorphs through hydrogenation of borophene. *Science* **2021**, *371*, 1143–1148. [CrossRef] [PubMed]
41. Niibe, M.; Cameau, M.; Cuong, N.T.; Sunday, O.I.; Zhang, X.; Tsujikawa, Y.; Okada, S.; Yubuta, K.; Kondo, T.; Matsuda, I. Electronic structure of a borophene layer in rare-earth aluminum/chromium boride and its hydrogenated derivative borophane. *Phys. Rev. Mater.* **2021**, *5*, 084007. [CrossRef]
42. Mannix, A.J.; Zhou, X.F.; Kiraly, B.; Wood, J.D.; Alducin, D.; Myers, B.D.; Liu, X.; Fisher, B.L.; Santiago, U.; Guest, J.R.; et al. Synthesis of borophenes: Anisotropic, two-dimensional boron polymorphs. *Science* **2015**, *350*, 1513–1516. [CrossRef]
43. Feng, B.; Zhang, J.; Zhong, Q.; Li, W.; Li, S.; Li, H.; Cheng, P.; Meng, S.; Chen, L.; Wu, K. Experimental realization of two-dimensional boron sheets. *Nat. Chem.* **2016**, *8*, 563–568. [CrossRef] [PubMed]
44. Li, W.; Kong, L.; Chen, C.; Gou, J.; Sheng, S.; Zhang, W.; Li, H.; Chen, L.; Cheng, P.; Wu, K. Experimental realization of honeycomb borophene. *Sci. Bull.* **2018**, *63*, 282–286. [CrossRef]

45. Kiraly, B.; Liu, X.; Wang, L.; Zhang, Z.; Mannix, A.J.; Fisher, B.L.; Yakobson, B.I.; Hersam, M.C.; Guisinger, N.P. Borophene synthesis on Au (111). *ACS Nano* **2019**, *13*, 3816–3822. [CrossRef]
46. Feng, B.; Sugino, O.; Liu, R.Y.; Zhang, J.; Yukawa, R.; Kawamura, M.; Iimori, T.; Kim, H.; Hasegawa, Y.; Li, H.; et al. Dirac Fermions in Borophene. *Phys. Rev. Lett.* **2017**, *118*, 096401. [CrossRef]
47. Liu, X.; Li, Q.; Ruan, Q.; Rahn, M.S.; Yakobson, B.I.; Hersam, M.C. Borophene synthesis beyond the single-atomic-layer limit. *Nat. Mater.* **2022**, *21*, 35–40. [CrossRef]
48. Buzea, C.; Yamashita, T. Review of the superconducting properties of MgB2. *Supercond. Sci. Technol.* **2001**, *14*, R115. [CrossRef]
49. Mori, T.; Borrmann, H.; Okada, S.; Kudou, K.; Leithe-Jasper, A.; Burkhardt, U.; Grin, Y. Crystal structure, chemical bonding, electrical transport, and magnetic behavior of $TmAlB_4$. *Phys. Rev. B* **2007**, *76*, 064404. [CrossRef]
50. Gschneidner, K.A.; Bunzli, J.C.G.; Pecharsky, V.K. *Handbook on the Physics and Chemistry of Rare Earths*; Elsevier: Amsterdam, The Netherlands, 2004.
51. Kistanov, A.A.; Cai, Y.; Zhou, K.; Srikanth, N.; Dmitriev, S.V.; Zhang, Y.W. Exploring the charge localization and band gap opening of borophene: a first-principles study. *Nanoscale* **2018**, *10*, 1403–1410. [CrossRef]
52. Kochaev, A.; Meftakhutdinov, R.; Sibatov, R.; Katin, K.; Maslov, M.; Efimov, V. Enhanced properties of covalently coupled borophene-graphene layers through fluorination and hydrogenation. *Appl. Surf. Sci.* **2021**, *562*, 150150. [CrossRef]
53. Pizzi, G.; Vitale, V.; Arita, R.; Blügel, S.; Freimuth, F.; Géranton, G.; Gibertini, M.; Gresch, D.; Johnson, C.; Koretsune, T.; et al. Wannier90 as a community code: New features and applications. *J. Phys. Condens. Matter* **2020**, *32*, 165902. [CrossRef]
54. Wu, Q.; Zhang, S.; Song, H.F.; Troyer, M.; Soluyanov, A.A. WannierTools: An open-source software package for novel topological materials. *Comput. Phys. Commun.* **2018**, *224*, 405–416. [CrossRef]
55. Lei, B.; Zhang, Y.Y.; Du, S.X. Band engineering of B2H2 nanoribbons. *Chin. Phys. B* **2019**, *28*, 046803. [CrossRef]
56. Fu, L.; Kane, C.L. Topological insulators with inversion symmetry. *Phys. Rev. B* **2007**, *76*, 045302. [CrossRef]
57. Niibe, M.; Mukai, M.; Miyamoto, S.; Shoji, Y.; Hashimoto, S.; Ando, A.; Tanaka, T.; Miyai, M.; Kitamura, H. Characterization of light radiated from 11 m long undulator. *AIP Conf. Proc. Am. Inst. Phys.* **2004**, *705*, 576–579.
58. Fujino, A.; Ito, S.I.; Goto, T.; Ishibiki, R.; Kondo, J.N.; Fujitani, T.; Nakamura, J.; Hosono, H.; Kondo, T. Hydrogenated borophene shows catalytic activity as solid acid. *ACS Omega* **2019**, *4*, 14100–14104. [CrossRef] [PubMed]
59. Fujino, A.; Ito, S.I.; Goto, T.; Ishibiki, R.; Osuga, R.; Kondo, J.N.; Fujitani, T.; Nakamura, J.; Hosono, H.; Kondo, T. Ethanol–ethylene conversion mechanism on hydrogen boride sheets probed by in situ infrared absorption spectroscopy. *Phys. Chem. Chem. Phys.* **2021**, *23*, 7724–7734. [CrossRef]
60. Rojas, K.I.M.; Cuong, N.T.; Nishino, H.; Ishibiki, R.; Ito, S.I.; Miyauchi, M.; Fujimoto, Y.; Tominaka, S.; Okada, S.; Hosono, H.; et al. Chemical stability of hydrogen boride nanosheets in water. *Commun. Mater.* **2021**, *2*, 1–8. [CrossRef]
61. Ito, S.I.; Hirabayashi, T.; Ishibiki, R.; Kawamura, R.; Goto, T.; Fujita, T.; Yamaguchi, A.; Hosono, H.; Miyauchi, M.; Kondo, T. Hydrogen Boride Sheets as Reductants and the Formation of Nanocomposites with Metal Nanoparticles. *Chem. Lett.* **2020**, *49*, 789–793. [CrossRef]
62. Grishakov, K.S.; Katin, K.P.; Kochaev, A.I.; Kaya, S.; Gimaldinova, M.A.; Maslov, M.M. Ab initio Study of Hydrogen Adsorption on Metal-Decorated Borophene-Graphene Bilayer. *Energies* **2021**, *14*, 2473. [CrossRef]
63. Kawamura, R.; Cuong, N.T.; Fujita, T.; Ishibiki, R.; Hirabayashi, T.; Yamaguchi, A.; Matsuda, I.; Okada, S.; Kondo, T.; Miyauchi, M. Photoinduced hydrogen release from hydrogen boride sheets. *Nat. Commun.* **2019**, *10*, 1–8. [CrossRef] [PubMed]
64. Ploysongsri, N.; Vchirawongkwin, V.; Ruangpornvisuti, V. Hydrogen boride nanotubes and their C, N, O decoration and doping derivatives as materials for hydrogen-containing gases storage and sensing: A SCC–DFTB study. *Vacuum* **2021**, *187*, 110140. [CrossRef]

molecules

Review

History of Cobaltabis(dicarbollide) in Potentiometry, No Need for Ionophores to Get an Excellent Selectivity

Anca-Iulia Stoica [1,2], Clara Viñas [2,*] and Francesc Teixidor [2,*]

1 Department of Water-Atmosphere Resources and Life Science, University of Natural Resources and Life Sciences, 1180 Vienna, Austria
2 Institut de Ciencia de Materials de Barcelona, ICMAB-CSIC, 08193 Bellaterra, Spain
* Correspondence: clara@icmab.es (C.V.); teixidor@icmab.es (F.T.)

Abstract: This work is a mini-review highlighting the relevance of the θ metallabis(dicarbollide) $[3,3'-Co(1,2-C_2B_9H_{11})_2]^-$ with its peculiar and differentiating characteristics, among them the capacity to generate hydrogen and dihydrogen bonds, to generate micelles and vesicles, to be able to be dissolved in water or benzene, to have a wide range of redox reversible couples and many more, and to use these properties, in this case, for producing potentiometric membrane sensors to monitor amine-containing drugs or other nitrogen-containing molecules. Sensors have been produced with this monoanionic cluster $[3,3'-Co(1,2-C_2B_9H_{11})_2]^-$. Other monoanionic boron clusters are also discussed, but they are much fewer. It is noteworthy that most of the electrochemical sensor species incorporate an ammonium cation and that this cation is the species to be detected. Alternatively, the detection of the borate anion itself has also been studied, but with significantly fewer examples. The functions of the borate anion in the membrane are different, even as a doping agent for polypyrrole which was the conductive ground on which the PVC membrane was deposited. Apart from these cases related to *closo* borates, the bulk of the work has been devoted to sensors in which the θ metallabis (dicarbollide) $[3,3'-Co(1,2-C_2B_9H_{11})_2]^-$ is the key element. The metallabis (dicarbollide) anion, $[3,3'-Co(1,2-C_2B_9H_{11})_2]^-$, has many applications; one of these is as new material used to prepare an ion-pair complex with bioactive protonable nitrogen containing compounds, $[YH]_x[3,3'-Co(1,2-C_2B_9H_{11})_2]_y$ as an active part of PVC membrane potentiometric sensors. The developed electrodes have Nernstian responses for target analytes, i.e., antibiotics, amino acids, neurotransmitters, analgesics, for some decades of concentrations, with a short response time, around 5 s, a good stability of membrane over 45 days, and an optimal selectivity, even for optical isomers, to be used also for real sample analysis and environmental, clinical, pharmaceutical and food analysis.

Keywords: Ion-Selective Electrodes; potentiometry; ion pair complexes; cobaltabis(dicarbollide); nitrogen containing compounds

Citation: Stoica, A.-I.; Viñas, C.; Teixidor, F. History of Cobaltabis(dicarbollide) in Potentiometry, No Need for Ionophores to Get an Excellent Selectivity. *Molecules* **2022**, *27*, 8312. https://doi.org/10.3390/molecules27238312

Academic Editors: Takahiro Kondo, Iwao Matsuda and Josep M. Oliva-Enrich

Received: 19 October 2022
Accepted: 22 November 2022
Published: 29 November 2022

1. Introduction, Objectives and Characteristics of *Closo*-Borates and Metallabis(dicarbollides)

This mini-review deals with the potentiometric application of metallabis(dicarbollides) in detecting basic nitrogen containing compounds, mostly in some pharmaceuticals. Thus it does not provide a summary of the organic chemistry of nitrogenous compounds, which is very extensive; according to Jonnalagadda et al. [1], there have been over 97,400 papers only dedicated to nitrogen heterocycles between 2009 and early 2020, nor on the wide list of top prescribed drugs containing nitrogen heterocycles that has been comprehensively reviewed [2], although a large proportion of these could be target compounds to be analyzed by the potentiometric method reported here; it does not deal on analytical techniques on drug analysis, that have been well reviewed [3], or more specifically on the application of electrochemical methods for pharmaceutical or drug analysis [4]. Concerning potentiometric sensors, more specifically Ion Selective Electrodes (ISEs), have been broadly employed as one of the most important electrochemical approaches for pharmaceutical drug analysis [5,6]. Since the advent of nanoscience, nanomaterial components and concepts are available that can improve the design of ISEs [7–9], thus it is expected that a new momentum for the fabrication of selective ISEs and nanomaterials-based potentiometric platforms for pharmaceutical drug analysis will take place. In this work, we will prove that by using metallabis(dicarbollides) it is possible to design and manufacture very selective, very stable, long-lasting, ionophore-free ISEs, without internal solutions and reference electrodes in the working electrode; all thanks to the unique characteristics of metallabis(dicarbollides).

Some years ago, little was known about the properties in a solution of the metallabis(dicarbollide) cobaltabis(dicarbollide) anion $[3,3'\text{-}Co(1,2\text{-}C_2B_9H_{11})_2]^-$ [10], also known as [*o*-COSAN]$^-$, which possess the two C atoms connected. Most efforts had been devoted to its synthesis and derivatization, although its redox reversibility was well known. There were three known redox couples, $Co^{4+/3+}$, $Co^{3+/2+}$, and $Co^{2+/1+}$, each of them reversible [11–13].

The molecule has a structure that, while not rigid, does not change volume or shape. Its shape resembles the theta letter, θ [14], which allows a non-free rotation around the cobalt atom, so that three types of conformers can be generated, *transoid*, *cisoid*, and *gauche* [15]. As recently demonstrated, the three conformers show different properties depending on the environment. If the environment is polar, water, or in the presence of ions, the *cisoid* conformer is prevalent. If it is in vacuum or non-polar solvents, the *transoid* conformer is dominant [16]. Therefore, depending on the environment, the properties in the solution are very different. Thus, in aqueous media, the space around the 4 $C_{cluster}$-Hs is hydrophobic while the space around the 18 B-Hs is hydrophilic [15]. Apart from the hydrophobic/hydrophilic interactions, non-covalent hydrogen and dihydrogen interactions play a key role [17]. Figure 1 displays the five conformers of the isomer $[3,3'\text{-}Co(1,2\text{-}C_2B_9H_{11})_2]^-$, abbreviated as [*o*-COSAN]$^-$, which are *cisoid*-1, *gauche*-1, *transoid*, *gauche*-2, and *cisoid*-2 (being *cisoid*-1 and *cisoid*-2 as well as *gauche*-1 and *gauche*-2 equivalent.

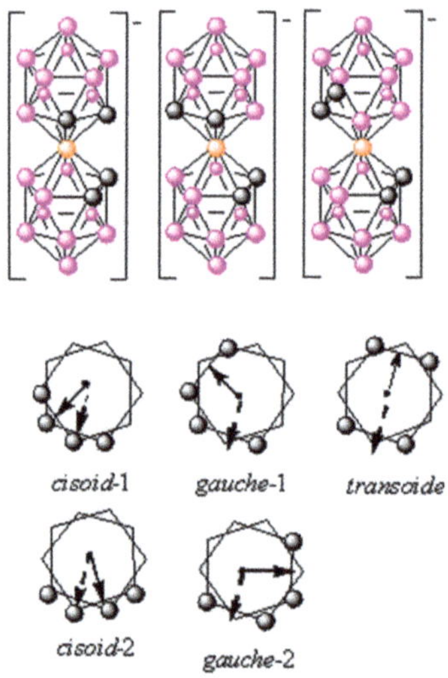

Figure 1. Schematic representation of the icosahedral anionic cobaltabis(dicarbollide) cluster$^-$ conformers (the arrows indicate the direction of dipole moments of compounds). Circles in grey represent the C_c-H vertices; the orange ones correspond to metal (M = Co^{3+}, Fe^{3+}), while the circles in pink correspond to B-H vertices.

These interactions explain why aggregates, vesicles, or micelles with monomers coexist in an aqueous solution [14]. This amphiphilic behavior also explains that [*o*-

COSAN]$^-$ can pass through cell membranes [18]. Apart from this type of interaction, the θ cobaltabis(dicarbollide) [*o*-COSAN]$^-$ interacts strongly with proteins as demonstrated by the interaction with Bovine Serum Albumin BSA [19]. In this case, there are about one hundred $[3,3'\text{-Co}(1,2\text{-}C_2B_9H_{11})_2]^-$ units per each BSA protein. This amount of cobaltabis(dicarbollide) per BSA coated the entire surface of the BSA that was interpreted by considering two phenomena: the anchoring capacity of [*o*-COSAN]$^-$ with amino acids whose residue contained amino groups, i.e., Lysine, Arginine, and Histidine, and the self-assembly capacity of the [*o*-COSAN]$^-$ anions. The cobaltabis(dicarbollide) anion can dissolve in very non-polar and very polar media depending on the cation, but in the case of H $[3,3'\text{-Co}(1,2\text{-}C_2B_9H_{11})_2]$, it is soluble from benzene to water. This makes it a unique compound.

2. Generalities of Ion Selective Electrodes (ISEs) and First Steps in the Use of *Closo* Borates and Metallabis(dicarbollides) as ISEs

Ion Selective Electrodes (ISEs) are transducers or sensors that convert the activity of a specific ion dissolved in a solution into an electrical potential, which can be measured by a voltmeter. ISEs have different applications in clinical, pharmaceutical, environmental, and food processing industries [6,20–24] due to their efficiency from an economical point of view, and analysis time. These sensors are related to low price, and following work due to Bloch, Simon, and Thomas on PVC=based membranes, their performance was improved a lot concerning the limit of detection and selectivity [25–27].

In 1999, $[3,3'\text{-M}(1,2\text{-}C_2B_9H_{11})_2]^-$ (M = Co^{3+}, Fe^{3+}, Ni^{3+}) compounds were implemented in PVC membranes to study their performance as Cs^+ sensors in ISEs. The three metallabis(dicarbollide) complexes displayed a similar behavior with a near-Nernstian response close to 51 mV decade^{-1} and, the $[3,3'\text{-Co}(1,2\text{-}C_2B_9H_{11})_2]^-$ anion was chosen as the parent on which C-substitution, both alkyl and aromatic were done. The species $[3,3'\text{-Co}(1\text{-}CH_3\text{-}2\text{-}(CH_2)_n\,OR\text{-}1,2\text{-}C_2B_9H_9)_2]^-$ ($[1]^-$: n = 3, R = $-CH_2CH_3$; $[2]^-$: n = 3, R = $-(CH_2)_2OCH_3$; $[3]^-$: n = 3, R = $-(CH_2)_3CH_3$; $[4]^-$: n = 6, R= $-(CH_2)_3CH_3$), and $[3,3'\text{-Co}(1\text{-}C_6H_5\text{-}1,2\text{-}C_2B_9H_{10})_2]^-$ ($[5]^-$) and $[3,3'\text{-Co}(1,7\text{-}(C_6H_5)_2\text{-}1,7\text{-}C_2B_9H_9)_2]^-$ ($[6]^-$) were tested for ^{137}Cs, ^{90}Sr, and ^{152}Eu in extraction as long as for potentiometric detection of Cs. In addition, permeability tests on supported liquid membranes with H [6], H [4], and H [6] showed that these compounds present the highest values reported so far for this sort of radionuclide transport experiment [28].

By the year 2006, when we initiated our potentiometric work on using metallabis (dicarbollides) to generate electroactive salts for the selective determination of amine-containing relevant drugs, two key experimental data were known for synthetic boron cluster chemists that were very relevant to start this research: the water insolubility of salts of Cs, used for Cs^+ sensing [28] and alkylammonium with anionic boron clusters and that these same salts were soluble in organic solvents like THF. This concept has been utilized with another series of *closo* boranes, e.g., tetradecylammonium triethylammonium-*closo*-dodecaborate as the electroactive species to determine $[B_{12}H_{11}N(C_2H_5)_3]^-$ [29], or the use of the sulfonium derivative of the *closo*-hydridodecaborate anion $[B_{10}H_9S(C_{18}H_{37})_2]^-$ as the active component for a potentiometric lidocaine-selective sensor [30]. Concerning the use of the *closo* borate clusters the 2, 3, 4, 5, 6, 7, 8, 9, 10, 11, 12-undecabromocarborane anion, $[1\text{-HCB}_{11}Br_{11}]^-$ has been studied as an alternative to the best lipophilic tetraphenylborate, 3,5-[bis(trifluoromethyl)phenyl]borate demonstrating a much higher persistence in the potentiometric membrane [31].

Common key components of a membrane ISE are an inner reference solution on one side of the membrane, a second reference electrode in contact with the analyte solution, and the membrane itself on which at each interface is established an ion-exchange equilibrium that results in charge separation at each interface producing a phase-boundary potential [32, 33]. When concentrations of the ion to be measured on both sides of the membrane are not equal, a membrane potential develops. The difference in potential is measured by the two reference electrodes. There are four major types of membrane: glass, crystalline, liquid, and polymer. The last two are also known as ion exchange membranes. In this mini-review, we mostly dedicate to the polymer type in which the selective membrane consists of three main components: ionophores, a polymer matrix, and a plasticizer. Figure 2 shows a schematic representation of the ISEs with PVC membrane in the solid-state, left, and on the right with a polypyrrol support [34] that performs as the conducting material on which stay the Ionophores, within the PVC membrane, which can be ionic or neutral; these are complexing agents capable of reversibly binding ions. Typically, solid membranes contain an ionophore, ion-exchanger additives (i.e., either alkylammonium salts for anion sensing or tetraphenylborates for cation sensing) [35,36], and a plasticizer which is the organic medium that is supposed to allow the transport of charges within the membranes. Instead of the ionophore, some membranes contain one ion pair complex, the anion part typically being tetraphenylborate and the cation part, the protonated analyte to be measured as the electroactive substance [37–41]. The ion pair complexes are charged ionophores, which bind ions in a more complex binding than mere electrostatic interaction, show increased selectivity towards a primary ion, and possess high strength of association constants [42,43].

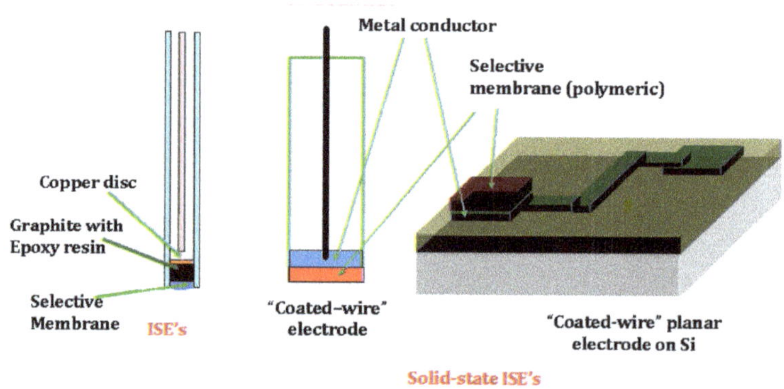

Figure 2. Schematic representation of solid ISEs (**left**) and the similar coated-wire or Si wafer (**right**).

Very relevant for this research on potentiometry based on *closo*-borate anions was that all PVC membranes for potentiometric sensors were prepared in THF. Therefore, the ammonium salts of $[3,3'\text{-Co}(1,2\text{-C}_2\text{B}_9\text{H}_{11})_2]^-$ and PVC had the same solubility requirements: very soluble in THF and insoluble in water. From the point of view of making the membrane, it seemed that all factors were pointing in the same direction. If now we restrict to the θ cobaltabis(dicarbollide) $[3,3'\text{-Co}(1,2\text{-C}_2\text{B}_9\text{H}_{11})_2]^-$ potentiometric electrodes and to make the construction simpler, even at the cost of having a lower limit of detection, LOD, we have moved away from the traditional electrode in which the membrane separated the solution with the analyte from an internal aqueous solution in which there was an inner reference electrode, see Figure 3 top. In 2006, the research addressed developing polymeric sensing membranes based on the principles of host-guest chemistry, as they allowed the selectivity of the sensor to be modulated. Many hundreds of receptors have been developed for this purpose. Typically, there was a lipophilic ion exchanger in addition to the ionophore, which was the gateway for the ions to be measured to enter the membrane. Commonly tetraphenylborate had been used as an ion exchanger, but also the boron cluster perbrominated *closo*-dodecacarborane anion, $[1\text{-HCB}_{11}\text{Br}_{11}]^-$ had been used for this purpose [31]. It was then considered that the emf response of such membranes was described in a simple way by the phase boundary model [44–46], which assumes a localized equilibrium across interfaces and does not consider changes in potential within the membrane or the sample solution [47,48]. A view is shown in Figure 3, bottom. The applied equation is:

$$E_{PB} = E_I^0 + \frac{RT}{z_I F} \ln \frac{a_1(aq)}{a_1(org)}$$

where $a_I(aq)$ and $a_I(org)$ are the activity of the ion I (with charge z_I) in the aqueous and organic phase boundaries, and are derived from the chemical standard potentials in either phase.

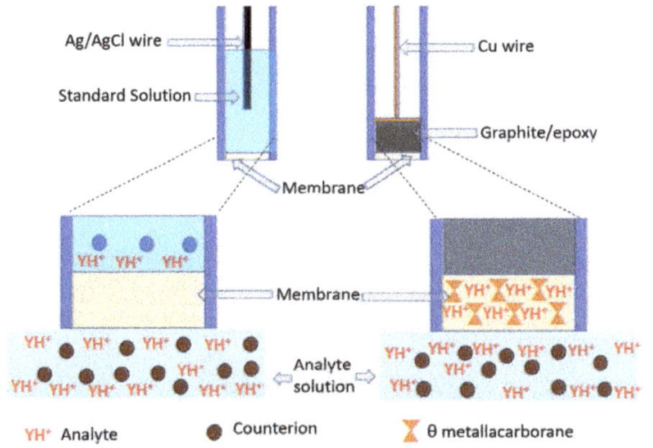

Figure 3. Schematic representation of conventional membrane liquid ISEs (**top left**) and membrane solid state (**top right**) with a magnification of their lower part displaying where the analytes are, and where the metallabis(dicarbollides) is.

The above equation is reduced to the Nernst equation if the activity of the ion to be studied is constant in the organic phase. This required the presence of a lipophilic ion exchanger in the membrane [44–46] otherwise, the membrane would lose its selectivity [47, 48].

$$emf = K + \frac{RT}{z_I F} \ln a_I(aq)$$

3. The Metallabis(dicarbollide) $[3,3'\text{-Co}(1,2\text{-C}_2\text{B}_9\text{H}_{11})_2]^-$ as an Active Component of Membrane Solid State ISE

Concerning the use of θ cobaltabis(dicarbollide) $[3,3'\text{-Co}(1,2\text{-C}_2\text{B}_9\text{H}_{11})_2]^-$ we assumed the basic concept of the rationale behind the recognition of ISE membranes, to suggest the build-up of a potential difference between the bulk of the membrane and the outer analyte aqueous phase (Figure 3 bottom right). The $[3,3'\text{-Co}(1,2\text{-C}_2\text{B}_9\text{H}_{11})_2]^-$ anion provides stability to all participating agents in the membrane. The novelty of this strategy is that the anion $[3,3'\text{-Co}(1,2\text{-C}_2\text{B}_9\text{H}_{11})_2]^-$ is not the sensing part, but the cation ([cation-NH]$^+$) that leads to the selectivity. But this cation is strongly interacting with $[3,3'\text{-Co}(1,2\text{-C}_2\text{B}_9\text{H}_{11})_2]^-$, unlike tetraphenylborate to illustrate with a relevant example. For the latter, only electrostatic interactions are expected, but for $[3,3'\text{-Co}(1,2\text{-C}_2\text{B}_9\text{H}_{11})_2]^-$ in addition to these, hydrogen and dihydrogen bonds occur. Therefore, ion-pair complexes of this type do not fit with the traditional definition of ionophore; hence the importance of this unique cobaltabis(dicarbollide) anion in (bio)sensors.

Therefore, we had an ion exchanger, $[3,3'\text{-Co}(1,2\text{-C}_2\text{B}_9\text{H}_{11})_2]^-$, the PVC, and the plasticizer, all within the membrane, and remarkably it was not necessary to design and synthesize complex molecules that would be selective for a given analyte Y. This would represent a major breakthrough as any protonable amine could be eligible as a candidate to be measured and certainly, it would represent a readily available source of electrochemical sensors while retaining selectivity as far as we were concerned. The presence of Y, in the membrane at the appropriate concentration, could already be achieved by adding the ion-pair complex $[\text{YH}][3,3'\text{-Co}(1,2\text{-C}_2\text{B}_9\text{H}_{11})_2]$, and there was confidence that YH^+ would not leak out of the membrane. Most biologically active compounds have in their structure one or more amino groups that are able to be protonated, thus, our target analytes were compounds with pharmaceutical and medical applications, i.e., antibiotics, amino acids, neurotransmitters, analgesics, etc. [49–56]. Figure 4 displays the schematic general synthetic procedure of the electro-active [cation-NH]$_x$[3,3′-Co(1,2-$C_2B_9H_{11}$)$_2$]$_y$ salt, as well as the ISE electrode assembling.

a)

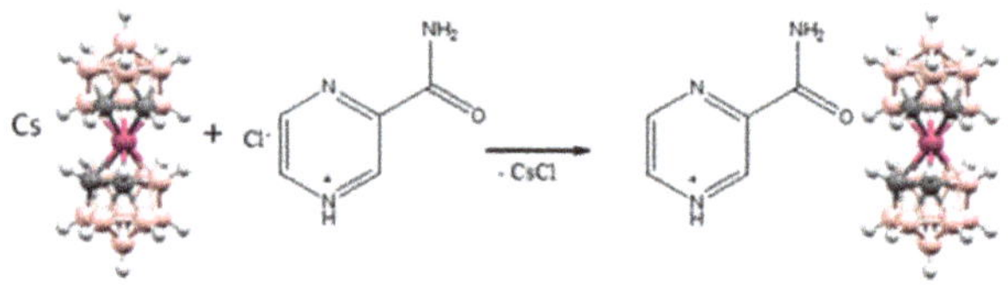

b)

Figure 4. (**a**) Synthesis of the electro-active [YH]$_x$[3,3Co(1,2-$C_2B_9H_{11}$)$_2$]$_y$ salt for pyrazinamide (PZA). (**b**) Photo showing the simplicity of the ISE electrode preparation.

It remained to be demonstrated whether this simplicity in the electrochemical sensor allowed for the detection of optically active species. Since the work of Simon et al. in 1975 [57], some papers and reviews have appeared dedicated to electrochemical enantioselective sensors and biosensors based on molecular chiral receptors, such as cyclodextrins, calixarenes, calixresorcinarenes, and crown-ethers, to form a complex preferentially with one of the enantiomers [58–63]. An important achievement of the [3,3′-Co(1,2-$C_2B_9H_{11}$)$_2$]$^-$ in the membrane preparation for ISEs was the possibility to prepare ion pair complexes for enantiomers that were able to differentiate with a good selectivity of one enantiomer in the presence of the second one without using a chiral receptor, and this turned out to be possible [50]. This unexpected result after comparing with the current techniques described earlier to differentiate enantiomers must be a consequence of the strong interactions displayed by [3,3′-Co(1,2-$C_2B_9H_{11}$)$_2$]$^-$ with the enantiomer in the membrane that prevents its fast rotation and mobility and therefore facilitates a better recognition. Recently chiral sensing systems based on chiral inorganic platforms have been reported for electrochemical recognition of enantiomers [64–66]. Also, [3,3′-Co(1,2-$C_2B_9H_{11}$)$_2$]$^-$ has been used in the development of ISEs for the analysis of tropane alkaloids (tropane, atropine, and scopolamine [67], the analysis of antipyrine and its metabolites/derivatives from environmental water monitoring, which are (besides their beneficial health effect) of growing concern based on their occurrence and fate in water and the environment [56] as well as for serotonin detection [52].

The chemical structures of bioactive nitrogen-containing compounds used in the ion-pair approach with [3,3′-Co(1,2-$C_2B_9H_{11}$)$_2$]$^-$ had the formula [cation-NH]$_x$[3,3′-Co(1,2-$C_2B_9H_{11}$)$_2$]$_y$ and some of the reported amines analyzed to date are presented in Figure 5. In this figure, the amino groups that were expected to be basic enough to generate the stable ion pair with [3,3′-Co(1,2-$C_2B_9H_{11}$)$_2$]$^-$ have been highlighted in red.

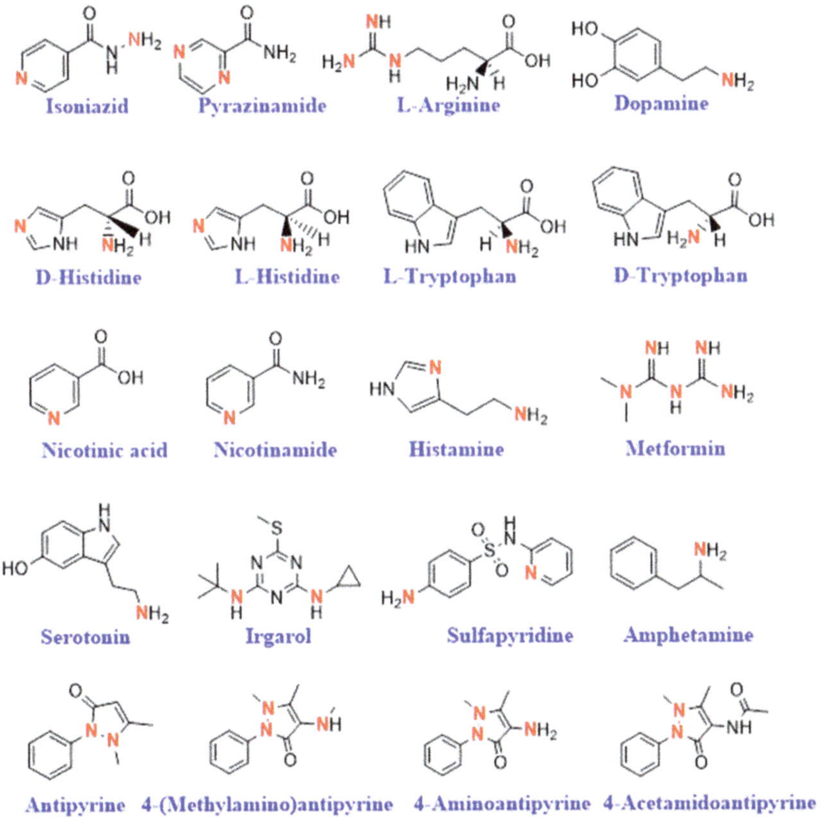

Figure 5. Chemical structures of bioactive compounds used to prepare $[YH]_x[3,3'\text{-}Co(1,2\text{-}C_2B_9H_{11})_2]_y$. In red color were the amino groups that were expected to be basic enough to generate stable ion-pairs with $[o\text{-COSAN}]^-$.

The x and y values in the formula of the ion-pair complex, $[\text{cation-NH}]_x[3,3'\text{-}Co(1,2\text{-}C_2B_9H_{11})_2]_y$ were established by ^{1}H-NMR upon integration because the $C_{cluster}$-H proton atoms of $[3,3'\text{-}Co(1,2\text{-}C_2B_9H_{11})_2]^-$ were easily identified (3.94 ppm in d_6-acetone) and were weighted with regard to singular proton atoms, equally well identified, from the cation. The electroactive ion-pair complex made this way was so simple that practically no other spectroscopic or elemental analysis technique was required, but in many cases, Nuclear Magnetic Resonance (^{1}H{^{11}B}, ^{11}B, ^{11}B{1H} ^{13}C{1H}) NMR, Fourier Transform Infrared Spectroscopy (FTIR), Elemental Analysis (EA), MALDI-TOF-MS spectroscopies were also used as further characterization (Table 1).

Table 1. Analytical performances of ISEs prepared for [cation-NH]$_x$[3,3′-Co(1,2-$C_2B_9H_{11}$)$_2$]$_y$. The plasticizer is abbreviated as: *o*-nitro phenyl octyl ether (NPOE); di-octyl phthalate (DOP); di-butyl phthalate (DBP); di-octyl sebacate (DOS), bis(2-ethylhexyl) phthalate (DEHP); tributyl phosphate (TBP).

Samples	Ion Pair Complex Formula	Plasticizer	Slope (mV·Decade^{-1})	Concentration Range (mol·dm^{-3})	Detection Limit (mol·dm^{-3})	Reference
Isoniazid	[H_3INH][*o*-COSAN]$_3$	NPOE	52.37	1.00×10^{-4}–1.00×10^{-1}	5.00×10^{-5}	[49]
Isoniazid	[H_3INH][*o*-COSAN]$_3$	DOP	47.80	1.00×10^{-4}–1.00×10^{-1}	5.80×10^{-5}	[49]
Pyrazinamide	H[HPZA]$_2$[*o*-COSAN]$_3$	NPOE	56.98	5.00×10^{-4}–1.00×10^{-1}	3.00×10^{-5}	[49]
Pyrazinamide	H[HPZA]$_2$[*o*-COSAN]$_3$	DOP	46.70	5.00×10^{-5}–1.00×10^{-1}	1.00×10^{-5}	[49]
L-Arginine	[HArg][*o*-COSAN]	DBP	45.80	5.00×10^{-6}–1.00×10^{-1}	3.00×10^{-6}	[50]
L-Arginine	[HArg][*o*-COSAN]	DEHP	37.70	1.00×10^{-5}–1.00×10^{-1}	5.00×10^{-5}	[50]
D-Histidine	[H_2His][*o*-COSAN]$_2$	DBP	36.50	1.00×10^{-5}–1.00×10^{-1}	8.00×10^{-6}	[50]
D-Histidine	[H_2His][*o*-COSAN]$_2$	DEHP	42.40	5.00×10^{-5}–1.00×10^{-1}	2.00×10^{-5}	[50]
L-Histidine	[H_2His][*o*-COSAN]$_2$	DBP	47.40	5.00×10^{-6}–1.00×10^{-1}	1.00×10^{-6}	[50]
L-Histidine	[H_2His][*o*-COSAN]$_2$	DEHP	48.50	5.00×10^{-4}–1.00×10^{-1}	1.00×10^{-4}	[50]
D-Tryptophan	[HTry][*o*-COSAN]	DBP	60.50	5.00×10^{-7}–1.00×10^{-1}	2.00×10^{-7}	[50]
L-Tryptophan	[HTry][*o*-COSAN]	DBP	62.60	5.00×10^{-7}–1.00×10^{-1}	1.00×10^{-7}	[50]
Dopamine	[HDA][*o*-COSAN]	Dibutylsebacate	44.97 ± 1.23	1.00×10^{-6}–1.00×10^{-2}	0.80×10^{-6}	[51]
Dopamine	[HDA][*o*-COSAN]	Bis (2-ethyl hexyl) phthalate	53.23 ± 1.75	1.00×10^{-5}–1.00×10^{-2}	7.20×10^{-6}	[51]
Dopamine	[HDA][*o*-COSAN]	NPOE	58.17 ± 1.44	5.00×10^{-6}–1.00×10^{-2}	1.00×10^{-6}	[51]
Dopamine	[HDA][*o*-COSAN]	DOP	55.96 ± 0.85	5.00×10^{-6}–1.00×10^{-2}	1.00×10^{-6}	[51]
Nicotinamide	[HNAmd][*o*-COSAN]	DOP	52.11 ± 1.17	5.00×10^{-6}–1.00×10^{-2}	1.00×10^{-6}	[51]
Nicotinic acid	[HNA][*o*-COSAN]	DOP	57.55 ± 0.88	1.00×10^{-6}–1.00×10^{-2}	0.70×10^{-6}	[51]
Histamine	[H_2His][*o*-COSAN]$_2$	NPOE	31.62 ± 0.43	5.00×10^{-6}–1.00×10^{-2}	0.80×10^{-6}	[51]
Metformin	[H_2Met][*o*-COSAN]$_2$	NPOE	25.82 ± 1.91	1.00×10^{-5}–1.00×10^{-2}	6.00×10^{-6}	[51]
Serotonin	[HSer][*o*-COSAN]	DBS	50.50 ± 0.50	2.25×10^{-5}–1.00×10^{-2}	4.51×10^{-6}	[52]
Serotonin	[HSer][*o*-COSAN]	DBP	50.60 ± 0.50	2.25×10^{-5}–1.00×10^{-2}	1.17×10^{-5}	[52]
Serotonin	[HSer][*o*-COSAN]	TBP	60.50 ± 0.30	2.25×10^{-5}–1.00×10^{-2}	1.56×10^{-5}	[52]
Serotonin	[HSer][*o*-COSAN]	NPOE	51.10 ± 0.10	2.25×10^{-5}–1.00×10^{-2}	1.70×10^{-5}	[52]
Irgarol	[Irg-H][*o*-COSAN]	DOP	56.67 ± 2.30	1.00×10^{-5}–1.00×10^{-1}	3.00×10^{-6}	[53]
Irgarol	[Irg-H][*o*-COSAN]	DOS	57.17 ± 1.70	1.00×10^{-5}–1.00×10^{-1}	2.00×10^{-6}	[53]
Irgarol	[Irg-H][*o*-COSAN]	NPOE	48.21 ± 6.40	1.00×10^{-5}–1.00×10^{-1}	4.00×10^{-6}	[53]

Table 1. *Cont.*

Samples	Ion Pair Complex Formula	Plasticizer	Slope (mV·Decade^{-1})	Concentration Range (mol·dm^{-3})	Detection Limit (mol·dm^{-3})	Reference
Sulfapyridine	A-H][*o*-COSAN]	NPOE	47.69	1.00×10^{-6}–1.00×10^{-3}	4.00×10^{-6}	[54]
Sulfapyridine	[A-H][*o*-COSAN]	DOS	61.29	1.00×10^{-6}–1.00×10^{-3}	1.00×10^{-6}	[54]
Sulfapyridine	[A-H][*o*-COSAN]	DOP	61.26	1.00×10^{-6}–1.00×10^{-3}	1.00×10^{-5}	[54]
Amphetamine	[Amph-H][*o*-COSAN]	DBP	60	1.00×10^{-5}–1.00×10^{-3}	12.00×10^{-6}	[55]
Amphetamine	[Amph-H][*o*-COSAN]	DOP	42	1.00×10^{-5}–1.00×10^{-3}	8.00×10^{-6}	[55]
Amphetamine	[Amph-H][*o*-COSAN]	DOS	53	1.00×10^{-5}–1.00×10^{-3}	4.00×10^{-5}	[55]
Amphetamine	[Amph-H][*o*-COSAN]	NPOE	45	1.00×10^{-5}–1.00×10^{-3}	2.00×10^{-5}	[55]
Antipyrine	[AP][*o*-COSAN]	NPOE	79.6 ± 4.9	1.00×10^{-5}–1.00×10^{-2}	$70.8 \times 10^{-6} \pm 9.3$	[56]
Antipyrine	[AP][*o*-COSAN]	DBS	57.0 ± 1.4	1.00×10^{-5}–1.00×10^{-2}	$29.8 \times 10^{-6} \pm 2.2$	[56]
4-(methylamino) antipyrine	[MAAP][*o*-COSAN]	NPOE	33.9 ± 1.0	1.00×10^{-5}–1.00×10^{-2}	$27.3 \times 10^{-6} \pm 1.5$	[56]
4-(methylamino)antipyrine	[MAAP][*o*-COSAN]	DBS	48.2 ± 1.0	1.00×10^{-5}–1.00×10^{-2}	$279.5 \times 10^{-6} \pm 7.0$	[56]
4-aminoantipyrine	[AAP][*o*-COSAN]	NPOE	54.6 ± 3.8	1.00×10^{-5}–1.00×10^{-2}	$88.2 \times 10^{-6} \pm 14.8$	[56]
4-aminoantipyrine	[AAP][*o*-COSAN]	DBS	71.2 ± 6.1	1.00×10^{-5}–1.00×10^{-2}	$342.0 \times 10^{-6} \pm 27.2$	[56]
4-acetamidoantipyrine	[AAAP][*o*-COSAN]	NPOE	57.0 ± 2.0	1.00×10^{-5}–1.00×10^{-2}	$252.2 \times 10^{-6} \pm 18.7$	[56]

4. Results

As is common theory, the performance of ISEs is given by several parameters: the slope of the calibration curve, the linear working range, the limit of detection (LOD), selectivity and response time. These parameters strongly depend on the composition, stability, and reproducibility of the membrane and these derived from the $[YH]_x[3,3'\text{-}Co(1,2\text{-}C_2B_9H_{11})_2]_y$ methodology are reported in Table 1.

The value of the slope of the calibration curve has to be in correlation with the number of charged species, proof of the Nernstian response of the prepared membranes. One of the most important components of the membrane composition is the nature (polar or nonpolar) and percentage of plasticizer, which ensure and improve the stability in time, the sensitivity, and selectivity.

pH is an important analytical parameter with a direct influence on sensitivity and selectivity. In our studies the influence of the pH on the ISEs answer in terms of Nernstian response and linear concentration range and for selectivity of the prepared membrane towards the analyte versus different interferences was done.

Commonly, the stability of the prepared membranes, in terms of slope was monitored for a period of 45 days. In this time, it was observed that the prepared membranes remain functional without degradation in performances for all of the studied analytes. The [*o*-COSAN]$^-$ ISEs correlate very well with those traditionally made with ionophores taking advantage of its ease of preparation and very favorably of the lack of ionophore design and hence synthetic simplicity.

Another important parameter for ISEs is the response time, the time that the electrode needs until reaching a stable potential, which is directly correlated with the membrane thickness. For most of the membranes prepared following the protocol described in these papers the response time was around 5 s.

From the analytical point of view, selectivity is one of the most important parameters, especially when it is necessary to determine the target compounds from a complex matrix. In these studies, it was observed that a controlled tuning of the chemical composition of the membrane, in terms of ratio between the ion pair complex and plasticizers and the nature of plasticizers was possible to improve the selectivity

The potentiometric selectivity coefficients ($Kpot_{A/B}$) were calculated based on the Nikolsky Eisenman equation [68,69] using the fixed interferences method (FIM). For each studied analyte, the selectivity of the prepared ISEs were tested for the possible inorganic and organic interference compounds and it was observed that the PVC membrane based on ion-pair complexes made between [*o*-COSAN]$^-$ and the protonated tested analyte assured a good selectivity. The advantages of using $[3,3'\text{-Co}(1,2\text{-}C_2B_9H_{11})_2]^-$ as an ion-pair generator compared with other anions are based on: its unique 3D aromaticity [70], its high chemical and thermal stability (withstanding strong acid, moderate base, high temperatures and intense radiation) [71,72], as well as its biological stability (neither degradation nor chemical modification compounds were identified after cells' uptake) [73,74], its lipophilicity [75], low-charge density [76], small volume molecules with a size of 1.1 × 0.6 nm [77], capacity to produce B–H···H–N dihydrogen bonds [19,77–79], and B–H···O, $C_{cluster}$–H···N and B-H···Na or B-H···K hydrogen bonds [80–82] as well as unconventional cooperative effect such as $C_{cluster}$–H···S–H···H–B hydrogen/dihydrogen bond interaction [83] that enhance the membrane stability over the time a fact that directly correlates with the sensitivity of the analyte determination. Further to these stability enhancing characteristics, it is our belief that the fact that the anion is redox reversible really influences in the good performances of these $[3,3'\text{-Co}(1,2\text{-}C_2B_9H_{11})_2]^-$ dependent electroactive ion-pair species.

5. Why Do These Membranes with Metallabis(dicarbollides), Being So Simple in Their Composition, Give Such Excellent Results?

Our explanation is simple; it is due to the θ-shape structure and chemical composition of the cobaltabis(dicarbollide) that give it unique properties. For instance, if we compare tetraphenylborate and $[1\text{-HCB}_{11}Br_{11}]^-$ described in the paper as lipophilic ion-exchangers with $[3,3'\text{-Co}(1,2\text{-}C_2B_9H_{11})_2]^-$ it is noted that in the case of borate there are no B-H bonds but B-Br bonds, and the same for the tetraphenylborate that have C-H bonds so these cannot generate hydrogen and dihydrogen bonds. We believe this is essential. Thereby neither the one nor the other will generate strong interactions with the ammonium cation or with the plasticizer solvent. Therefore, they have high mobility in the membrane. This is not the case with $[3,3'\text{-Co}(1,2\text{-}C_2B_9H_{11})_2]^-$, which does make these strong interactions, and therefore in our view, generates more reticulate, and therefore more stability and a higher fixation of the concentration of the analyte in the membrane. The cobaltabis(dicarbollide) has dimensions of 1.1 nm in length and 0.6 nm in width and is surrounded by hydrogens, the mentioned B-H bonds, that have considerable hydride character but not enough to be unstable in protic solvents. This sufficient hydride character of the B-H groups enables it to interact strongly with H-N units. The non-bonding interactions are weak, but if there are many, they become a strong interaction. Surely, this is what makes it unnecessary to have ionophores in these membranes and that, on the other hand, the common ion to be detected that is present on both sides of the interface remains constant within the membrane giving the appropriate stability and sensitivity. In this case, we do not measure the anion, the cobaltabis(dicarbollide); we measure the cation.

6. Conclusions

Typically, ISEs for cations or anions require a ligand to complex them and thus achieve target selectivity. This ligand required a design and synthetic process that was usually laborious because it required the formation constant with the target ion to be much higher than an interfering one. This selectivity was also associated with a low dissociation constant, which could lead to difficulties in transferring information between the analyte phase and the internal phase. Our knowledge and understanding of the characteristics of [*o*-

COSAN]$^-$ and its high appetence for protonated amines as well as its solubility properties in aqueous media and in THF indicated that it could generate very efficient and easy to produce ionic pairs to detect amine cations, which made it very interesting for the determination of pharmaceuticals. The different potentials for the different redox couples in which [*o*-COSAN]$^-$ could participate also made us believe that this system could facilitate information transfer between the analyte phase and the internal phase of the electrode. We could not observe the latter property as we did not exceed the 10^{-6} M detection limit, but we did find a very versatile membrane with a cationic electroactive substance, which was the one we wanted to detect. This cation was compensated by the anionic cobaltabis(dicarbollides), [cation-NH]$_x$[3,3′-Co(1,2-$C_2B_9H_{11}$)$_2$]$_y$, which allowed us to easily adapt it to the target we wanted to investigate. The strong interactions between the electroactive cation and [*o*-COSAN]$^-$ and between [*o*-COSAN]$^-$ and the plasticizer solvent favored a highly stable system very suitable for detecting amines as indicated in this work. The system is highly extrapolated to different amines, including enantiomers, which it detects in a clearly discriminatory way with respect to their optical isomers.

Thus, we developed membranes based on ion pair complexes between metallabis (dicarbollides), [3,3′-Co(1,2-$C_2B_9H_{11}$)$_2$]$^-$, and bioactive protonable nitrogen containing compounds, [cation-NH]$_x$[3,3′-Co(1,2-$C_2B_9H_{11}$)$_2$]$_y$ that have proven that the properties of this anion open new directions for using these ISEs in environmental, clinical, pharmaceutical and food application and for miniaturization and mass production for routine analysis.

Author Contributions: C.V. and F.T. developed the idea. A.-I.S., C.V. and F.T. wrote the draft. F.T. and C.V. assembled the article and, gave it in its final form. All authors have read and agreed to the published version of the manuscript.

Funding: This work research was funded by the European Union's Horizon 2020 grant number 768686, the Spanish Ministerio de Economía y Competitividad grant number PID2019-106832RB-I00 and, the Generalitat de Catalunya grant number 2017SGR1720.

Institutional Review Board Statement: Not applicable.

Informed Consent Statement: Not applicable.

Data Availability Statement: Not applicable.

Acknowledgments: This mini review is dedicated to Nicole Jaffrezic-Renault for her great contribution to the field of sensors and biosensors.

Conflicts of Interest: There are no conflict on interests between the authors.

References

1. Kerru, N.; Gummidi, L.; Maddila, S.; Gangu, K.K.; Jonnalagadda, S.B. A Review on Recent Advances in Nitrogen-Containing Molecules and Their Biological Applications. *Molecules* **2020**, *25*, 1909. [CrossRef] [PubMed]
2. Heravi, M.M.; Zadsirjana, V. Prescribed Drugs Containing Nitrogen Heterocycles: An Overview. *RSC Adv.* **2020**, *10*, 44247–44311. [CrossRef] [PubMed]
3. Siddiqui, M.R.; AlOthman, Z.A.; Rahman, N. Analytical techniques in pharmaceutical analysis: A review. *Arab. J. Chem.* **2017**, *10*, S1409–S1421. [CrossRef]
4. Xu, Q.; Yuan, A.J.; Zhang, R.; Bian, X.; Chen, D.; Hu, X. Application of Electrochemical Methods for Pharmaceutical and Drug Analysis. *Curr. Pharm. Anal.* **2009**, *5*, 144–155. [CrossRef]
5. Mostafa, I.M.; Meng, C.; Dong, Z.; Lou, B.; Xu, G. Potentiometric Sensors for the Determination of Pharmaceutical Drugs. *Anal. Sci.* **2022**, *38*, 23–37. [CrossRef] [PubMed]
6. Kharitonov, S.V. Ion-Selective Electrodes in Medicinal Drug Determination. *Russ. Chem. Rev.* **2007**, *76*, 361. [CrossRef]
7. Tong, H.Y.; Meng, J.; Liang, J.Y.; Li, J.P. Molecularly Imprinted Electrochemical Luminescence Sensor Based on Core-Shell Magnetic Particles with ZIF-8 Imprinted Material. *Sens. Actuators B Chem.* **2021**, *330*, 129405. [CrossRef]
8. Li, X.H.; Kuang, X.J.; Sun, J.L. Rare Earth Elements Based Oxide Ion Conductors. *Inorg. Chem. Front.* **2021**, *8*, 1374. [CrossRef]
9. Yao, M.M.; Huang, J.K.; Deng, Z.H.; Jin, W.Y.; Yuan, Y.L.; Nie, J.F.; Wang, H.; Du, F.Y.; Zhang, Y. Transforming Glucose into Fluorescent Graphene Quantum Dots Via Microwave Radiation for Sensitive Detection of Al^{3+} Ions Based on Aggregation-Induced Enhanced Emission. *Analyst* **2020**, *145*, 6981. [CrossRef]
10. Hawthorne, M.F.; Andrews, T.D. Carborane Analogues of Cobalticinium Ion. *Chem. Commun.* **1965**, *19*, 443–444. [CrossRef]

11. Lever, A.B.P. Electrochemical Parametrization of Metal Complex Redox Potentials, using the Ruthenium(III)/Ruthenium(II) Couple to Generate a Ligand Electrochemical Series. *Inorg. Chem.* **1990**, *29*, 1271–1285. [CrossRef]
12. Morris, J.H.; Gysling, H.J.; Reed, D. Electrochemistry of Boron Compounds. *Chem. Rev.* **1985**, *85*, 51–76. [CrossRef]
13. Núñez, R.; Tarrés, M.; Ferrer-Ugalde, F.; de Biani, F.F.; Teixidor, F. Electrochemistry and Photoluminescence of Icosahedral Carboranes, Boranes, Metallacarboranes, and their Derivatives. *Chem. Rev.* **2016**, *116*, 14307–14378. [CrossRef]
14. Bauduin, P.; Prevost, S.; Farras, P.; Teixidor, F.; Diat, O.; Zemb, T. A Theta-Shaped Amphiphilic Cobaltabisdicarbollide Anion: Transition from Monolayer Vesicles to Micelles. *Angew. Chem. Int. Ed.* **2011**, *50*, 5298–5300. [CrossRef]
15. Malaspina, D.C.; Viñas, C.; Teixidor, F.; Faraudo, J. Atomistic Simulations of COSAN: Amphiphiles without a Head-and-Tail Design Display "Head and Tail" Surfactant Behavior. *Angew. Chem. Int. Ed.* **2020**, *59*, 3088–3092. [CrossRef]
16. Juárez-Pérez, E.J.; Núñez, R.; Viñas, C.; Sillanpää, R.; Teixidor, F. The Role of C–H···H–B Interactions in Establishing Rotamer Configurations in Metallabis (dicarbollide) Systems. *Eur. J. Inorg. Chem.* **2010**, *16*, 2385–2392. [CrossRef]
17. Farràs, P.; Juárez-Pérez, E.J.; Lepšík, M.; Luque, R.; Núñez, R.; Teixidor, F. Metallacarboranes and their Interactions: Theoretical Insights and their Applicability. *Chem. Soc. Rev.* **2012**, *41*, 3445–3463. [CrossRef]
18. Verdiá-Báguena, C.; Alcaraz, A.; Aguilella, V.M.; Cioran, A.M.; Tachikawa, S.; Nakamura, H.; Teixidor, F.; Viñas, C. Amphiphilic COSAN and I2-COSAN Crossing Synthetic Lipid Membranes: Planar Bilayers and Liposomes. *Chem. Commun.* **2014**, *50*, 6700–6703. [CrossRef]
19. Fuentes, I.; Pujols, J.; Viñas, C.; Ventura, S.; Teixidor, F. Dual Binding Mode of Metallacarborane Produces a Robust Shield on Proteins. *Chem. Eur. J.* **2019**, *25*, 12820–12829. [CrossRef]
20. de Marco, R.; Clarke, G.; Pejcic, B. Ion-Selective Electrode Potentiometry in Environmental Analysis. *Electroanalysis* **2007**, *19*, 1987–2001. [CrossRef]
21. Mikhelson, K.N. *Ion-Selective Electrodes*; Springer: Berlin/Heidelberg, Germany, 2013.
22. Bakker, E.; Pretsch, E. The New Wave of Ion-Selective Electrodes. *Anal. Chem.* **2007**, *26*, 46–51.
23. Bakker, E.; Pretsch, E. Modern Potentiometry. *Angew. Chem. Int. Ed.* **2007**, *46*, 5660–5668. [CrossRef]
24. Lewenstam, A. Routines and Challenges in Clinical Application of Electrochemical Ion-Sensors. *Electroanalysis* **2014**, *26*, 1171–1181. [CrossRef]
25. Bloch, R.; Shatkay, A.; Saroff, H.A. Fabrication and Evaluation of Membranes as Specific Electrodes for Calcium Ions. *Biophys. J.* **1967**, *7*, 865–877. [CrossRef]
26. Moody, G.J.; Oke, R.B.; Thomas, J.D.R. A Calcium-Sensitive Electrode Based on a Liquid Ion Exchanger in a Poly (Vinyl Chloride) Matrix. *Analyst* **1970**, *95*, 910–918. [CrossRef]
27. Stefanac, Z.; Simon, W. Ion Specific Electrochemical Behavior of Macrotetrolides in Membranes. *Microchem. J.* **1967**, *12*, 125–132. [CrossRef]
28. Viñas, C.; Gómez, S.; Bertran, J.; Barron, J.; Teixidor, F.; Dozol, J.-F.; Rouquette, H.; Kivekkäs, R.; Sillanpää, R. C-substituted bis (dicarbollide) metal compounds as sensors and extractants of radionuclides from nuclear wastes. *J. Organomet. Chem.* **1999**, *581*, 188–193. [CrossRef]
29. Kopytin, A.V.; Zhizhin, K.Y.; Urusov, Y.I.; Mustyatsa, V.N.; Kokunov, Y.V.; Kuznetsov, N.T. Potentiometric Sensors with Membranes Based on Ionic Liquid Tetradecylammonium Triethylammonio-closo-Dodecaborate. *J. Anal. Chem.* **2012**, *67*, 168–171. [CrossRef]
30. Kubasov, A.S.; Turishev, E.S.; Kopytin, A.V.; Shpigun, L.K.; Zhizhin, K.Y.; Kuznetsov, N.T. Sulfonium closo-hydridodecaborate Anions as Active Components of a Potentiometric Membrane Sensor for Lidocaine Hydrochloride. *Inorg. Chim. Acta.* **2021**, *514*, 119992. [CrossRef]
31. Peper, S.; Telting-Diaz, M.; Almond, P.; Albrecht-Schmitt, T.; Bakker, A. Perbrominated closo-Dodecacarborane Anion, [1-$HCB_{11}Br_{11}]^-$, as an Ion Exchanger in Cation-Selective Chemical Sensors. *Anal. Chem.* **2002**, *74*, 1327–1332. [CrossRef]
32. Bakker, E.; Nägele, M.; Schaller, U.; Pretsch, E. Applicability of the Phase Boundary Potential Model to the Mechanistic Understanding of Solvent Polymeric Membrane-Based Ion-Selective Electrodes. *Electroanalysis* **1995**, *7*, 817–822. [CrossRef]
33. Karpfen, F.M.; Randles, J.E.B. Ionic Equilibria and Phase-Boundary Potentials in Oil-Water Systems. *Trans. Faraday Soc.* **1953**, *49*, 823–831. [CrossRef]
34. Núñez, R.; Romero, I.; Teixidor, F.; Viñas, C. Icosahedral Boron Clusters: A Perfect Tool for the Enhancement of Polymer Features. *Chem. Soc. Rev.* **2016**, *45*, 5147–5173. [CrossRef] [PubMed]
35. Peper, S.; Qin, Y.; Almond, P.; McKee, M.; Telting-Diaz, M.; Albrecht-Schmitt, T.; Bakker, E. Ion-Pairing Ability, Chemical Stability, and Selectivity Behavior of Halogenated Dodecacarborane Cation Exchangers in Neutral Carrier-Based Ion-Selective Electrodes. *Anal. Chem.* **2003**, *75*, 2131–2139. [CrossRef]
36. Bakker, E.; Pretsch, E. Lipophilicity of Tetraphenylborate Derivatives as Anionic Sites in Neutral Carrier-Based Solvent Polymeric Membranes and Lifetime of Corresponding Ion-Selective Electrochemical and Optical Sensors. *Anal. Chim. Acta* **1995**, *309*, 7–17. [CrossRef]
37. Ortuno, J.A.; Rodenas, V.; Garcıa, M.S.; Albero, M.I.; Sanchez-Pedreno, C.A. New Tiapride Selective Electrode and Its Clinical Application. *Sensors* **2007**, *7*, 400–409. [CrossRef]
38. Malongo, K.; Blankert, B.; Kambu, O.; Amighi, K.; Nsangu, J.; Kauffmann, J.M. Amodiaquine Polymeric Membrane Electrode. *J. Pharm. Biomed. Anal.* **2006**, *41*, 70–76. [CrossRef]
39. Aboul-Enein, H.Y.; Sun, X.X.; Sun, C.J. Ion Selective PVC Membrane Electrode for the Determination of Methacycline Hydrochloride in Pharmaceutical Formulation. *Sensors* **2002**, *2*, 424–431. [CrossRef]

40. Kulapina, E.G.; Barinova, O.V. Ion-Selective Electrodes for the Determination of Nitrogen-Containing Medicinal Substances. *J. Anal. Chem.* **2001**, *56*, 457–460. [CrossRef]
41. Ortuño, J.A.; Hernández, J.; Sánchez-Pedreño, C. Ion-Selective Electrode for the Determination of Some Multidrug Resistance Reversers. *Sens. Actuators B Chem.* **2006**, *119*, 282–287. [CrossRef]
42. Bakker, E.; Pretsch, E.; Bühlmann, P. Selectivity of Potentiometric Ion Sensors. *Anal. Chem.* **2000**, *72*, 1127–1133. [CrossRef] [PubMed]
43. Freiser, H. *Ion-Selective Electrodes in Analytical Chemistry*; Springer Science & Business Media: Berlin, Germany, 2012.
44. Bakker, E.; Chumbimuni-Torres, K. Modern Directions for Potentiometric Sensor. *J. Braz. Chem. Soc.* **2008**, *19*, 621–629. [CrossRef]
45. Guggenheim, E.A. The Conceptions of Electrical Potential Difference between Two Phases and the Individual Activities of Ions. *J. Phys. Chem.* **1929**, *33*, 842–849. [CrossRef]
46. Guggenheim, E.A. On the Conception of Electrical Potential Difference between two Phases. II. *J. Phys. Chem.* **1930**, *34*, 1540–1543. [CrossRef]
47. Theorell, T. An Attempt to Formulate a Quantitative Theory of Membrane Permeability. *Proc. Soc. Exp. Biol. Med.* **1935**, *33*, 282–285. [CrossRef]
48. Meyer, K.H.; Sievers, J.F. La Perméabilité des Membranes I. Théorie de la Perméabilité Ionique. *Helv. Chim. Acta* **1936**, *19*, 649–664. [CrossRef]
49. Stoica, A.I.; Vinas, C.; Teixidor, F. Application of the Cobaltabisdicarbollide Anion to the Development of Ion Selective PVC Membrane Electrodes for Tuberculosis Drug Analysis. *Chem. Commun.* **2008**, *48*, 6492–6494. [CrossRef]
50. Stoica, A.I.; Vinas, C.; Teixidor, F. Cobaltabisdicarbollide Anion Receptor for Enantiomer-Selective Membrane Electrodes. *Chem. Commun.* **2009**, *33*, 4988–4990. [CrossRef]
51. Stoica, A.I.; Kleber, C.; Vinas, C.; Teixidor, F. Ion Selective Electrodes for Protonable Nitrogen Containing Analytes: Metallacarboranes as Active Membrane Components. *Electrochim. Acta* **2013**, *113*, 94–98. [CrossRef]
52. Bliem, C.; Fruhmann, P.; Stoica, A.I.; Kleber, C. Development and Optimization of an Ion-selective Electrode for Serotonin Detection. *Electroanalysis* **2017**, *29*, 1635–1642. [CrossRef]
53. Saini, A.; Gallardo-Gonzalez, J.; Baraket, A.; Fuentes, I.; Viñas, C.; Zine, N.; Bausells, J.; Teixidor, F.; Errachid, A. A Novel Potentiometric Microsensor for Real-Time Detection of Irgarol using the Ion-Pair Complex $[\text{Irgarol-H}]^+[\text{Co}(C_2B_9H_{11})_2]^-$. *Sens. Actuators B Chem.* **2018**, *268*, 164–169. [CrossRef]
54. Saini, A.; Fuentes, I.; Viñas, C.; Zine, N.; Bausells, J.; Errachid, A.; Teixidor, F. A Simple Membrane with the Electroactive $[\text{Sulfapyridine-H}]^+[\text{Co}(C_2B_9H_{11})_2]^-$ for the Easy Potentiometric Detection of Sulfonamides. *J. Organomet. Chem.* **2019**, *893*, 32–38. [CrossRef]
55. Gallardo-Gonzalez, J.; Saini, A.; Baraket, A.; Boudjaoui, S.; Alcácer, A.; Strelas, A.; Teixidor, F.; Zine, N.; Bausells, J.; Errachid, A. A Highly Selective Potentiometric Amphetamine Microsensor Based on All-Solid-State Membrane Using a New Ion-Pair Complex, $[3,3'\text{-Co}(1,2\text{-closo-}C_2B_9H_{11})_2]^-[C_9H_{13}NH]^+$. *Sens. Actuators B Chem.* **2018**, *266*, 823–829. [CrossRef]
56. Mayerhuber, L.; Trattner, S.; Luger, S.; Weigelhofer, G.; Hametner, C.; Fruhmann, P. Development of Ion-Selective Electrodes for Antipyrine and its Derivatives as Potential Tool for Environmental Water Monitoring. *J. Electroanal. Chem.* **2021**, *886*, 115110–115117. [CrossRef]
57. Thoma, A.P.; Cimerman, Z.; Fiedler, U.; Bedekovic, D.; Güggi, M.; Jordan, P.; May, K.; Pretsch, E.; Prelog, V.; Simon, W. Enantiomerenselektives Verhalten in Membranen Eines Chiralen Elektrisch Neutralen Ionophores. *Chimia* **1975**, *29*, 344–346.
58. Trojanowicz, M. Enantioselective Electrochemical Sensors and Biosensors: A Mini-Review. *Electrochem. Commun.* **2014**, *38*, 47–52. [CrossRef]
59. Trojanowicz, M.; Kaniewska, M. Electrochemical Chiral Sensors and Biosensors. *Electroanalysis* **2009**, *21*, 229–238. [CrossRef]
60. Yasaka, Y.; Yamamoto, T.; Kimura, K.; Shono, T. Simple Evaluation of Enantiomer-Selectivity of Crown Ether Using Membrane Electrode. *Chem. Lett.* **1980**, *9*, 769–772. [CrossRef]
61. Maruyama, K.; Sohmiya, H.; Tsukube, H. New Chiral Host Molecules Derived from Naturally Occurring Monensin Ionophore. *J. Chem. Soc. Chem. Commun.* **1989**, 864–865. [CrossRef]
62. Maruyama, K.; Sohmiya, H.; Tsukube, H. Enantiomer Recognition of Organic Ammonium Salts by Podand and Crown-e Monensin Amides: New Synthetic Strategy for Chiral Receptors. *Tetrahedron* **1992**, *48*, 805–818. [CrossRef]
63. Kaniewska, M.; Sikora, T.; Kataky, R.; Trojanowicz, M. Enantioselectivity of Potentiometric Sensors with Application of Different Mechanisms of Chiral Discrimination. *J. Biochem. Biophys. Methods* **2008**, *70*, 1261–1267. [CrossRef] [PubMed]
64. Ji, J.; Qu, L.; Wang, Z.; Li, G.; Feng, W.; Yang, G. A facile electrochemical chiral sensor for tryptophan enantiomers based on multiwalled carbon nanotube/hydroxypropyl-β-cyclodextrin functionalized carboxymethyl cellulose. *Microchem. J.* **2022**, *175*, 107133. [CrossRef]
65. Liua, N.; Yang, B.; Yin, Z.-Z.; Cai, W.; Li, J.; Kong, Y. A chiral sensing platform based on chiral metal-organic framework for enantiodiscrimination of the isomers of tyrosine and tryptophan. *J. Electroanal. Chem.* **2022**, *918*, 116445. [CrossRef]
66. Lu, Q.; Chen, L.; Meng, Q.; Jiang, Y.; Xie, L. A biomolecule chiral interface base on BSA for electrochemical recognition of amine enantiomers. *Chirality* **2021**, *33*, 385–396. [CrossRef] [PubMed]
67. Luger, S.; Mayerhuber, L.; Weigelhofer, G.; Hein, T.; Holzer, B.; Hametner, C.; Fruhmann, P. Development of Ion-selective Electrodes for Tropane, Atropine, and Scopolamine—A Concept for the Analysis of Tropane Alkaloids. *Electroanalysis* **2022**, *34*, 1579–1586. [CrossRef]

68. Analytical Chemistry Division, Commission on Analytical Nomenclature. Recommendations for Nomenclature of Ion-Selective Electrodes, "Recommendations–1975". *Pure Appl. Chem.* **1976**, *48*, 127–132. [CrossRef]
69. Buck, R.P.; Lindner, E. Recommendations for Nomenclature of Ion-Selective Electrodes IUPAC Recommendations 1994. *Pure Appl. Chem.* **1994**, *66*, 2527–2536. [CrossRef]
70. Poater, J.; Viñas, C.; Bennour, I.; Gordils, S.E.; Sola, M.; Teixidor, F. Too Persistent to Give Up: Aromaticity in Boron Clusters Survives Radical Structural Changes. *J. Am. Chem. Soc.* **2020**, *142*, 9396–9407. [CrossRef]
71. Grimes, R.N. *Carboranes*, 3rd ed.; Academic Press: New York, NY, USA, 2016.
72. Plesek, J. Potential applications of the boron cluster compounds. *Chem. Rev.* **1992**, *92*, 269–278. [CrossRef]
73. Tarrés, M.; Canetta, E.; Paul, E.; Forbes, J.; Azzouni, K.; Viñas, C.; Teixidor, F.; Harwood, A.J. Biological interaction of living cells with COSAN-based synthetic vesicles. *Sci. Rep.* **2015**, *5*, 7804. [CrossRef]
74. Fuentes, I.; García-Mendiola, T.; Sato, S.; Pita, M.; Nakamura, H.; Lorenzo, E.; Teixidor, F.; Marques, F.; Viñas, C. Metallacarboranes on the Road to Anticancer Therapies: Cellular Uptake, DNA Interaction, and Biological Evaluation of Cobaltabisdicarbollide [COSAN]−. *Chem. Eur. J.* **2018**, *24*, 17239–17254. [CrossRef] [PubMed]
75. Bennour, I.; Ramos, M.N.; Nuez-Martínez, M.; Xavier, J.A.M.; Buades, A.B.; Sillanpää, R.; Teixidor, F.; Choquesillo-Lazarte, D.; Romero, I.; Martinez-Medina, M.; et al. Water soluble organometallic small molecules as promising antibacterial agents: Synthesis, physical–chemical properties and biological evaluation to tackle bacterial infections. *Dalton Trans.* **2022**, *51*, 7188–7209. [CrossRef]
76. Masalles, C.; Llop, J.; Vinas, C.; Teixidor, F. Extraordinary overoxidation resistance increase in self-doped polypyrroles by using non-conventional low charge-density anions. *Adv. Mater.* **2002**, *14*, 826–829. [CrossRef]
77. Goszczynski, T.M.; Fink, K.; Kowalski, K.; Lesnikowski, Z.J.; Boratynski, J. Interactions of Boron Clusters and their Derivatives with Serum Albumin. *Sci. Rep.* **2017**, *7*, 9800. [CrossRef] [PubMed]
78. Cıgler, P.; Kozısek, M.; Rezacova, P.; Brynda, J.; Otwinowski, Z.; Pokorna, J.; Plesek, J.; Grüner, B.; Doleckova-Maresova, L.; Masa, M.; et al. From nonpeptide toward noncarbon protease inhibitors: Metallacarboranes as specific and potent inhibitors of HIV protease. *Proc. Natl. Acad. Sci. USA* **2005**, *102*, 15394–15399. [CrossRef]
79. Fanfrlık, J.; Brynda, J.; Rezac, J.; Hobza, P.; Lepsık, M. Interpretation of Protein/Ligand Crystal Structure using QM/MM Calculations: Case of HIV-1 Protease/Metallacarborane Complex. *J. Phys. Chem. B* **2008**, *112*, 15094–15102. [CrossRef]
80. Matějíček, P.; Zedník, J.; Ušelová, K.; Pleštil, J.; Fanfrlík, J.; Nykänen, A.; Ruokolainen, J.; Hobza, P.; Procházka, K. Stimuli-Responsive Nanoparticles Based on Interaction of Metallacarborane with Poly (ethylene oxide). *Macromolecules* **2009**, *42*, 4829–4837. [CrossRef]
81. Tarrés, M.; Viñas, C.; Gonzalez-Cardoso, P.; Hänninen, M.M.; Sillanpää, R.; Dordovic, V.; Uchman, M.; Teixidor, F.; Matejicek, P. Aqueous Self-Assembly and Cation Selectivity of Cobaltabisdicarbollide Dianionic Dumbbells. *Chem. Eur. J.* **2014**, *20*, 6786–6794. [CrossRef]
82. Teixidor, F.; Pedrajas, J.; Rojo, I.; Viñas, C.; Kivekäs, R.; Sillanpää, R.; Sivaev, I.; Bregadze, V.; Sjöberg, S. Chameleonic Capacity of [3,3′-Co(1,2-C2B9H11)2]-in Coordination. Generation of the Highly Uncommon S (thioether)-Na Bond. *Organometallics* **2003**, *22*, 3414–3423. [CrossRef]
83. Planas, J.G.; Viñas, C.; Teixidor, F.; Comas-Vives, A.; Ujaque, G.; Lledós, A.; Light, M.E.; Hursthouse, M.B. Self-Assembly of Mercaptane−Metallacarborane Complexes by an Unconventional Cooperative Effect: A C−H···S−H···H−B Hydrogen/Dihydrogen Bond Interaction. *J. Am. Chem. Soc.* **2005**, *127*, 15976–15982. [CrossRef]

molecules

MDPI

Article

Electrochemistry of Cobalta Bis(dicarbollide) Ions Substituted at Carbon Atoms with Hydrophilic Alkylhydroxy and Carboxy Groups

Lukáš Fojt [1,*], Bohumír Grüner [2], Jan Nekvinda [2], Ece Zeynep Tűzűn [2], Luděk Havran [1] and Miroslav Fojta [1]

1 Department of Biophysical Chemistry and Molecular Oncology, Institute of Biophysics of the Czech Academy of Sciences, Královopolská 135, 612 65 Brno, Czech Republic; raven@ibp.cz (L.H.); fojta@ibp.cz (M.F.)

2 Department of Synthesis, Institute of Inorganic Chemistry of the Czech Academy of Sciences, Hlavní 1001, 250 68 Řež, Czech Republic; gruner@iic.cas.cz (B.G.); nekvinda@iic.cas.cz (J.N.); tuzun@iic.cas.cz (E.Z.T.)

* Correspondence: fojt@ibp.cz

Abstract: In this study we explore the effect on the electrochemical signals in aqueous buffers of the presence of hydrophilic alkylhydroxy and carboxy groups on the carbon atoms of cobalta bis(dicarbollide) ions. The oxygen-containing *exo*-skeletal substituents of cobalta bis(dicarbollide) ions belong to the perspective building blocks that are considered for bioconjugation. Carbon substitution provides wider versatility and applicability in terms of the flexibility of possible chemical pathways. However, until recently, the electrochemistry of compounds substituted only on boron atoms could be studied, due to the unavailability of carbon-substituted congeners. In the present study, electrochemistry in aqueous phosphate buffers is considered along with the dependence of electrochemical response on pH and concentration. The compounds used show electrochemical signals around −1.3 and +1.1 V of similar or slightly higher intensities than in the parent cobalta bis(dicarbollide) ion. The signals at positive electrochemical potential correspond to irreversible oxidation of the boron cage (the C2B9 building block) and at negative potential correspond to the reversible redox process of (CoIII/CoII) at the central atom. Although the first signal is typically sharp and its potential can be altered by a number of substituents, the second signal is complex and is composed of three overlapping peaks. This signal shows sigmoidal character at higher concentrations and may be used as a diagnostic tool for aggregation in solution. Surprisingly enough, the observed effects of the site of substitution (boron or carbon) and between individual groups on the electrochemical response were insignificant. Therefore, the substitutions would preserve promising properties of the parent cage for redox labelling, but would not allow for the further tuning of signal position in the electrochemical window.

Keywords: metallacarborane; cobalta bis(dicarbollide) ions; glassy carbon electrode; differential pulse voltammetry

Citation: Fojt, L.; Grüner, B.; Nekvinda, J.; Tűzűn, E.Z.; Havran, L.; Fojta, M. Electrochemistry of Cobalta Bis(dicarbollide) Ions Substituted at Carbon Atoms with Hydrophilic Alkylhydroxy and Carboxy Groups. *Molecules* **2022**, *27*, 1761. https://doi.org/10.3390/molecules27061761

Academic Editors: Takahiro Kondo, Iwao Matsuda and Josep M. Oliva-Enrich

Received: 17 February 2022
Accepted: 6 March 2022
Published: 8 March 2022

1. Introduction

The bis(icosahedral) cobalta bis(dicarbollide)(1-) ion, discovered by M. F. Hawthorne in 1965 [1–5], contains two η^5 coordinated dicarbollide ligands that sandwich a central Co(III) atom [4–6]. Due to its unique properties, such as aromaticity [7], space-filling properties, high thermal and chemical stability, low toxicity, hydrophobic interactions due to the hydridic character of B-H bonds [4,8–10], the formation of dihydrogen bonds [9,11], and easy penetration across phospholipid bilayers [12–15] and cell membranes [15–20], this ion has been applied in different areas of contemporary materials research [4,21] and medicinal chemistry [4,22–24]. The scope of potential applications in drug design includes enzyme inhibitors such as HIV-protease [25–27], carbonic anhydrase IX [16,28,29], anticancer compounds [16,29,30], antibiotics and antimycotics [31–35] and components for modulating the hydrophobic interactions of biomolecules [36–38].

The electronic structure in the inner orbital corresponds to a fully delocalized aromatic system [7], and along with the presence of a Co(III) central atom, generates interesting electrochemical properties [39]. Previous studies [24,39–43] and our recent results [44] suggest that the cobalta bis(dicarbollide) ion can be considered a good candidate for the tunable electrochemical labelling of biomolecules. Indeed, some icosahedral metallacarboranes have already been used as redox labels for the electrochemical detection of biomolecules [45,46]. The cobalta bis(dicarbollide) ion has been used in bioconjugation studies [3,4,40,41]. In addition, a study on the closely related ferra bis(dicarbollide) ion appeared recently in the literature [40]. According to the published results, derivatives of cobalta bis(dicarbollide) proved to have favorable electrochemical properties when modified boronated nucleotides were studied alone [44] and when such building blocks were incorporated into DNA [40]. The design of all these systems is based on the cleavage of cyclic ether rings bound to the B(8) boron atom of the cobalta bis(dicarbollide) ion by a nucleophilic attack of amino functions on the α-carbon position adjacent to the oxonium atom of the ring [3,4,41,47]. This truly universal chemical method, which may be called the "boron click approach", was first reported by Plešek in 1991 [48]. Thus, only compounds substituted on boron atoms have hitherto been studied. The boron cluster is attached to the biomolecule by a six-atom-long linker terminated by a protonated ammonium group. This has some further implication, not due to the chain length, but rather due to the presence within it of three donor atoms. This type of ethyleneglycol connection is susceptible to forming crown-ether-like arrangements with sodium ions (or K^+ and Ca^{2+} ions) [49,50]. This results in increased hydrophobic properties and a tendency to self-assemble into micelles [51] or undergo compartmentation in solution. It is also well known from carborane chemistry that simple protonatable groups on boron atoms have distinctly different physicochemical properties than those on carbons.

Herein we therefore focus on a new alternative system based on carbon substitutions, presenting the first electrochemical screening study over a series of compounds containing hydroxyalkyl [52] and carboxyl [53] groups on the cobalt bis(dicarbollide) ion. This is possible only now as synthetic pathways to these substitutions have appeared only recently. In general, -COOH, -SH, and -OH groups attached on carbon atoms in icosahedral carboranes [4,54] are distinctively more acidic and better comparable in properties to organic compounds than those on boron atoms. The compounds selected for this study thus contain polar oxygen atoms in terminal units, which, according to our previous results on other clusters [55–58], increased the solubility of boron compounds in phosphate buffers and aqueous media and improved electrochemical response. The selection was also made with the aim of using these compounds as biolabels. Herein we compare the electrochemical properties of carbon-substituted derivatives with similar boron-substituted compounds available by contemporary synthetic routes, which contain methoxy groups at B(8,8′) sites [42,59]. All compounds provided an electrochemical response that was sufficiently distinct from typical windows for biomolecules. The results are a first step in the development of direct site-specific labelling of biomolecules, following in situ bioconjugation schemes using known methods. Thus, the -COOH group could be used for attachment to a biomolecule by an established procedure of ester bond [60] or amide bond formation, or procedures for esterification reactions when considering the studied alcohols.

2. Results and Discussion

All samples were sandwich complexes consisting of a central cobalt atom coordinated between two C_2B_9 ligands. The samples were modified by different *exo*-skeletal substituents. Within this series, the substituents comprised alkylhydroxy or carboxy residues. The systematic formulae of metal bis(dicarbollide) ions salts investigated in this study are given in Table 1. For the structural formulae, please see Figure 1. In contrast to our previous results obtained with metallacarboranes and their *exo*-skeletal derivatives [44], all *exo*-skeletal substituents used in this study were bound to the carborane cage at the carbon atom (previously only at the boron atom). Our aim was to determine to what extent the presence and position of a hydrophilic substituent can alter the electrochemical behavior of metallacarborane derivatives. This would reflect differences of electronic densities resulting from the contribution of substituents attached at carbon or boron atoms.

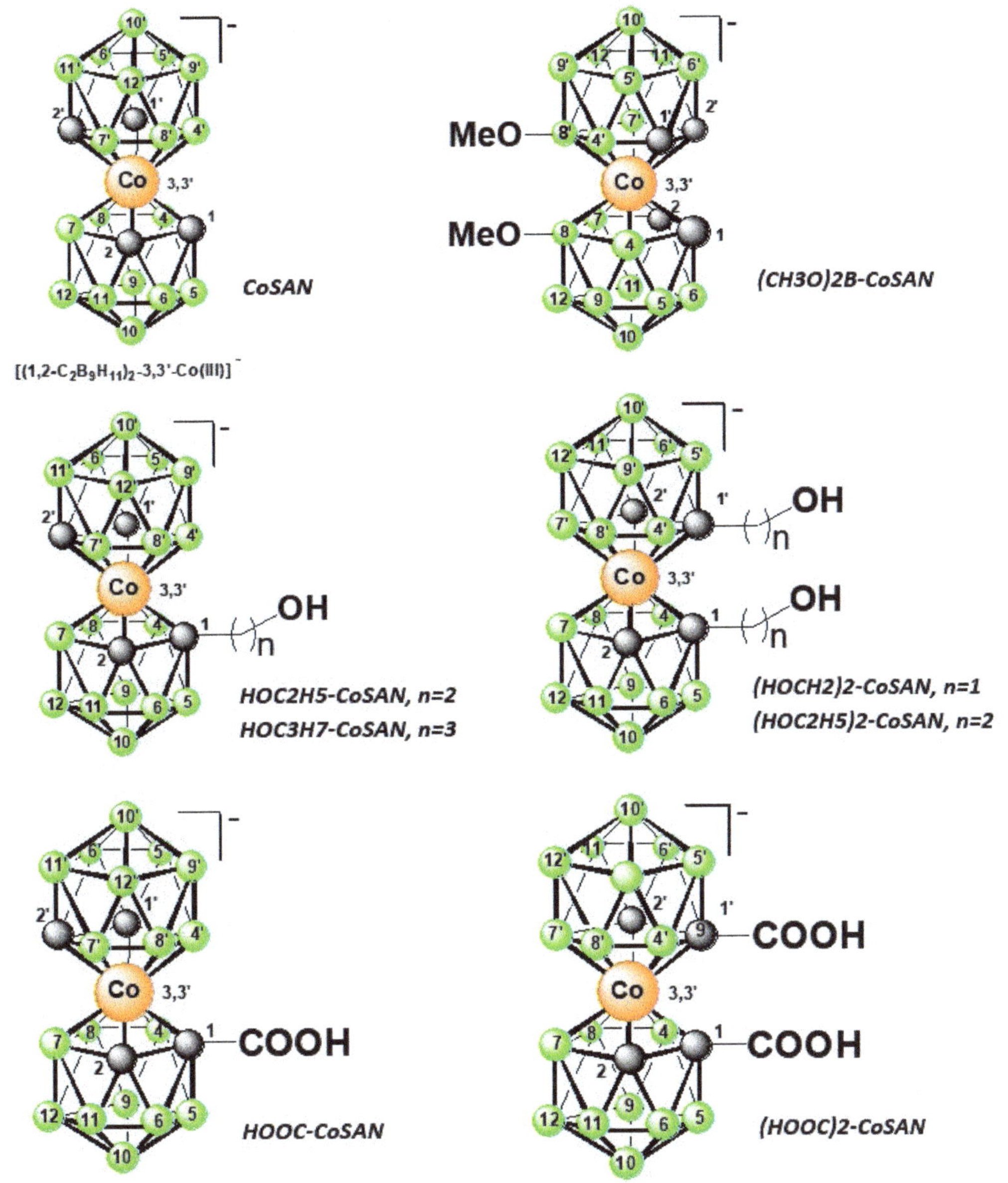

Figure 1. Schematic structural formulas of the metallacarboranes and all used *exo*-skeletal substituents. For all cases, black ball—CH group; green ball—BH group.

Table 1. Peak positions and heights with simple characteristics of used BCCs on GCE, pH = 8, concentration of all metallaborates 500 μM. For the BCCs marked * the highest achievable concentration of 500 μM was used in the electrolyte solution.

	BCC Charge	Label in the Text	MW	Peak Position/V	Peak Height/μA·cm^{-2}
closo-[(1,2-$C_2B_9H_{11}$)$_2$-3,3′-Co)]Cs *	1-	*CoSAN*	323.74	−1.30; 1.14; 1.29; 1.45;	121.0; 55.5; 217.0; 160.0;
[(1-HOC_2H_5-1,2-$C_2B_9H_{10}$)(1′,2′-$C_2B_9H_{11}$)-3,3′-Co)]Cs	1-	*HOC2H5-CoSAN*	367.79	−1.24; 1.16; 1.47;	8.7;191.2; 65.7;
[(1-HOC_3H_7-1,2-$C_2B_9H_{10}$)(1′,2′-$C_2B_9H_{11}$)-3,3′-Co)] Me_3NH	1-	*HOC3H7-CoSAN*	382.83	−1.26; 1.17; 1.30; 1.58;	50.6; 75.9; 114.1; 30.2;
[1,1′-($HOCH_2$1,2-$C_2B_9H_{10}$)$_2$-3,3′-Co)] Me_3NH	1-	*(HOCH2)2-CoSAN*	377.73	−1.13; 1.16; 1.38; 1.44;	31.7; 297.9; 209.7; 228.6;
[1,1′-(HOC_2H_5-1,2-$C_2B_9H_{10}$)$_2$-3,3′-Co)] Me_3NH	1-	*(HOC2H5)2-CoSAN*	411.74	−1.15; 1.17; 1.26; 1.49;	57.0; 231.8; 375.6; 160.7;
[(1-HOOC-1,2-$C_2B_9H_{10}$)(1′,2′-$C_2B_9H_{11}$)-3,3′-Co)] Me_4N	1-	*HOOC-CoSAN*	367.75	−1.31; 1.16;1.32;	44.5; 192.5; 355.6;
[1,1′-$(HOOC)_2$-(1,2-$C_2B_9H_{10}$)$_2$-3,3′-Co)] Me_4N	1-	*(HOOC)2-CoSAN*	411.76	−1.21; 1.17; 1.32;	51.1; 339.0; 406.5;
[8,8′-$(CH_3O)_2$-(1,2-$C_2B_9H_{10}$)$_2$-3,3′-Co)] Me4N * [42]	1-	*(CH3O)2B-CoSAN*	377.73	−1.30; 1.15; 1.32; 1.45; 1.68;	94.3; 410.0; 371.0; 377.0; 14.1;

2.1. Electrochemical Behavior of Hydroxy- and Carboxy-Substituted Metallacarboranes

The electrochemical behavior of all measured samples is displayed in Figure 2. The electrochemical data, systematic formulae and abbreviations used in the text are presented in Table 1. All compounds studied give a single symmetrical peak in the negative part of the potential window of the used GCEs. This peak is ascribed to the Co(III/II) redox process. We can observe oxidation (the measurement starts at E = −1.7 V) after corresponding electrochemical reduction to Co(II). According to the CV measurements (see Figure 3 for the *HOOC-CoSAN* and *(HOOC)2-CoSAN*, the situation for the other samples is similar, see Figure S1 in ESI. We can deduce the reversible character of this negatively situated electrochemical response (the scan rate–peak height dependence (not shown) suggest that the electrochemical reaction is driven by diffusion).

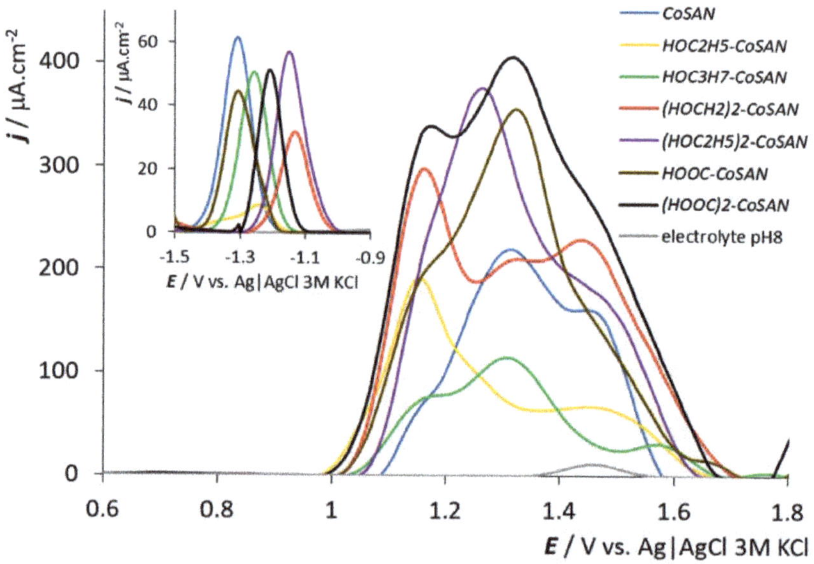

Figure 2. DPV of all used compounds, PB, pH = 8, 500 μM concentrations of all samples, GCE. See caption in the panel for sample identification.

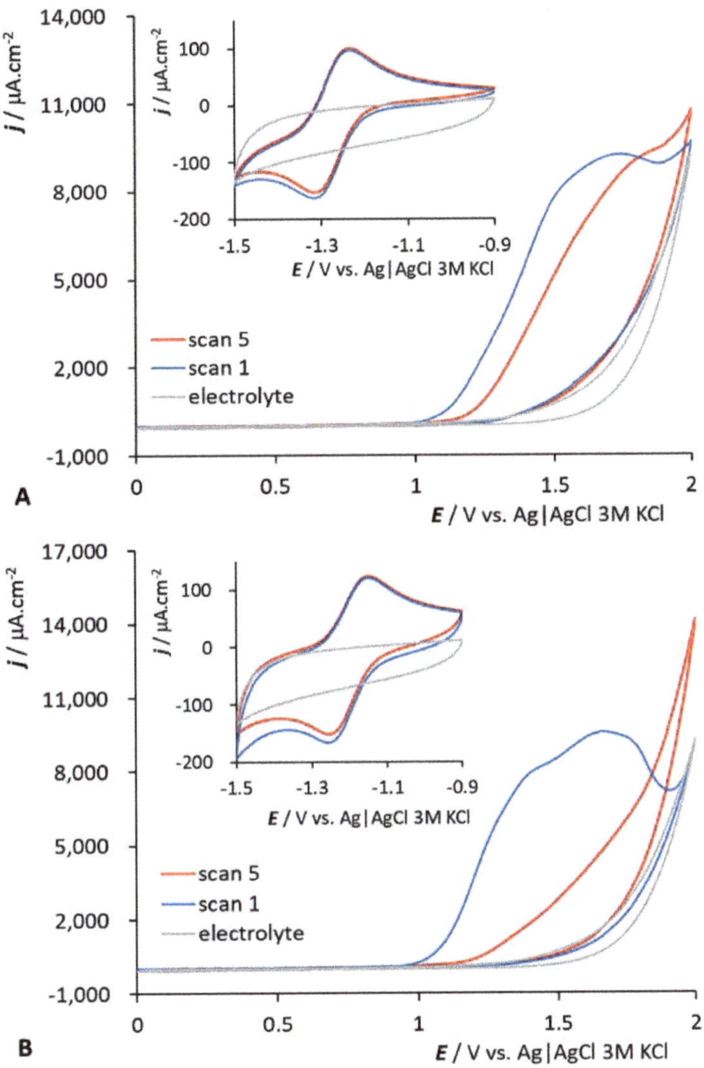

Figure 3. Cyclic voltammogram of *HOOC-CoSAN* (**A**) and *(HOOC)2-CoSAN* (**B**) in PB, 1000 μM concentration, GCE, pH = 8, scan rate v = 100 $mV \cdot s^{-1}$.

The peak position for the reversible electrochemical reaction has an interesting trend for all measured samples: the peak height is similar for almost all samples, about 50 $\mu A \cdot cm^{-2}$, except for the ill-developed peak for *HOC2H5-CoSAN*. Starting at E = −1.30 V for the parent *CoSAN* and a similar value for the simplest substitution, *HOOC-CoSAN*, the compounds with a single *exo*-skeletal substituent exhibit generally positively shifted electrochemical oxidation potential. This continues to grow up to E = −1.10 V for the disubstituted dicarboxylic acid and dialkylhydroxy derivatives *(HOCH2)2-CoSAN* and *(HOC2H5)2-CoSAN*. The reason for this shift can be seen in the perturbation of the electronic and geometric structure of the parent *CoSAN* by the hydrophilic substituents, which seem to facilitate the electrochemical reaction.

However, the situation in the positive part of the potential window is completely different. In accordance with our previous observations [44,55–58], the electrochemical response is irreversible in this region and could be ascribed to the electrochemical oxidation of the carborane cage in the structure of the metallacarborane sandwich (in this case the two C_2B_9 dicarbollide ligands).

If the *CoSAN* is considered as a template, we can observe one main broad peak situated at E = +1.32 V. Focusing on the negative potential, a small shoulder can be observed. Moving up to positive potentials, one overlapping peak appears at a potential of around E = +1.45 V, which is near to the potential of the electrode–electrolyte surface reaction [44] (see the peak for the electrolyte at the same potential). All studied samples exhibit similar, but quite complex electrochemical behavior. The electrochemical signals comprise three overlapping peaks situated at similar potentials. Generally, a similar line shape is observed for samples with identical substituents that differ in their numbers (most clearly for *(HOOC)2-CoSAN* and *HOOC-CoSAN*). However, peak height and peak separation differ markedly. The reason for this behavior may be similar to the situation at the negative potential part of the electrochemical window. As is known from the literature [61], the cobalta bis(dicarbollide) ion has three energy minima, resulting in discrete rotamers—cisoid, transoid and gauche—identified by chemical calculations and found in solid-state structures [2–4]. The three peaks may thus originate in energy minima corresponding to the higher stability of the three different rotamers. As expected, the terminal hydroxy and carboxy groups may be involved in hydrogen bonding, and furthermore, the longer chains may contribute to steric clashes that may a play role in the distribution of different rotamers within a time scale, which could be reflected in the position of observed maxima and the shape of electrochemical responses. This hypothesis is further supported by the recent electrochemical study of *CoSAN* substituted with a rigidifying bridge interconnecting two C_2B_9 ligands and that led to the induction of a conformationally restricted cisoid conformation. Unlike the compounds presented herein, those compounds showed only one major electrochemical peak in the respective range [62]. Furthermore, even if it is known that the speed of rotation is quite high in solution (as described recently for the closely related iron sandwich) [63], the situation at the electrode interface is different. The compounds are slightly stabilized when accessing the electrode surface, where often at least slight adsorption takes place (which may hinder the rotation feasibility). Secondly, the compounds are asymmetrically substituted and correspond to racemic mixtures. The electrochemistry represents an average over all forms of molecules; thus, we are able to observe all three rotamers simultaneously. The reason for the shift in electrochemical potential may lie in the different accessibility of the electrode surface to boron and carbon atoms in given rotamers. Coming out from our previously published results on metallacarboranes and their *exo*-skeletal derivatives [44,56] and newly obtained knowledge, we can assume that higher potential shift and/or the emergence of new peaks strongly depends on the physicochemical properties of the *exo*-skeletal substituents. In the case of *CoSAN*, a significant oxidation peak shift occurs in the case of iodine and chlorine *exo*-skeletal substituents [44] (iodine has bulky valence orbitals, chlorine has ahigh electronegativity). The present results corroborate these findings.

2.2. *pH Dependencies of Selected BCCs*

The pH dependence of selected anions is shown in Figure 4; the results for samples at 200 μM concentration are presented and discussed. We selected samples with one and two carboxylic groups for this study, which sit directly on the carbon atom and can be easily protonated/deprotonated—compounds corresponding to *(HOOC)2-CoSAN* and *HOOC-CoSAN*. The situation is simple at the positive potentials (the region of the electrochemical oxidation of dicarbollide ligands). For both selected samples, only one dominant peak is observed. *HOOC-CoSAN* has peak height maxima for pH = 6, *(HOOC)2-CoSAN* for pH = 8. At lower and higher pH, the peak height gradually decreases with only a small shift in the observed potentials. The situation for *HOC3H7-CoSAN* and *(HOC2H5)2-CoSAN* is similar, and differs for *(HOCH2)2-CoSAN* and *HOC2H5-CoSAN*—for the three highest pH values (6, 8 and 10) we can observe the same peak heights, but with a shift of electrochemical potential (to more positive values with the decrease of pH; see Figure S2 in ESI). In the negative potential region of the Co(III) redox process, we can discern a similar trend for all samples. At pH = 2, the signal becomes distorted or highly elevated. Both these effects have their origin in the protonation of the groups and the formation of hydrogen bonds. This may result (in dynamic time scales) in a slower rotation of ligands or in partially restrained conformation, particularly in the case of disubstituted species, where a formation of bridging -OH-H_3O^+-HO- arrangement between two dicarbollide ligands can be expected.

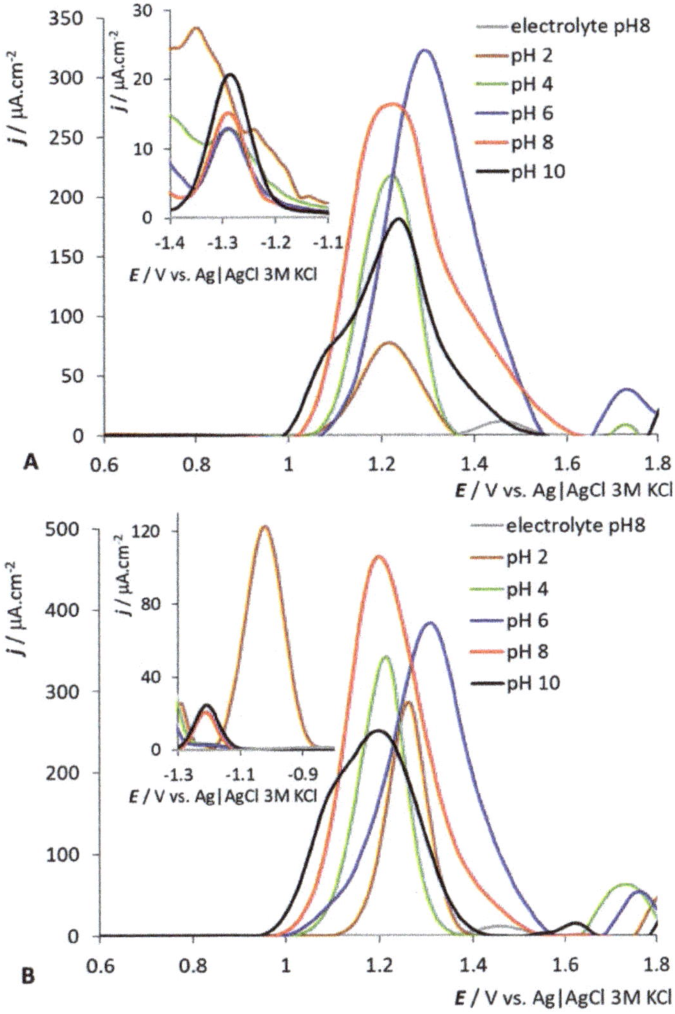

Figure 4. pH dependencies of 200 µM *HOOC-CoSAN* (**A**) and *(HOOC)2-CoSAN* (**B**), GCE. See the captions in the panels for pH value identification.

2.3. *Concentration Dependence*

The results (Figures 5 and S3 in ESI) show a similar trend for all samples, with the exception of HOC2H5-CoSAN. For higher concentrations, sigmoidal types of dependence are generally observed at the positive potential area of the polyhedral carborane cage oxidation; for higher sample concentrations, the signal does not increase linearly. This behavior has already been observed in other metallacarboranes [44]. Based on current and previous results we can assume that these effects may originate in the tendency of the metallacarborane anions to aggregate in aqueous solution. This behavior is well known from recent studies using other techniques [49,51], which further support the electrochemical results. In the positive potential area and at high concentrations (approximately above 250 μM), the predominant peak is located at approximately +1.30 V. When the concentrations decrease, this peak slightly decreases, and its maximum is transformed into another peak located around +1.15 V. The exception is HOC2H5-CoSAN, where the peak at +1.15 V remains at the same potential even after increasing the concentration to 1 mM (see ESI, Figure S3). When considering the emerging potential of metallacarboranes in medicine and biochemistry [8,27,39], rapid information about aggregation and/or micelle formation in water-based media is of particular concern. From the results described in this paragraph, it follows that electrochemical methods can provide a new tool for easy inspection if a substituted boron anion tends to aggregate in aqueous solution under particular experimental conditions. In particular, this provides for a ready estimation of the concentration at which the formation of aggregates or micelles starts to occur.

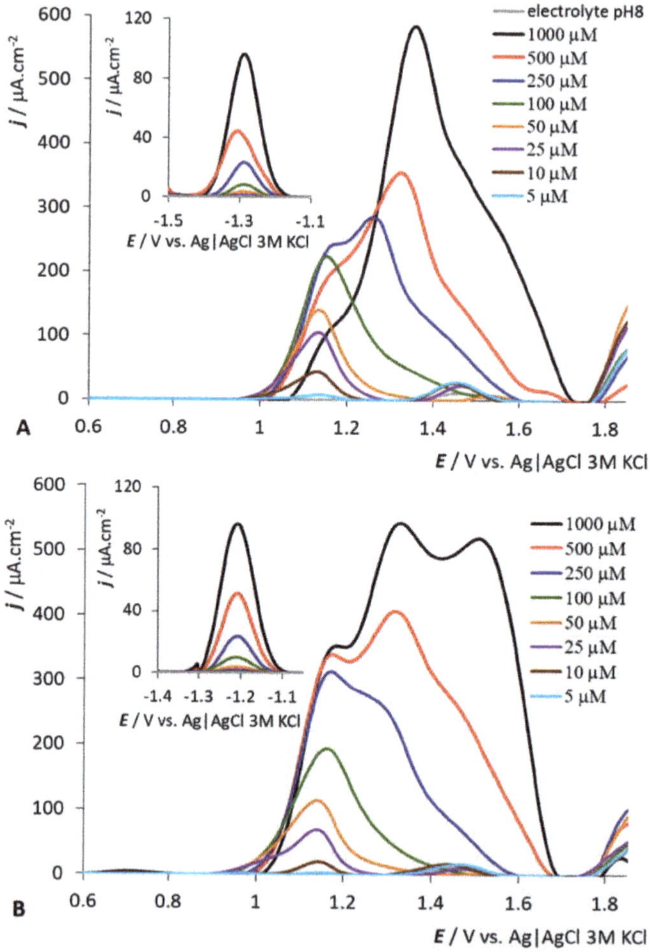

Figure 5. Concentration dependencies of *HOOC-CoSAN* (**A**) and *(HOOC)2-CoSAN* (**B**), at pH 8, GCE. See the captions in the panels for concentration identification.

Interestingly, the concentration dependence in the region of the Co(III) redox process, on which most electrochemical studies are centered, does not follow this trend and the dependence on concentration is almost linear. Indeed, as can be expected, the reversible reaction may be not so highly affected by this behavior. Therefore, only the signal of the irreversible oxidation of the carborane polyhedral can be used for diagnostics of the aggregation phenomena. On the other hand, this signal has a complex character and its deconvolution is needed.

2.4. Comparison of Different Substituent Positions (B or C Atom)

It is reasonable that the most important changes in electron density can be introduced by substitution at the carbon and boron atoms in ligand planes that bind in η-fashion to the central cobalt atom in *CoSAN* polyhedra. We compare here the compound with two $(HOCH_2)_2$ substituents on the carbon atom *(HOCH2)2-CoSAN* with a similar compound that has two $(O\text{-}CH_3)_2$ groups on boron B(8,8′) sites in *(CH3O)2B-CoSAN*. The latter compound should be taken for comparison due to the limited progress of the cage modification. Unfortunately, no fully identical congeners that would have the same substitution on the B or C atom are currently available by synthesis. The electrochemical responses of these two samples in the positive potential area are complex (see Figures 2 and 6) and consist of several overlapping peaks. For this reason, we used Fytik software for the deconvolution of the electrochemical curves in this potential region. The deconvolution was performed using a Gaussian distribution with an average fit error of 10%. For better clarity we also included the electrochemical response of the parent *CoSAN*. The results are displayed in Figure 6. Surprisingly enough, there is no distinguishable difference between those two substitutions except for the peak height in the range of positive potentials ascribed to cage oxidation. The character of the peak is quite complex, and is referred to in the discussion in Section 2.1. The deconvolution provided four peaks for each compound, in which two positive peaks are only slightly shifted in the case of disubstituted compounds; only the most positive signal significantly differs in its position. Nevertheless, this peak might originate from the electrode surface–electrolyte reaction and cannot be considered sufficiently reliable, as mentioned above. Thus, the potential of the fitted peaks does not change, only the peak heights. The corresponding number of the three peaks may give indirect evidence that the peaks are associated with the presence of rotamers. This is because the B(8)-H bond with the highest electron density known to be easily activated in an EINS-type reaction [2] cannot be involved in the redox process in *(CH3O)2B-CoSAN*. If the complex character of the signal is associated with the three most electron-rich sites H-B(8,9,12), the deconvolution would inherently provide a simpler pattern for *(CH3O)2B-CoSAN* because most of the reactive sites B(8,8′) are blocked by substitution.

2.5. General Remarks

Herein we present the electrochemistry of hydroxy and carboxy *exo*-skeletal derivatives of CoSAN. Present knowledge on the electrochemistry of metallacarboranes indicates that the responses at the positive potentials could be denoted as the electrochemical oxidation of the core carborane structure. The peaks located in the negative potential part are connected with the reversible redox processes of the central Co atom. Our results indicate that the oxidation of the polyhedral boron cage is completely irreversible. Unfortunately, the electrochemical methods that would allow for the theoretical evaluation of these processes are available only for reversible or quasi-reversible reactions (e.g., the Pourbaix diagrams). Therefore, the mechanism of the studied compounds' electrochemical responses cannot be understood fully. The reason for the uncommon electrochemical behavior of the boron cluster compounds may lie in their specific electronic properties. Considering the irreversible oxidation process, the electronic density is probably removed from the B-H [44]. Due to electron delocalization over the boron cages, we can assume that the exact place of the electron exchange cannot be defined.

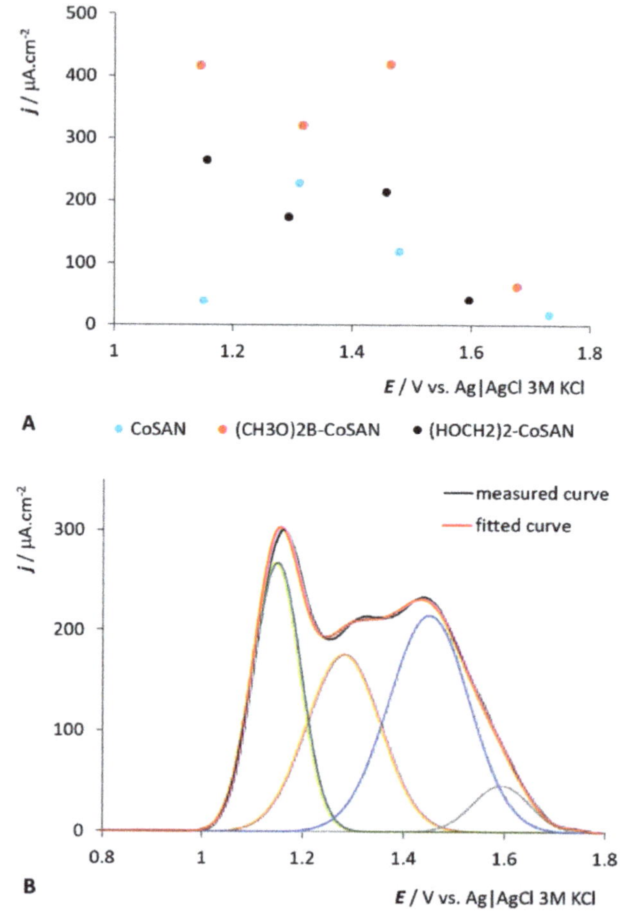

Figure 6. Deconvolution results (peak heights and position of fitted peaks) for the *CoSAN*, *(HOCH2)2-CoSAN* and *(CH3O)2B-CoSAN* [44] (**A**). Example of measured, sum of fitted peaks and individual fitted peaks for *(HOCH2)2-CoSAN* (**B**). All fitted curves were measured in PB of pH = 8, 500 μM concentration.

3. Materials and Methods

3.1. Synthesis

All compounds used in this study were prepared in accordance with previously published procedures. The carbon-substituted hydroxyalkyl derivatives of the general formula $[(1\text{-}HO(CH_2)_n\text{-}1,2\text{-}C_2B_9H_{10})(1',2'\text{-}C_2B_9H_{11})\text{-}3,3'\text{-}Co)]Cs$ and $[1,1'\text{-}(1\text{-}HO(CH_2)_n\text{-}1,2\text{-}C_2B_9H_{10})_2\text{-}3,3'\text{-}Co]$ were prepared by low-temperature lithiation reaction of the $Cs[(C_2B_9H_{11})2\text{-}3,3'\text{-}Co]$ with BuLi in DME followed by reaction with *para*-formaldehyde, oxirane or trimethylene oxide [52]. The products were isolated by chromatography and crystallization. In the case of dihydroxyalkyl compounds, only the racemic diastereoisomer was used in this study, and was isolated according to the procedure described in the literature [16]. The carboxylic acids were prepared by reactions of the carbon lithiated ion with carbon dioxide [53]. The boron-substituted compound was isolated from reaction mixtures obtained from acid-catalyzed reactions of cobalta bis(dicarbollide) with para-formaldehyde, as described in the literature [59]. The compounds were characterized by NMR, MS and HPLC methods, with results matching data published in previous papers [52,53,59].

3.2. Electrochemistry

An Autolab 302 potentiostat (Ecochemie, The Netherlands, and now Metrohm) connected to a conventional electrochemical cell with a three-electrode system was used for the electrochemical experiments. A platinum wire (1 mm diameter, purity 99.99%, Safina, Czech Republic) counter electrode and an Ag | AgCl | 3 M KCl reference electrode (Metrohm, Switzerland) were used. A glassy carbon electrode (GCE, 2 mm diameter, Metrohm, Switzerland) was used as the working electrode. Differential pulse voltammetry (DPV, pulse amplitude of 25 mV, pulse width of 50 ms, and scan rate of 8 $mV \cdot s^{-1}$) and cyclic voltammetry (step 5 mV, scan rate 100 $mV \cdot s^{-1}$) were used for the measurements. All measurements were performed at room temperature (295 K). DPV curves were baseline corrected before further processing. Current values were normalized to the geometrical surface area of the used electrodes. The GCE was pretreated by mechanical polishing with silicon carbide papers (SiC polishing papers, Struers, Denmark) and the polishing was finalized using 1 μm diamond particles in spray-on Lecloth B polishing cloth (both Leco, St. Joseph, MI, USA). As a final pretreatment step the GCE was sonicated in tri-distilled water. For electrochemical measurements, phosphate buffers (PBs) of various pH values were mixed from NaH_2PO_4 and Na_2HPO_4. The concentration of phosphate anions was kept at 0.2 M in all solutions. The systematic formulas of metal bis(dicarbollide) ion salts investigated in this study are given in Table 1. Cations in the salts are specified in Table 1, and served for anion precipitation in the last synthesis step, exchanged for sodium cation using Amberlite CG-120 (Fluka) to increase their solubility in water. Samples used in this study were prepared as 1 mM aqueous solutions according to weighting (in some cases, the solubility in water was lower—see Table 1). All other chemicals were purchased from Sigma-Aldrich and were of the highest available purity, and triple-distilled water was used as the solvent.

4. Conclusions

This study presents the electrochemistry of different carbon-substituted cobalta bis(dicarbollide) ion derivatives, substituted with groups containing oxygen atoms, in phosphate buffer. Our results indicate that the presence of different substituents containing terminal hydroxyalkyl or carboxylic residues do not markedly alter the high electrochemical response of the parent *CoSAN*, and that some compounds show even slightly enhanced signals. Therefore, these derivatives can be considered good building blocks with groups suitable for the redox labelling of biomolecules. On the other hand, the electrochemical potential in the negative interval corresponding to the reversible Co(III)/Co(II) redox couple seems to be affected more by the increasing number of substituents on the boron cage than by their nature or the site of substitution. A second potential area of electrochemical response corresponding to an electrochemical oxidation of the cage was also observed. This signal is located in the positive range of potentials from +1.1 to +1.3 V and shows complex character composed of four overlapping peaks, as follows from the deconvolution of selected electrochemical records. The number of peaks did not change even in the case of the boron-substituted compound used for comparison. However, the mutual intensities of the three observed peaks indicate pH and concentration dependency. In the first case this is connected with hydrogen bond formation between substituents in the acidic range, which seems to result in partially limited rotation of the ligands around the central atom. The sigmoidal character of the peak observed for higher concentrations in the latter case might be used as a rapid diagnostic tool for aggregation of ions in aqueous solution.

Supplementary Materials: The following are available online at https://www.mdpi.com/article/10.3390/molecules27061761/s1, Figure S1: Cyclic voltammogram of *HOC2H5-COSAN*, *HOC3H7-COSAN*, *(HOCH2)2-COSAN* and *(HOC2H5)2-COSAN (D)* in PB, 1000 μM concentration, GCE, pH = 8, scan rate ν = 100 $mV \cdot s^{-1}$. Figure S2: pH dependencies of *HOC2H5-COSAN*, *HOC3H7-COSAN*, *(HOCH2)2-COSAN* and *(HOC2H5)2-COSAN* for 200 μM concentration, GCE. Figure S3: Concentration dependencies of *HOC2H5-COSAN, HOC3H7-COSAN, (HOCH2)2-COSAN* and *(HOC2H5)2-COSAN* at pH 8, GCE.

Author Contributions: Conceptualization, L.F. and B.G.; methodology, L.F. and B.G.; validation, L.H.; formal analysis, L.F. and B.G.; investigation, L.F., B.G., J.N. and E.Z.T.; resources, L.F. and B.G.; writing—original draft preparation, L.F. and B.G.; writing—review and editing, L.H. and M.F.; visualization, L.F.; funding, L.F. and B.G. All authors have read and agreed to the published version of the manuscript.

Funding: This research was funded by the Czech Science Foundation, Project No. 19-04630S.

Institutional Review Board Statement: Not applicable.

Informed Consent Statement: Not applicable.

Data Availability Statement: All data used to support the findings of this study are included within the article and are also available from the corresponding author upon request.

Conflicts of Interest: The authors declare no conflict of interest.

Sample Availability: Samples of the compounds are available from the authors.

References

1. Hawthorne, M.F.; Andrews, T.D. Carborane analogues of cobalticinium ion. *Chem. Commun.* **1965**, *19*, 443–444. [CrossRef]
2. Sivaev, I.B.; Bregadze, V.I. Chemistry of cobalt bis(dicarbollides). A review. *Collect. Czech. Chem. Commun.* **1999**, *64*, 783–805. [CrossRef]
3. Dash, B.P.; Satapathy, R.; Swain, B.R.; Mahanta, C.S.; Jena, B.B.; Hosmane, N.S. Cobalt bis(dicarbollide) anion and its derivatives. *J. Organomet. Chem.* **2017**, *849–850*, 170–194. [CrossRef]
4. Grimes, R.N. *Carboranes*, 3rd ed.; Academic Press Ltd-Elsevier Science Ltd.: London, UK, 2016; pp. 1–1041.
5. Saxena, A.K.; Hosmane, N.S. Recent Advances in the Chemistry of Carborane Metal-Complexes Incorporating D-Block and F-Block Elements. *Chem. Rev.* **1993**, *93*, 1081–1124. [CrossRef]
6. Hawthorne, M.F. Chemistry of polyhedral species derived from transition metals and carboranes. *Accounts Chem. Res.* **1968**, *1*, 281–288. [CrossRef]
7. Junqueira, G.M.A. Remarkable aromaticity of cobalt bis(dicarbollide) derivatives: A NICS study. *Theor. Chem. Acc.* **2018**, *137*, 7. [CrossRef]
8. Gozzi, M.; Schwarze, B.; Hey-Hawkins, E. Preparing (Metalla)carboranes for Nanomedicine. *ChemMedChem* **2021**, *16*, 1533–1565. [CrossRef]
9. Viñas, C.; Tarres, M.; Gonzalez-Cardoso, P.; Farras, P.; Bauduin, P.; Teixidor, F. Surfactant behaviour of metallacarboranes. A study based on the electrolysis of water. *Dalton Trans.* **2014**, *43*, 5062–5068. [CrossRef]
10. Plešek, J. Potential Applications of the Boron Cluster Compounds. *Chem. Rev.* **1992**, *92*, 269–278. [CrossRef]
11. Fanfrlík, J.; Lepšík, M.; Horinek, D.; Havlas, Z.; Hobza, P. Interaction of Carboranes with Biomolecules: Formation of Dihydrogen Bonds. *ChemPhysChem* **2006**, *7*, 1100–1105. [CrossRef]
12. Dubey, R.D.; Sarkar, A.; Shen, Z.Y.; Bregadze, V.I.; Sivaev, I.B.; Druzina, A.A.; Zhidkova, O.B.; Shmal'ko, A.V.; Kosenko, I.D.; Sreejyothi, P.; et al. Effects of Linkers on the Development of Liposomal Formulation of Cholesterol Conjugated Cobalt Bis(dicarbollides). *J. Pharm. Sci.* **2021**, *110*, 1365–1373. [CrossRef] [PubMed]
13. Bregadze, V.I.; Sivaev, I.B.; Dubey, R.D.; Semioshkin, A.; Shmal'ko, A.V.; Kosenko, I.D.; Lebedeva, K.V.; Mandal, S.; Sreejyothi, P.; Sarkar, A.; et al. Boron-Containing Lipids and Liposomes: New Conjugates of Cholesterol with Polyhedral Boron Hydrides. *Chem.-Eur. J.* **2020**, *26*, 13832–13841. [CrossRef] [PubMed]
14. Assaf, K.I.; Begaj, B.; Frank, A.; Nilam, M.; Mougharbel, A.S.; Kortz, U.; Nekvinda, J.; Grüner, B.; Gabel, D.; Nau, W.M. High-Affinity Binding of Metallacarborane Cobalt Bis(dicarbollide) Anions to Cyclodextrins and Application to Membrane Translocation. *J. Org. Chem.* **2019**, *84*, 11790–11798. [CrossRef] [PubMed]
15. Verdia-Baguena, C.; Alcaraz, A.; Aguilella, V.M.; Cioran, A.M.; Tachikawa, S.; Nakamura, H.; Teixidor, F.; Vinas, C. Amphiphilic COSAN and I2-COSAN crossing synthetic lipid membranes: Planar bilayers and liposomes. *Chem. Commun.* **2014**, *50*, 6700–6703. [CrossRef]
16. Gruner, B.; Brynda, J.; Das, V.; Sicha, V.; Stepankova, J.; Nekvinda, J.; Holub, J.; Pospisilova, K.; Fabry, M.; Pachl, P.; et al. Metallacarborane Sulfamides: Unconventional, Specific, and Highly Selective Inhibitors of Carbonic Anhydrase IX. *J. Med. Chem.* **2019**, *62*, 9560–9575. [CrossRef]
17. Grzelczak, M.P.; Danks, S.P.; Klipp, R.C.; Belic, D.; Zaulet, A.; Kunstmann-Olsen, C.; Bradley, D.F.; Tsukuda, T.; Vinas, C.; Teixidor, F.; et al. Ion Transport across Biological Membranes by Carborane-Capped Gold Nanoparticles. *ACS Nano* **2017**, *11*, 12492–12499. [CrossRef]
18. Tarres, M.; Canetta, E.; Vinas, C.; Teixidor, F.; Harwood, A.J. Imaging in living cells using nu B-H Raman spectroscopy: Monitoring COSAN uptake. *Chem. Commun.* **2014**, *50*, 3370–3372. [CrossRef]

19. Gona, K.B.; Zaulet, A.; Gomez-Vallejo, V.; Teixidor, F.; Llop, J.; Vinas, C. COSAN as a molecular imaging platform: Synthesis and "in vivo" imaging. *Chem. Commun.* **2014**, *50*, 11415–11417. [CrossRef]
20. Teixidor, C.V.; Teixidor, F.; Harwood, A.J. *Cobaltabisdicarbollide-Based Synthetic Vesicles: From Biological Interaction to In Vivo Imaging*; John Wiley & Sons Ltd.: Chichester, UK, 2018; pp. 159–173.
21. Zhu, Y.H.; Hosmane, N.S. Advanced carboraneous materials. *J. Organomet. Chem.* **2017**, *849–850*, 286–292. [CrossRef]
22. Lesnikowski, Z.J. Challenges and Opportunities for the Application of Boron Clusters in Drug Design. *J. Med. Chem.* **2016**, *59*, 7738–7758. [CrossRef]
23. Issa, F.; Kassiou, M.; Rendina, M. Boron in Drug Discovery: Carboranes as Uniquie Pharmacophores in Biologically Active Compounds. *Chem. Rev.* **2011**, *111*, 5701–5722. [CrossRef] [PubMed]
24. Olejniczak, A.B.; Nawrot, B.; Lesnikowski, Z.J. DNA Modified with Boron-Metal Cluster Complexes M(C2B9H11)(2) Synthesis, Properties, and Applications. *Int. J. Mol. Sci.* **2018**, *19*, 3501. [CrossRef] [PubMed]
25. Cigler, P.; Kozisek, M.; Rezacova, P.; Brynda, J.; Otwinowski, Z.; Pokorna, J.; Plesek, J.; Gruner, B.; Doleckova-Maresova, L.; Masa, M.; et al. From nonpeptide toward noncarbon protease inhibitors: Metallacarboranes as specific and potent inhibitors of HIV protease. *Proc. Natl. Acad. Sci. USA* **2005**, *102*, 15394–15399. [CrossRef] [PubMed]
26. Kožíšek, M.; Cígler, P.; Lepšík, M.; Fanfrlík, J.; Řezáčová, P.; Brynda, J.; Pokorná, J.; Plešek, J.; Grüner, B.; Grantz-Šašková, K.; et al. Inorganic polyhedral metallacarborane inhibitors of HIV protease: A new approach to overcoming antiviral resistance. *J. Med. Chem.* **2008**, *59*, 4839–4843. [CrossRef]
27. Řezáčová, P.; Pokorná, J.; Brynda, J.; Kožíšek, M.; Cígler, P.; Lepšík, M.; Fanfrlík, J.; Řezáč, J.; Šašková, K.G.; Sieglová, I.; et al. Design of HIV Protease Inhibitors Based on Inorganic Polyhedral Metallacarboranes. *J. Med. Chem.* **2009**, *52*, 7132–7141. [CrossRef]
28. Gruner, B.; Kugler, M.; El Anwar, S.; Holub, J.; Nekvinda, J.; Bavol, D.; Ruzickova, Z.; Pospisilova, K.; Fabry, M.; Kral, V.; et al. Cobalt Bis(dicarbollide) Alkylsulfonamides: Potent and Highly Selective Inhibitors of Tumor Specific Carbonic Anhydrase IX. *Chempluschem* **2021**, *86*, 352–363. [CrossRef]
29. Kugler, M.; Nekvinda, J.; Holub, J.; El Anwar, S.; Das, V.; Šícha, V.; Pospíšilová, K.; Fábry, M.; Král, V.; Brynda, J.; et al. Inhibitors of CA IX Enzyme Based on Polyhedral Boron Compounds. *ChemBioChem* **2021**, *22*, 2741–2761. [CrossRef]
30. Fuentes, I.; Garcia-Mendiola, T.; Sato, S.; Pita, M.; Nakamura, H.; Lorenzo, E.; Teixidor, F.; Marques, F.; Viñas, C. Metallacarboranes on the Road to Anticancer Therapies: Cellular Uptake, DNA Interaction, and Biological Evaluation of Cobaltabisdicarbollide COSAN (-). *Chem.-Eur. J.* **2018**, *24*, 17239–17254. [CrossRef]
31. Popova, T.; Zaulet, A.; Teixidor, F.; Alexandrova, R.; Vinas, C. Investigations on antimicrobial activity of cobaltabisdicarbollides. *J. Organomet. Chem.* **2013**, *747*, 229–234. [CrossRef]
32. Vankova, E.; Lokocova, K.; Matatkova, O.; Krizova, I.; Masak, J.; Gruner, B.; Kaule, P.; Cermak, J.; Sicha, V. Cobalt bis-dicarbollide and its ammonium derivatives are effective antimicrobial and antibiofilm agents. *J. Organomet. Chem.* **2019**, *899*, 8. [CrossRef]
33. Swietnicki, W.; Goldeman, W.; Psurski, M.; Nasulewicz-Goldeman, A.; Boguszewska-Czubara, A.; Drab, M.; Sycz, J.; Goszczynski, T.M. Metallacarborane Derivatives Effective against Pseudomonas aeruginosa and Yersinia enterocolitica. *Int. J. Mol. Sci.* **2021**, *22*, 6762. [CrossRef] [PubMed]
34. Kvasnickova, E.; Masak, J.; Cejka, J.; Matatkova, O.; Sicha, V. Preparation, characterization, and the selective antimicrobial activity of N-alkylammonium 8-diethyleneglycol cobalt bis-dicarbollide derivatives. *J. Organomet. Chem.* **2017**, *827*, 23–31. [CrossRef]
35. Fink, K.; Uchman, M. Boron cluster compounds as new chemical leads for antimicrobial therapy. *Coord. Chem. Rev.* **2021**, *431*, 10. [CrossRef]
36. Fuentes, I.; Pujols, J.; Vinas, C.; Ventura, S.; Teixidor, F. Dual Binding Mode of Metallacarborane Produces a Robust Shield on Proteins. *Chem.-Eur. J.* **2019**, *25*, 12820–12829. [CrossRef]
37. Fink, K.; Boratynski, J.; Paprocka, M.; Goszczynski, T.M. Metallacarboranes as a tool for enhancing the activity of therapeutic peptides. *Ann. N. Y. Acad. Sci.* **2019**, *1457*, 128–141. [CrossRef]
38. Goszczynski, T.M.; Fink, K.; Boratynski, J. Icosahedral boron clusters as modifying entities for biomolecules. *Expert Opin. Biol. Ther.* **2018**, *18*, 205–213. [CrossRef]
39. Núñez, R.; Tarres, M.; Ferrer-Ugalde, A.; de Biani, F.F.; Teixidor, F. Electrochemistry and Photoluminescence of Icosahedral Carboranes, Boranes, Metallacarboranes, and Their Derivatives. *Chem. Rev.* **2016**, *116*, 14307–14378. [CrossRef]
40. Kodr, D.; Yenice, C.P.; Simonova, A.; Saftic, D.P.; Pohl, R.; Sykorova, V.; Ortiz, M.; Havran, L.; Fojta, M.; Lesnikowski, Z.J.; et al. Carborane- or Metallacarborane-Linked Nucleotides for Redox Labeling. Orthogonal Multipotential Coding of all Four DNA Bases for Electrochemical Analysis and Sequencing. *J. Am. Chem. Soc.* **2021**, *143*, 7124–7134. [CrossRef]
41. Olejniczak, A.B.; Lesnikowski, Z.J. *Boron Clusters as Redox Labels for Nucleosides and Nucleic Acids*; World Scientific Publ Co Pte Ltd.: Singapore, 2019; pp. 1–13.
42. Gonzμlez-Cardoso, P.; Stoica, A.I.; Farràs, P.; Pepiol, A.; Vinas, C.; Teixidor, F. Additive Tuning of Redox Potential in Metallacarboranes by Sequential Halogen Substitution. *Chem.-Eur. J.* **2010**, *16*, 6660–6665. [CrossRef]
43. Rudakov, A.; Shirokii, V.L.; Knizhnikov, V.A.; Bazhanov, A.V.; Vecher, E.I.; Maier, N.A.; Potkin, V.I.; Ryabtsev, A.N.; Petrovskii, P.V.; Sivaev, I.B.; et al. Electrochemical synthesis of halogen derivatives of bis(1,2-dicarbollyl)cobalt(III). *Russ. Chem. Bull. Int. Ed.* **2004**, *53*, 2554–2557. [CrossRef]
44. Fojt, L.; Gruner, B.; Sicha, V.; Nekvinda, J.; Vespalec, R.; Fojta, M. Electrochemistry of icosahedral cobalt bis(dicarbollide) ions and their carbon and boron substituted derivatives in aqueous phosphate buffers. *Electrochim. Acta* **2020**, *342*, 11. [CrossRef]

45. Ziolkowski, R.; Olejniczak, A.B.; Gorski, L.; Janusik, J.; Lesnikowski, Z.J.; Malinowska, E. Electrochemical detection of DNA hybridization using metallacarborane unit. *Bioelectrochemistry* **2012**, *87*, 78–83. [CrossRef] [PubMed]
46. Palecek, E.; Bartosik, M. Electrochemistry of Nucleic Acids. *Chem. Rev.* **2012**, *112*, 3427–3481. [CrossRef] [PubMed]
47. Semioshkin, A.A.; Sivaev, I.B.; Bregadze, V.I. Cyclic oxonium derivatives of polyhedral boron hydrides and their synthetic applications. *Dalton Trans.* **2008**, *2008*, 977–992. [CrossRef]
48. Selucky, P.; Plesek, J.; Rais, J.; Kyrs, M.; Kadlecova, L. Extraction of Fission-Products into Nitrobenzene with Dicobalt Tris-Dicarbollide and Ethyleneoxy-Substituted Cobalt Bis- Dicarbollide. *J. Radioanal. Nucl. Chem.-Artic.* **1991**, *149*, 131–140. [CrossRef]
49. Zaulet, A.; Teixidor, F.; Bauduin, P.; Diat, O.; Hirva, P.; Ofori, A.; Vinas, C. Deciphering the role of the cation in anionic cobaltabisdicarbollide clusters. *J. Organomet. Chem.* **2018**, *865*, 214–225. [CrossRef]
50. Olejniczak, A.B.; Milecki, J.; Schroeder, G. The Effect of Stereochemistry on Sodium Ion Complexation in Nucleoside-Metallacarborane Conjugates. *Bioinorg. Chem. Appl.* **2010**, *2010*, 196064. [CrossRef]
51. Dorďovic, V.; Tošner, Z.; Uchman, M.; Zhigunov, A.; Reza, M.; Ruokolainen, J.; Pramanik, G.; Cígler, P.; Kalikova, K.; Gradzielski, M.; et al. Stealth Amphiphiles: Self-Assembly of Polyhedral Boron Clusters. *Langmuir* **2016**, *32*, 6713–6722. [CrossRef]
52. Grüner, B.; Švec, P.; Šícha, V.; Padělková, Z. Direct and facile synthesis of carbon substituted alkylhydroxy derivatives of cobalt bis(1,2-dicarbollide), versatile building blocks for synthetic purposes. *Dalton Trans.* **2012**, *41*, 7498–7512. [CrossRef]
53. Nekvinda, J.; Sicha, V.; Hnyk, D.; Gruner, B. Synthesis, characterisation and some chemistry of C- and B-substituted carboxylic acids of cobalt bis(dicarbollide). *Dalton Trans.* **2014**, *43*, 5106–5120. [CrossRef]
54. Scholz, M.; Hey-Hawkins, E. Carbaboranes as Pharmacophores: Properties, Synthesis, and Application Strategies. *Chem. Rev.* **2011**, *111*, 7035–7062. [CrossRef] [PubMed]
55. Fojt, L.; Fojta, M.; Grüner, B.; Vespalec, R. Electrochemistry of closo-dodecaborate dianion and its simple exo-skeletal derivatives at carbon electrodes in aqueous phosphate buffers. *J. Electroanal. Chem.* **2013**, *707*, 38–42. [CrossRef]
56. Fojt, L.; Fojta, M.; Gruner, B.; Vespalec, R. Electrochemistry of parent and exo-skeletally substituted icosahedral monocarba and dicarbaboranes and their derivatives at the graphite carbon electrode in aqueous phosphate buffers. *J. Electroanal. Chem.* **2014**, *730*, 16–19. [CrossRef]
57. Fojt, L.; Nekvinda, J.; El Anwar, S.; Gruner, B.; Havran, L.; Fojta, M. Simple Electrochemical Characterization of ortho-carborane and Some of its Exo-skeletal Derivatives. *Electroanalysis* **2020**, *32*, 1859–1866. [CrossRef]
58. Fojt, L.; Grüner, B.; Holub, J.; Havran, L.; Fojta, M. Electrochemistry of icosahedral metal full and half sandwich metallacarboranes in phosphate buffers. *J. Electroanal. Chem.* **2022**, *910*, 116165. [CrossRef]
59. Plešek, J.; Grüner, B.; Šícha, V.; Bőhmer, V.; Císařová, I. [(8,8′-μ-CH2O(CH3)-(1,2-C2B9H10)2-3,3′-Co]0 zwitterion as a versatile building block for introduction of the cobalt bis(dicarbollide) ion into organic molecules. *Organometallics* **2012**, *31*, 1703–1715. [CrossRef]
60. Dong, Y.; Liu, D.; Yang, Z. A brief review of methods for terminal functionalization of DNA. *Methods* **2014**, *67*, 116. [CrossRef] [PubMed]
61. Bühl, M.; Hnyk, D.; Macháček, J. Computational study of structures and properties of metallaboranes: Cobalt bis(dicarbollide). *Chem.-Eur. J.* **2005**, *11*, 4109–4120. [CrossRef] [PubMed]
62. El Anwar, S.; Pazderová, L.; Bavol, D.; Bakardjiev, M.; Růžičková, Z.; Horáček, O.; Fojt, L.; Kučera, R.; Grűner, B. Structurally rigidified cobalt bis(dicarbollide) derivatives, a chiral platform for labelling of biomolecules and new materials. *Chem. Commun.* **2022**, *58*, 2572–2575. [CrossRef] [PubMed]
63. Bednarska-Szczepaniak, K.; Dziedzic-Kocurek, K.; Przelazly, E.; Stanek, J.; Lesnikowski, Z.J. Intramolecular rotations and electronic states of iron in the iron bis(dicarbollide) complex Fe (C2B9H11)(2) studied by a Fe-57 nuclear probe and computational methods. *Chem. Commun.* **2022**, *58*, 391–394. [CrossRef]

molecules

MDPI

Article

Electrochemical Characterization and Voltammetric Determination of Methylisothiazolinone on a Boron-Doped Diamond Electrode

Magdalena Jakubczyk [1,*], Slawomir Michalkiewicz [1], Agata Skorupa [1] and Kinga Krajcarz [2]

[1] Institute of Chemistry, Jan Kochanowski University, PL-25406 Kielce, Poland
[2] Holy Cross Cancer Center, Tumors Markers Department, PL-25734 Kielce, Poland
* Correspondence: magdalena.jakubczyk@ujk.edu.pl

Abstract: The electrochemical properties of methylisothiazolinone (MIT), the most widely used preservative, were investigated by cyclic (CV) and differential pulse voltammetry (DPV) to develop a new method for its determination. To our knowledge, this is the first demonstration of a voltammetric procedure for the determination of MIT on a boron-doped diamond electrode (BDDE) in a citrate–phosphate buffer (C-PB) environment. The anodic oxidation process of methylisothiazolinone, which is the basis of this method, proved to be diffusion-controlled and proceeded with an irreversible two-electron exchange. The radical cations, as unstable primary products, were converted in subsequent chemical reactions to sulfoxides and sulfones, and finally to more stable final products. Performed determinations were based on the DPV technique. A linear calibration curve was obtained in the concentration range from 0.7 to 18.7 mg L^{-1}, with a correlation coefficient of 0.9999. The proposed procedure was accurate and precise, allowing the detection of MIT at a concentration level of 0.24 mg L^{-1}. It successfully demonstrated its suitability for the determination of methylisothiazolinone in household products without the need for any separation steps. The proposed method can serve as an alternative to the prevailing chromatographic determinations of MIT in real samples.

Keywords: methylisothiazolinone; boron-doped diamond electrode; voltammetry

Citation: Jakubczyk, M.; Michalkiewicz, S.; Skorupa, A.; Krajcarz, K. Electrochemical Characterization and Voltammetric Determination of Methylisothiazolinone on a Boron-Doped Diamond Electrode. *Molecules* **2022**, 27, 9013. https://doi.org/10.3390/molecules27249013

Academic Editors: Takahiro Kondo, Iwao Matsuda and Josep M. Oliva-Enrich

Received: 25 November 2022
Accepted: 14 December 2022
Published: 17 December 2022

1. Introduction

Many groups of water-based products, including cosmetics, personal care products, household products, and pharmaceuticals, require protection from microorganisms (fungi and bacteria) to ensure their properties, suitability, and safety for users and to extend their shelf life. Such functions are performed by preservatives. These are the compounds that, when added at relatively low concentrations to protected objects, block, destroy, inactivate and prevent the action of harmful organisms by chemical and/or biological means. Preservatives are characterized by their diverse chemical structure. Literature data show that the most numerous group of preservatives comprises isothiazolinones and other nitrogen compounds. Isothiazolinone derivatives, such as methylisothiazolinone (2-methyl-4-isothiazolin-3-one, MIT), methylchloroisothiazolinone (5-chloro-2-methyl-4-isothiazolin-3-one, CMIT), and 4,5-dichloro-2-octyl-4-isothiazolin-3-one (DCOIT) (Scheme 1), are used in various types of consumer products such as cosmetics (moisturizers, eye shadows, and make-up removers), hair and skin care products [1–3], adhesives [4], water-based paints [5], hydraulic fracturing fluids [6], biodiesels [7], reverse osmosis systems for water desalination [8], cooling water treatment [9], household products [3], and in the textile and paper industries [10]. This widescale use is due to their excellent biocidal properties at low concentrations, which in turn, are related to the active and oxidation-prone sulfur molecule.

In the past, a mixture of methylisothiazolinone and methylchloroisothiazolinone in a ratio of 1:3 (trade names: Kathon CG, Euxyl 400) was mainly used, but it was not until MIT started to be used in higher concentrations due to its allergenic effect that it was supposed to be a better-tolerated chemical compound [11–13]. However, it has also been shown in the literature to be highly allergenic [14,15]. Occupational cases of allergy to this compound involve medical personnel, painters, turners, mechanics, catering workers, cleaning staff, hairdressers and beauticians [16,17]. In 2013, MIT was announced as allergen of the year by the American Contact Dermatitis Society [18]. For this reason, the use of MIT in leave-on products was banned in 2016 [19]. In turn, in 2017, the European Commission published a new regulation limiting the use of MIT in rinse-off products to a maximum concentration of 0.0015% [20].

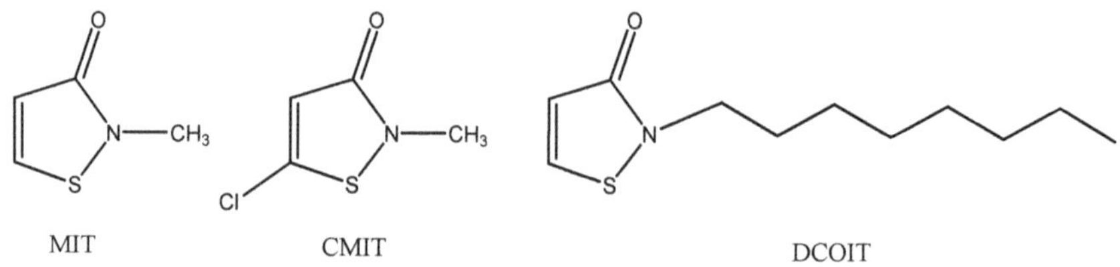

Scheme 1. Chemical structures of selected isothiazolinones.

The widespread use of isothiazolinones, as well as numerous reports in the literature on their harmful effects on the human body, make it important to quantify them in various matrices. The determination of isothiazolinones is most commonly performed by high-performance liquid chromatography (HPLC) [2,3,5,21–28] and sometimes by ultra-high-performance liquid chromatography (UHPLC) [1,10,29,30]. Gas chromatography (GC) is a much less commonly used technique for this purpose [28]. Mass spectrometry (MS) [25,28] and tandem mass spectrometry (MS/MS) [1,3,5,10,24,26,27,31] predominate as detection methods in the chromatographic analysis of these biocides. Spectrophotometric detectors in the UV range are also frequently used [2,21–23,29,30,32]. To our knowledge, only one method for the determination of MIT by HPLC coupled with electrochemical detection (ECD) has been developed to date. Abad-Gil et al. [33] used their previous studies on the voltammetric determination of MIT with a the gold electrode [34] to develop a procedure for the simultaneous chromatographic determination of antimicrobial agents in cosmetics. The *LOD* value of 30 μg L^{-1} obtained for MIT is lower than those previously reported for HPLC with UV detectors [2,21,29,30] and comparable for MS/MS [1]. The advantage of chromatographic techniques is excellent selectivity and sensitivity, allowing the simultaneous determination of many analytes and achieving low detection limits. Their serious drawbacks are expensive and complex equipment and the consumption of many reagents necessary to prepare the sample for analysis (clean-up, extraction or/and derivatization) and elution. Analytical methods based on voltammetry may be an attractive alternative for the determination of isothiazolinones. They do not require expensive equipment, and sample preparation is often limited to dissolving a sample in a suitable medium. They are equal in sensitivity and precision to chromatographic techniques [34,35].

To our knowledge, only four papers published to date deal with the voltammetric determination of isothiazolinones [34–37]. According to Abad-Gil et al. [34], MIT can be successfully determined on a gold anode in phosphate buffer solutions at pH 6 using square wave voltammetry (SWV) with a linear response up to 53 mg L^{-1}. Using adsorptive stripping voltammetry (SWAdSV), it was possible to analyze samples containing MIT at very low concentrations ranging from 0.027 to 0.12 mg L^{-1}. The *LOD* and *LOQ* values obtained were 2.8 and 9.4 mg L^{-1} for SWV and reached significantly lower values of up 0.008 and 0.027 mg L^{-1} for SWAdSV. The same group of researchers developed a voltammetric method for the determination of MIT in cosmetic and water samples using a screen-printed electrode (SPCE) modified with poly(diallyldimethylammonium) (PDDA) nanocomposite membranes containing a gold nanoparticles (AuNp) [35]. The experiments were performed using the cyclic voltammetry (CV) technique in 0.1 M NaOH solutions. This method allowed the determination of MIT in a concentration range from 8.7 to 36 mg L^{-1} and with the *LOD* and *LOQ* values of 2.6 and 8.7 mg L^{-1}, respectively. No interferences were observed in the presence of many cations and organic compounds.

A simple differential pulse voltammetric (DPV) method for the quantification of MIT and CMIT in cosmetics using the standard additional method developed by Wang et al. [36]. The determinations were preceded by the extraction of the analytes with dichloromethane. It was shown that the best results were obtained using carbon fiber (CF) microelectrode in solutions of 0.1 M $LiClO_4$ (pH 6.04). The linearity of the method was found over the range of 2–260 mg L^{-1} and 4–230 mg L^{-1} for MIT and CMIT, respectively.

An interesting indirect DPV method based on the interaction of cysteine with MIT and CMIT for their voltammetric determination on a glassy carbon electrode (GCE) in phosphate buffer solutions was proposed by Montoya et al. [37]. The specificity of these interactions and low peak potential values minimized the interference from matrix components. The determinations were based on the decrease in the cysteine oxidation signal with increasing biocide concentration.

The limited number of procedures developed for the voltammetric determination of isothiazolinones is probably due to little knowledge of their electrochemical properties in different environments. As far as we know, no systematic studies of the electrochemical properties of isothiazolinones have been conducted to find the optimal conditions for their voltammetric determination in real samples. The anodic oxidation of these biocides, most commonly MIT, CMIT, and DCOIT, has mainly been studied during the development of methods for their electrochemical degradation [38–41] or determination [34,37]. The working electrodes made of various materials such as Ti/SnO_2-Sb/PbO_2 [38], Sb_2O_3/α, β-PbO_2 [39], boron-doped diamond (BDD) [40], and carbon fiber felt (CFF) [41] have been used. All the oxidative degradation pathways described involved a ring-opening reaction and the breaking of the weak sulfur–nitrogen bond in MIT molecules is postulated [38,40], but cleavage of the carbon–carbon double bond is also considered [39,41]. Subsequent oxidation, hydrolysis, and loss of sulfur, as well as its conversion to SO_4^{2-} lead to the formation of many compounds, including N-containing products with fewer carbon atoms, organic acids (acetic and formic), which can eventually be slowly converted into CO_2, NO_3^-, H_2O, and additionally into HCl during CMIT oxidation [38–40]. In the case of a CFF-based flow-through electrode system (FES) the electrochemical degradation of MIT was via direct anodic oxidation, where the organic sulfur was oxidized to an unstable sulfoxide or sulfone structure [41]. The possibility of direct oxidation of isothiazolinones on the surface of a BDD anode was demonstrated by Kandavelu et al. [40]. The CV curves recorded in the presence of these compounds consisted of a well-defined anodic peak at 1.72 V and a shoulder at 1.58 V vs. SCE. The shape of the signals resulted from the overlapping of the two characteristic oxidation peaks of CMIT and MIT, respectively. The absence of any cathodic peaks indicated irreversible oxidation of the biocides. The same voltammetric technique was used to study the electrochemical properties of MIT on a gold electrode in phosphate buffer solutions in the pH range of 2–10 by Abad-Gil et al. [34]. The anodic oxidation of this compound was shown to be an irreversible and diffusion-controlled process in which two electrons and two protons were exchanged to yield sulfoxides and, ultimately, sulfones. Unlike other cited researchers [38–41], the authors of [34] did not include the ring-opening reaction in the proposed MIT oxidation mechanism. The results obtained were applied to the voltammetric determination of MIT [34] and to the construction of an electrochemical detector in HPLC for the detection of various antimicrobial agents, including MIT in cosmetic products [33]. The two-electron anodic oxidation processes of MIT, CMIT, and DCOIT in phosphoric acid and phosphate buffer solutions (pH 2–11) on GCE with their ring-opening were postulated by Montoya et al. [37]. The CV curves recorded in the studied solutions showed a poorly shaped MIT oxidation peak at a potential of about 1.5 V vs. Ag/AgCl. When CMIT was anodically oxidized, two overlapping peaks appeared at the same potential as the MIT peak. The oxidation of DCOIT in this environment occurs at potentials higher than those characteristic for MIT and CMIT (about 1.7 V vs. Ag/AgCl).

Considerably better-shaped peaks were obtained when the DPV technique was used. The anodic peak potentials of all three biocides remain constant up to pH 6 and decrease slightly above this pH value. This indicates that protons are not involved in the electrode reaction. The results show that all the biocides studied undergo a two-electron anodic ring-opening oxidation. The oxidation process is of *ECE* type in which the rate-determining step is a chemical reaction placed after the first partially reversible electron transfer.

To our knowledge, there has been no work to date on the use of a boron-doped diamond electrode (BDDE) for the voltammetric determination of isothiazolinones. No have citrate-phosphate buffer solutions been used for this purpose. Therefore, this study aims to investigate the electrochemical properties of MIT—the major isothiazolinone on BDDE in aqueous buffer solutions—to test the possibility of developing a voltammetric method for its determination in real samples.

2. Results and Discussion

2.1. Selection of the Best Voltammetric Conditions

In order to obtain an optimal environment for the investigation of the electrochemical properties of MIT, as well as for its voltammetric determination, the DPV curves were recorded in various types of buffer solutions in buffer-specific pH ranges (Figure 1). Britton-Robinson (B-RB), acetate (AcB), phosphate (PB), citrate (CB), and citrate–phosphate (McIlvaine, C-PB) aqueous buffers were tested. The position and shape of the curves, as well as the peak currents, were found to be closely dependent on the type of buffer and its pH. In each of the buffer solutions tested, it was found that there was a pH at which the peak current, I_p reached a maximum value. The curves corresponding to such conditions are shown in Figure 1B.

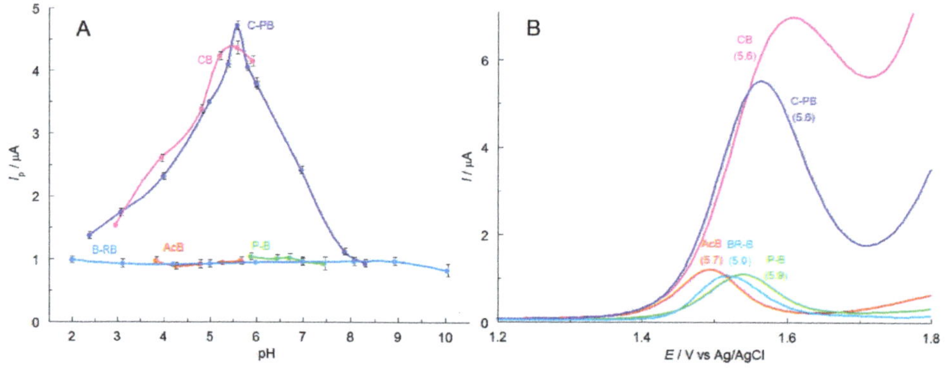

Figure 1. (**A**) Relationship between MIT (10.38 mg L^{-1}) oxidation peak current, I_p and the pH obtained in buffer solutions: B-RB, AcB, PB, CB, and C-PB on BDDE. (**B**) DPV curves (dE = 20 mV) recorded in tested solutions at pH (given in parenthesis) guaranteeing the maximum value of the peak current.

The best-shaped curves with the highest current intensity to guarantee high sensitivity of future determinations were observed in the McIlvaine buffer. In all other buffer solutions, asymmetric curves (CB) or much lower peak currents, I_p (B-RB, AcB, PB) were obtained. The data presented in Figure 1 indicate that optimal conditions can be obtained in C-PB solutions at pH 5.6. The oxidation of the analyte in this environment proceeds in one step giving a well-defined peak at a potential of about 1.535 V vs. Ag/AgCl. The DPV peak width at half height, $W_{1/2}$ of 0.130 V is well above the theoretical predicted for reversible exchange of electron, n ($W_{1/2} = 0.0904/n$ V at 25 °C [42]). This indicates an irreversible anodic oxidation of MIT. The same was observed in other buffers. Therefore, the McIlvaine buffer, pH 5.6, was chosen for further studies. It should be noted that the concentration of CP-B buffer components has no significant effect on the magnitude and shape of the MIT signal.

The next part of the experiments was the selection of the optimal working electrode material. For this purpose, DPV curves were recorded in the chosen composition of the solution on electrodes made of gold (Au), platinum (Pt), glassy carbon, and boron-doped diamond (Figure 2). When the electrode material was gold, to prevent its oxidation, the curves were recorded at 1.4 V. The only signal appeared at a potential of about 1 V. However, it cannot be associated with MIT because it also occurs in the case of the DPV curve recorded for the supporting electrolyte itself. This signal may be related to the formation of an oxide layer on the surface of the gold electrode. In the presence of MIT, only a slight increase in this signal was observed, which may suggest that oxidation of methylisothiazolinone takes place on the oxide layer. Another reason for the lack of a clear signal for MIT oxidation may be the blocking of the electrode surface by the analyte caused by the affinity of sulfur to gold. The signals recorded on GCE and platinum at potentials of about 1.7 and 1.8 V, respectively, are poorly shaped, unreproducible, and thus unsuitable for analytical purposes. The best-shaped and symmetrical MIT oxidation curve with the lowest residual currents was obtained on a boron-doped diamond electrode (Figure 2). In addition, this material provides the best repeatability of successively recorded curves. Therefore, BDDE with a diameter of 3 mm was used in further studies.

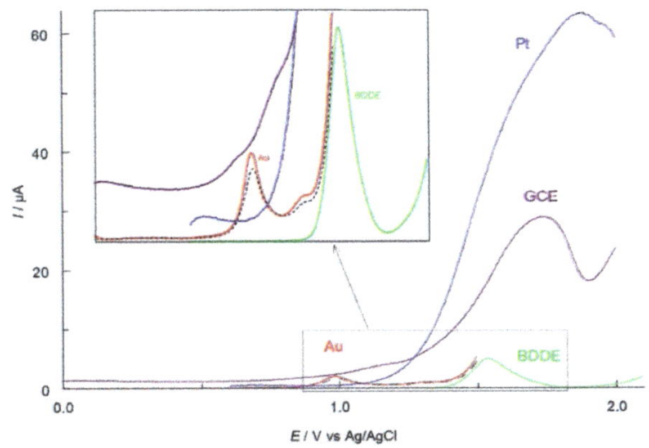

Figure 2. Comparison of DPV curves (dE = 20 mV) recorded in C-PB buffer (pH 5.6) containing MIT (10.38 mg L^{-1}) on different electrode materials (each Φ = 3 mm, given in the curves). The dashed line shows the curve recorded for the supporting electrolyte on the gold electrode.

In order to optimize the measurement conditions, the influence of the operating parameters of the DPV technique was also studied. They should ensure the best-shaped curves obtained, high resolution of signals, and the maximum peak current guaranteeing the highest sensitivity and, thus, achieving low detection limits. The operating parameters of the DPV technique and their test values are shown in Table 1. It was noted that I_p increased gradually as the DPV amplitude (dE) increased. Since values of dE greater than 50 mV cause a sharp increase in $W_{1/2}$, and thus reduce the resolution of the signals, an amplitude of 50 mV was chosen for further study. Considering the shape and MIT oxidation peak current, the potential step of 5 mV and the pulse width of 80 ms were selected.

Table 1. Optimization of DPV parameters.

Parameter	Tested Values	Optimal Value
Amplitude, dE/mV	5–150	50
Potential step, E_s/mV	2–13	5
Pulse width/ms	20–140	80

2.2. Electrochemical Properties of MIT in C-PB

The electrochemical properties of MIT on BDDE were performed in an experimentally selected C-PB buffer (pH 5.6). For this purpose, cyclic voltammetry was used. This voltammetric technique is usually applied to investigate the reduction and oxidation processes of both organic and inorganic species and to study their mechanisms. The analysis of CV curves allows obtaining information on the characteristics of electrode processes and accompanying homogeneous reactions, electrochemical properties of the analyzed substances, as well as their reduction or oxidation products.

The preliminary investigations indicate that MIT undergoes anodic oxidation in the applied environment giving a well-shaped, single peak at a potential above 1.5 V vs. Ag/AgCl. The selected CV curves of MIT recorded in the potential window from −1.2 to 2.0 V are shown in Figure 3. The position of the signal is similar to those reported in the literature and observed in other buffer solutions [36,37,40]. A characteristic feature of the CV curves recorded on BDDE, both in the absence and presence of MIT, were very small residual currents. The observed signal position corresponds to the DPV curves recorded in the same buffer solution (E_p = 1.535 ± 0.005 V vs. Ag/AgCl; Figure 1). The unlimited current increase above 1.8 V is related to the oxidation of the buffer components. As Figure 3 shows, no corresponding reduction peak is observed over a wide cathodic potential range. The cathodic signal did not appear even when the scan rate increased, and the polarization direction, E_λ was reversed near the anodic peak (Figure 3, inset). The reason for this can be due to a completely irreversible electrode reaction or instability of the primary MIT oxidation product, which may participate in the following homogeneous reaction near the electrode surface. This chemical reaction can transform it into non-electroactive products or those that were not reducible in the potential range available in the environment studied. Other researchers have also observed the absence of a cathodic peak on the CV curves recorded in the presence of these compounds [37,40].

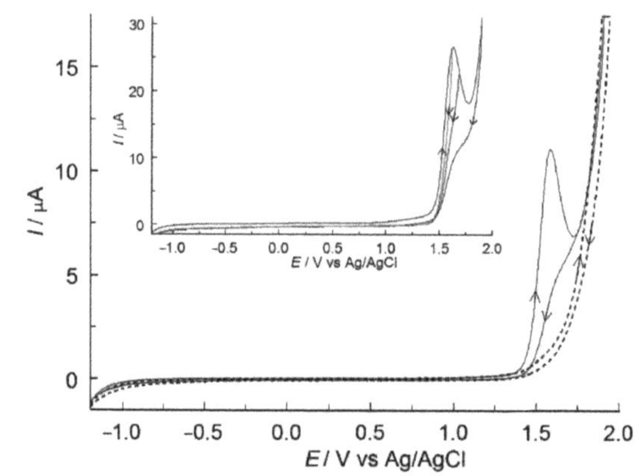

Figure 3. CVs (v = 50 mV s^{-1}) for MIT oxidation (10.38 mg L^{-1}) recorded on a BDDE in C-PB solution (pH 5.6). The dashed line is the residual current. Inset: CVs (v = 500 mV s^{-1}). The direction of electrode polarization was reversed from anodic to cathodic at potentials E_λ = 1.6; 1.7; 1.9 V.

In the next stage of the experiments, the influence of the scan rate, v in the range of 0.0062 to 0.5 V s^{-1}, on the MIT peak current and its peak potential was investigated. The CV curves shown in Figure 4 indicate the peak currents increase as the potential scan rate increases. According to the Randles-Sevcik equation, a linear relationship was observed between the anodic peak current and the square root of the scan rate (I_p (μA) = 38.26 $v^{1/2}$ (V s^{-1})$^{1/2}$ + 3.34, r = 0.9999, Figure 4, inset A). This indicates that the electrode process is controlled by the diffusion of the analyte onto the surface of the working electrode. This conclusion is confirmed by the plot of log I_p vs. log v (log I_p (μA) = 0.41 log v (V s^{-1}) + 1.60,

r = 0.9995, Figure 4, inset B), whose slope of 0.41 is close to the theoretical value of 0.5 characterizing diffusion-controlled processes [42,43].

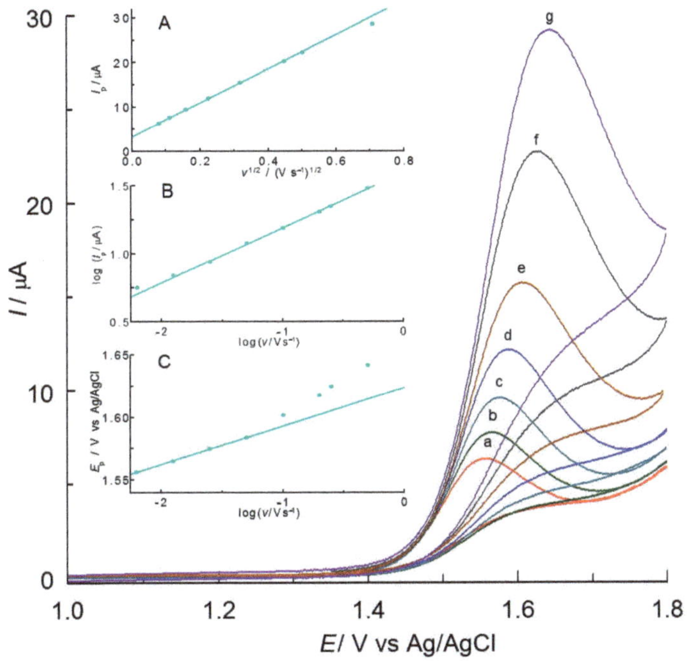

Figure 4. CVs for MIT oxidation (10.36 mg L^{-1}) on BDDE in C-PB solution (pH 5.6) at different scan rates: (a) 6.2; (b) 12.5; (c) 25; (d) 50; (e) 100; (f) 250; (g) 500 mV s^{-1}. Insets: (**A**) the relationship between anodic peak currents, I_p, and the square root of scan rate, $v^{1/2}$; (**B**) the dependence log I_p = f(log v); (**C**) the dependence E_p = f(log v).

When recording successive scans without cleaning the electrode surface, a small and gradual decrease in peak currents was observed. The largest changes were noticed between the first and second cycles (a 15% decrease in the peak current). Subsequently, the peak currents decreased by 8% and 2% between cycles 2 and 3, and 3 and 4, respectively. This indicates that the oxidation of MIT was accompanied by passivation of the electrode and, thus, a reduction in its active surface area. The peaks in successive cycles reached a stable height. Since the anodic oxidation of MIT is diffusion-controlled, the decrease in the peak current can be attributed to the partial adsorption of the final products of the electrode process. The reproducibility of the successively recorded curves was restored by applying cathodic polarization in the region of the hydrogen evolution potential (−1.2 V by 30 s). A similar decrease in the anodic peak current, corresponding to the oxidation of MIT and CMIT on BDDE in aqueous Na_2SO_4 solutions, was observed by Kandavelu et al. [40]. However, the electrode activity was restored in the region of water decomposition potential (>2.3 V vs. SCE).

The CV curves shown in Figure 4 indicate that the peak potential of the MIT oxidation, E_p, slightly shifted toward more positive with an increase in the potential scan rate. This indicates that the anodic oxidation of MIT is not reversible. To check the reversibility of the electrode process, a criterion based on the difference between the peak potential and the potential corresponding to 1/2 the peak current ($E_p - E_{p/2}$) was applied. The theoretical value characteristic of the reversible process is $0.0564/n$ V [42] at 298 K. The experimental values are higher than the theoretical ones, and they increased with an increasing potential scan rate from 0.078 V to 0.093 V (for v = 0.025 and 0.500 V s^{-1}, respectively). This confirms that the anodic oxidation of MIT is irreversible. In this case, the relationship between E_p and $E_{p/2}$ is described by the equation $E_p - E_{p/2} = 0.0477/n_e\alpha$ V, where α is the transfer coefficient, and n_e—the number of electrons exchanged in a rate-determining step [42,43]. Since most often $n_e = 1$, the value of α can be easily obtained from this equation [43]. An α value in the range of 0.61 to 0.51 was obtained for scan rates of 0.025 to 0.500 V s^{-1}. The irreversibility of this electrode reaction is confirmed by the relationship between the peak potential and the log of the scan rate shown in Figure 4, inset C. In this case, the peak potential should be a linear function of v and described by the equation: E_p (V) = $(2.303RT/2\alpha nF)\log[v\ (\mathrm{V\ s^{-1}})]$ + const. [42], where n is the total number of electrons transferred. The theoretical value of the slope of this relationship at 298 K should be $0.030/\alpha n$ V per decade change in scan rate. The experimentally obtained dependence (Figure 4, inset C) is described by the equation: E_p (V) = 0.03093 log v + 1.624 (r = 0.9997). The slope of this curve is close to the theoretical one predicted for a 2-electron irreversible electrode reaction at the experimental value of the transfer coefficient, α close to 0.5. The anodic oxidation of MIT thus proceeds with an irreversible exchange of two electrons.

The contribution of protons in the electrode process was checked for the effect of pH on the peak potentials of the CV curves in the range from 2.4 to 6.0. The obtained results indicate that an increase in pH caused only a slight shift in the oxidation peak potentials toward less positive values (data not shown). The obtained plot of peak potential, E_p vs. pH is linear and is expressed by the equation: E_p (V) = 1.605 − 0.0041 pH (r = 0.998). A slight slope of this relationship (0.0041 V/pH) deviates significantly from the theoretical one expected for an equal number of protons and electrons involved in the electrode reaction (0.0591 V/pH). These results indicate that the anodic oxidation of MIT in the tested pH range proceeds without proton exchange. They are consistent with the results obtained by Montoya et al. [37] on GCE in phosphate buffer solution.

2.3. *Mechanism*

The results obtained indicate that the anodic oxidation of MIT on BDDE in a citrate-phosphate buffer solution is diffusion-controlled and proceeds with an irreversible exchange of two electrons. Protons, however, do not participate in this heterogeneous reaction. In addition, the primary product is unstable and undergoes an irreversible chemical reaction near the surface of the working electrode. This is evidenced by the absence of cathodic peaks in the CV curves (Figures 3 and 4). Taking into account the results obtained, a probable mechanism for the anodic oxidation of MIT in C-PB was proposed, which is shown in Scheme 2. Given the susceptibility of sulfur to oxidation and its easy access to the electrode surface, the electrode process can proceed with its participation. The irreversible loss of the first electron (E_{i1}) leads to the formation of an unstable radical cation. This primary product can react with a water molecule, leading to the formation of sulfoxide. The sulfoxide thus formed can undergo a second irreversible stage of one-electron anodic oxidation (E_{i2}) to a radical cation, which is then transformed into a sulfone molecule under the influence of water. The oxidation of MIT to sulfoxides and sulfones has been suggested in studies of their direct electrochemical degradation on carbon fiber felt [41] and on a gold electrode [33,34]. Both sulfoxides and sulfones are unstable [41] and can undergo subsequent chemical reactions, including breaking the weakest sulfur–nitrogen bond and opening the ring [38,40]. These homogeneous chemical reactions can lead to the loss of a sulfur atom and its conversion to SO_4^{2-} ions and the formation of many stable compounds, such as N-containing products with fewer carbon atoms, NO_3^- ions, and organic acids (acetic and formic) [38–41]. It is worth noting that the presented mechanism is only probable and consistent with the literature data. Our laboratory has not been able to identify the primary and final products.

2.4. *Validation of the MIT Determination Method*

The applicability of the proposed voltammetric method for the determination of methylisothiazolinone was examined by measuring the peak current as a function of analyte concentration using the optimized parameters of the DPV technique. Figure 5 and inset A display the DPV curves at different MIT concentrations. It was found that the peak current increases linearly with increasing its concentration in the range of 0.7–18.7 mg L^{-1} (Figure 5B). The calibration curve was described by the equation: I_p (μA) = (1.147841 ± 0.032) × c (mg L^{-1}) − (0.095063 ± 0.087), r = 0.9999 (n = 25). Based on the standard deviation of the intercept S_b, and the slope a of the calibration curve, the detection limit ($LOD = 3.29\ S_b/a$) of MIT was determined to be 0.24 mg L^{-1}. This is well below the legal level allowed in cosmetics [20]. The *LOD* value is lower than those obtained by HPLC-DAD [2,21], square wave [34], cyclic voltammetry [35], comparable to the value for UHPLC-PDA [30] and higher for GC-MS [28], HPLC-MS/MS [27] and adsorptive stripping voltammetry [34]. Additionally, McIlvaine buffer pH 5.6 provides one of the wider useful concentration ranges compared to the literature data [21,29,34,35].

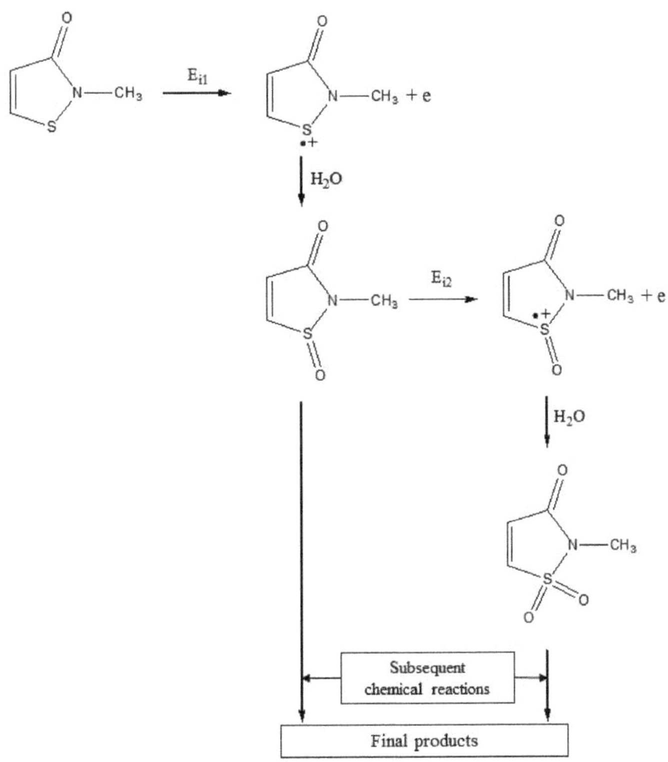

Scheme 2. Proposed mechanism of MIT oxidation.

The intra-day reproducibility was determined by recording DPV curves (n = 10) in 0.75 mg L^{-1} MIT solution. The obtained relative standard deviation (*RSD*) of the peak current of 0.6% demonstrates the excellent repeatability of the developed method. The inter-day precision of the results was determined by measuring I_p over a period of 5 consecutive days using the same solution, and the *RSD* value obtained was 1.1%.

To evaluate the selectivity of the procedure, the influence of possible interfering species, such eugenol, methyl paraben, 4-hydroxybenzoic acid, and also inorganic ions K^+, Na^+, Mg^{2+}, Ca^{2+}, SO_4^{2-}, NO_3^- which can be found in cosmetics and household products along with MIT, were investigated. The impact of the 10-fold excess of interferences concentration on the peak currents of MIT did not exceed 5%. These results indicate that it is possible to determine methylisothiazolinone in the presence of these interferents.

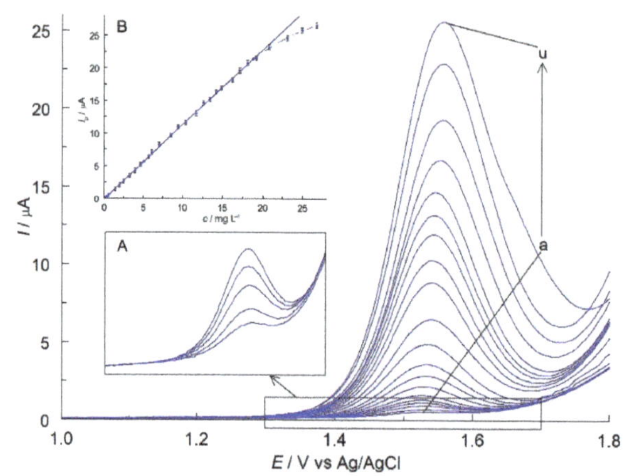

Figure 5. DPVs (dE = 50 mV) recorded on BDDE in C-PB solution (pH 5.6) containing different concentrations of MIT ranging from (*a*) 0.24–(*u*) 20.74 mg L^{-1}. Insets: (**A**) enlarging part of the main picture for small concentrations (0.24–0.98 mg L^{-1}); (**B**) calibration plot for MIT.

The developed method was verified by control determinations (the procedure described in Section 3.3). The methylisothiazolinone content of a selected household product was determined in the same way. The example curves are shown in Figure 6A. The content of the analyte in the solution was determined using mEAlab 2.1 software based on the dependence of the peak current on the concentration of the standard (Figure 6B). The resulting concentration in the real sample was then converted to the content in 1 g of product and statistically analyzed (Table 2).

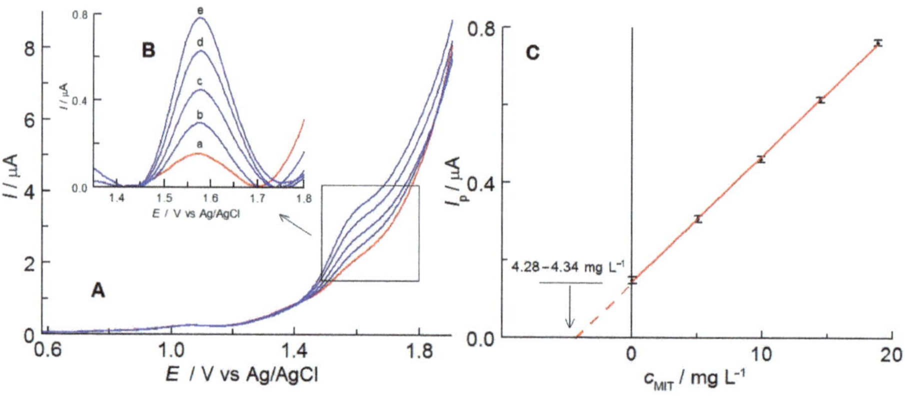

Figure 6. (**A**) DPVs recorded in a solution containing (a) Perwoll (V_0 = 2.0 mL) and after addition of the MIT standard solution (207.4 mg L^{-1}) with the volume of: (*b*) 50 μL, (*c*) 100 μL, (*d*) 150 μL, (*e*) 200 μL. (**B**) DPV curves from (**A**) after subtracting the background current. (**C**) Calibration plots for five determinations in the standard addition method.

Based on the analysis of the results of control determinations and those obtained in the solution of the selected household product, it was concluded that the developed procedure for the voltammetric determination of methylisothiazolinone was accurate (recovery, R = 99.0–99.5%, without Perwoll matrix) and precise ($RSD \leq 1.0\%$). The MIT content in control solutions with the matrix is consistent with the sum of the amounts introduced and determined in the matrix.

Table 2. Results of the methylisothiazolinone determination in control and in household products by DPV.

Sample	MIT Taken/ mg L^{-1}	1) MIT Found/ mg L^{-1}	% (w/w)	2) RSD/% (n = 5)	3) R/%
Control	0.96	0.95 ± 0.02	-	1.0	99.0
Control	4.98	4.96 ± 0.03	-	0.5	99.5
Control with Perwoll matrix	0.96	5.17 ± 0.04	0.0126 ± 0.001	0.7	538.5
Control with Perwoll matrix	4.98	9.15 ± 0.04	0.0224 ± 0.001	0.4	183.7
Perwoll	-	4.31 ± 0.03	0.0103 ± 0.0001	0.6	-

1) $x = x_{av} \pm t_{0.95} S_{av}$ for $n = 5$ and $t_{0.95} = 2.776$ (tabulated), S_{av}—denote standard deviation of mean, 2) relative standard deviation, 3) Recovery, $R = (x_{av}/\text{concentration involved}) \times 100\%$.

3. Materials and Methods

3.1. Reagents

All the reagents were of high quality and were used as received: 2-methyl-1,2-thiazol-3(2H)-one (MIT, ≥95%, Sigma-Aldrich, St. Louis, MO, USA), sodium acetate (CH_3COONa, AcNa, anhydrous, >99.0%, Merck, Darmstadt, Germany), sodium citrate ($Na_3C_6H_5O_7 \times 2H_2O$, ≥99.0%, Sigma-Aldrich, St. Louis, MO, USA), disodium hydrogen phosphate (Na_2HPO_4, anhydrous, ACS, Reag. Ph Eur, Merck, Darmstadt, Germany), potassium dihydrogen phosphate (KH_2PO_4, anhydrous, ACS, Reag. Ph Eur, Merck, Darmstadt, Germany), sodium hydroxide (NaOH, p.a., 98.8%, Pol-Aura, Olsztyn, Poland), glacial acetic acid (CH_3COOH, AcH, p.a. ACS, Merck, Darmstadt, Germany), phosphoric acid (H_3PO_4, 85 wt.%, Sigma-Aldrich, St. Louis, MO, USA), citric acid ($H_3C_6H_5O_7$, ≥99.5%, Sigma-Aldrich, St. Louis, MO, USA), sulfuric (VI) acid (H_2SO_4, p.a., 95 wt.%, Chempur, Piekary Slaskie, Poland). Perwoll washing liquid (Henkel, Wien, Austria) was used as an example of the MIT determination in the real sample.

Britton–Robinson buffer (B-RB) solutions were prepared by mixing phosphoric acid, boric acid, and acetic acid (all at a concentration of 0.04 mol L^{-1}). Their pH ranged from 2.0 to 10.0 and was adjusted with 0.2 mol L^{-1} NaOH. A series of McIlvaine buffers (citrate-phosphate buffers, C-PB) with pH in the range of 2.4–8.3 was obtained by mixing different volumes of 0.4 mol L^{-1} disodium hydrogen phosphate and 0.2 mol L^{-1} citric acid. Citrate buffers (CB) were prepared by mixing 0.2 mol L^{-1} solutions of citric acid and sodium citrate in different volumes. The pH range of these buffer solutions was from about 3.0 to 5.9. Phosphate buffers (PB) were prepared by mixing 0.2 mol L^{-1} solutions of disodium hydrogen phosphate and potassium dihydrogen phosphate in various amounts. The pH range of these buffer solutions was from about 5.9 to 7.5. Acetate buffers (AcB) were obtained by mixing 0.2 mol L^{-1} solutions of acetic acid and sodium acetate to obtain the desired pH in the range of 3.8 to 5.7. Double-distilled water was used to prepare the buffer solutions.

3.2. Apparatus

The voltammetric experiments were performed using a computer-controlled Model M161 electrochemical analyzer cooperating with EALab 2.1. software (mtm-anko, Cracow, Poland). A three-electrode glass measuring cell with a volume of 10 mL consisted of a boron-doped diamond electrode, BDDE with a diameter of 3 mm (BioLogic Science Instruments, Seyssinet-Pariset, France), a platinum wire (BASi, West Lafayette, IN, USA), and Ag/AgCl with 3 mol L^{-1} KCl (Mineral, Warsaw, Poland) as the working, counter and reference electrode, respectively. Some experiments were carried out using working electrodes of platinum, gold, and glassy carbon with a diameter of 3 mm (each BASi, West Lafayette, IN, USA).

The pH of the buffer solutions was measured using a CX-732 multifunction meter with a sensor consisting of a glass indicator electrode and an Ag/AgCl reference electrode (Elmetron, Zabrze, Poland).

All experiments were conducted at a constant temperature of 25 ± 1 °C.

3.3. Electrochemical Measurements

The EC behavior of MIT on BDDE was studied by cyclic and differential pulse voltammetry. Prior to all the voltammetric experiments, the BDDE surface was mechanically polished with 0.01 μm alumina powder slurry on a polishing cloth, then sonicated in distilled water and dried. In the next stage, the electrode surface was cathodically activated in 1 mol L^{-1} H_2SO_4, at a potential of −2.4 V, for a period of 5 min [44,45]; this procedure was repeated each day before starting all voltammetric measurements. The BDDE surface did not require additional treatment between the registrations of the voltammetric curves. This pre-treatment method guaranteed the highest activity of the electrode surface and, thus, the highest MIT oxidation currents, and the best reproducibility of voltammetric curves. The CV curves were recorded at scan rates from 6.2 to 500 mV s^{-1} in the potential range of −1.2 to 2.0 V vs. Ag/AgCl. To improve the electrode response, the DPV technique parameters, such as a potential step, the pulse width, and the pulse amplitude, were optimized. Their values are shown in Table 1. The DPV curves were recorded in the potential range of 1.0 V (beginning) to 1.8 V (end) vs. Ag/AgCl. The reproducibility of the successively recorded curves was restored by applying cathodic polarization in the area of the hydrogen evolution potential (−1.2 V by 30 s).

The main voltammetric measurements were carried out on BDDE in an experimentally selected citrate-phosphate buffer solution at pH 5.6. The calibration procedure was based on the DPV curves of MIT oxidation in the concentration range of 0.24 to 27.1 mg L^{-1}. Test solutions with the desired amount of MIT were obtained by dissolution in an appropriately defined pH buffer.

The quantitative analysis was carried out by the multiple standard addition method. Optimal determination conditions were obtained using the MIT standard at a concentration that guaranteed a significant increase in the analytical signal. The appropriate standard was added in portions of 50 μL to 2.0 mL to the solution of the test preparations. The sample preparation procedure was limited only to dissolving the product in a solution of citrate–phosphate buffer at pH 5.6. The test product was introduced in such an amount as to obtain an analyte signal suitable for quantitative analysis.

The reliability of the procedure was checked by control determinations at two MIT concentration levels of 0.96 and 4.98 mg L^{-1}, without and with the addition of a matrix (about 1 g of the Perwoll product containing MIT). The DPV curves were recorded in the stock solution and after the successive additions of the methylisothiazolinone standard. The constructed relationship $I_p = f(c_{MIT})$ was used to determine the concentration of the analyte. The procedure was repeated five times, and the results were subject to statistical analysis.

4. Conclusions

The obtained results indicate that citrate-phosphate buffer solution at pH 5.6 is a suitable medium for studying the electrochemical properties of methylisothiazolinone and its determination. The process of the anodic oxidation of MIT is characterized as diffusion-controlled and proceeds with an irreversible exchange of two electrons. Based on this electrode reaction, a new voltammetric method was developed using DPV on a boron-doped diamond electrode for the determination of MIT. To the best of our knowledge, this is the first procedure using C-PB and BDDE for this purpose. A linear relationship was obtained between the MIT concentration and the current response in the range of 0.7 to 18.7 mg L^{-1}, with a detection limit of 0.24 mg L^{-1}, which is lower than those obtained by square wave and cyclic voltammetry presented in the literature. The developed procedure is accurate and precise, and its utility was successfully demonstrated in the determination of methylisothiazolinone in household products. Its advantage is the simplicity of sample preparation, which does not require any separation steps, thus reducing the time of analysis. In addition, the use of aqueous buffer solutions and small sample volumes make the proposed method compliant with the principles of green chemistry. Consequently, the developed procedure can be a useful tool for quality control analysis of products containing MIT.

Author Contributions: Conceptualization, S.M.; methodology, M.J., S.M. and A.S.; validation, M.J.; formal analysis, M.J., S.M. and A.S.; investigation, M.J., S.M. and K.K.; writing—original draft preparation, M.J. and S.M.; writing—review and editing, M.J., S.M., A.S. and K.K.; visualization, M.J., A.S. and K.K.; supervision, S.M. All authors have read and agreed to the published version of the manuscript.

Funding: This work was financed from UJK Rector's GRANT no. SUPS.RN.22.021.

Institutional Review Board Statement: Not applicable.

Informed Consent Statement: Not applicable.

Data Availability Statement: The data presented in this study are available on request from the corresponding author.

Conflicts of Interest: The authors declare no conflict of interest.

Sample Availability: A sample of methylisothiazolinone is available from the authors.

References

1. Wittenberg, J.B.; Canas, B.J.; Zhou, W.; Wang, P.G.; Rua, D.; Krynitsky, A.J. Determination of methylisothiazolinone and methylchloroisothiazolinone in cosmetic products by ultra high performance liquid chromatography with tandem mass spectrometry. *J. Sep. Sci.* **2015**, *38*, 2983–2988. [CrossRef] [PubMed]
2. Hefnawy, M.; Al-Majed, A.; Mohammed, M.; Al-Ghusn, A.; Al-Musallam, A.; Al-Sowidan, N.; Al-Hamid, M.; Al-Homoud, A. Fast and Sensitive Liquid Chromatography Method for Simultaneous Determination of Methylisothiazolinone, Salicylic Acid and Parabens in Cosmetic Products. *Curr. Anal. Chem.* **2017**, *13*, 430–438. [CrossRef]
3. Alvarez-Rivera, G.; Dagnac, T.; Lores, M.; Garcia-Jares, C.; Sanchez-Prado, L.; Lamas, J.P.; Llompart, M. Determination of isothiazolinone preservatives in cosmetics and household products by matrix solid-phase dispersion followed by high-performance liquid chromatography-tandem mass spectrometry. *J. Chromatogr. A* **2012**, *1270*, 41–50. [CrossRef]
4. Goodier, M.C.; Zang, L.Y.; Siegel, P.D.; Warshaw, E.M. Isothiazolinone Content of US Consumer Adhesives: Ultrahigh-Performance Liquid Chromatographic Mass Spectrometry Analysis. *Dermatitis* **2019**, *30*, 129–134. [CrossRef] [PubMed]
5. Lundov, M.D.; Kolarik, B.; Bossi, R.; Gunnarsen, L.; Johansen, J.D. Emission of isothiazolinones from water-based paints. *Environ. Sci. Technol.* **2014**, *48*, 6989–6994. [CrossRef] [PubMed]
6. Kahrilas, G.A.; Blotevogel, J.; Stewart, P.S.; Borch, T. Biocides in hydraulic fracturing fluids: A critical review of their usage, mobility, degradation, and toxicity. *Environ. Sci. Technol.* **2015**, *49*, 16–32. [CrossRef] [PubMed]
7. Luz, G.V.S.; Sousa, B.A.S.M.; Guedes, A.V.; Barreto, C.C.; Brasil, L.M. Biocides used as additives to biodiesels and their risks to the environment and public health: A review. *Molecules* **2018**, *23*, 2698. [CrossRef] [PubMed]
8. Isaksson, M.; Gruvberger, B.; Bruze, M. Patch testing with serial dilutions of various isothiazolinones in patients hypersensitive to methylchloroisothiazolinone/methylisothiazolinone. *Contact Dermat.* **2014**, *70*, 270–275. [CrossRef]
9. Critchley, M.; Bentham, R. The efficacy of biocides and other chemical additives in cooling water systems in the control of amoebae. *J. Appl. Microbiol.* **2009**, *106*, 784–789. [CrossRef]
10. Lin, Q.B.; Wang, T.J.; Song, H.; Li, B. Analysis of isothiazolinone biocides in paper for food packaging by ultra-high-performance liquid chromatography-tandem mass spectrometry. *Food Addit. Contam.-Part A* **2010**, *27*, 1775–1781. [CrossRef]
11. Sukakul, T.; Kanchanapenkul, D.; Bunyavaree, M.; Limphoka, P.; Kumpangsin, T.; Boonchai, W. Methylchloroisothiazolinone and/or methylisothiazolinone in cosmetic products—A market survey. *Contact Dermat.* **2019**, *80*, 110–113. [CrossRef] [PubMed]
12. Aerts, O.; Meert, H.; Goossens, A.; Janssens, S.; Lambert, J.; Apers, S. Methylisothiazolinone in selected consumer products in Belgium: Adding fuel to the fire? *Contact Dermat.* **2015**, *73*, 142–149. [CrossRef]
13. Nielsen, H. Occupational exposdure to isothializolinones. A study based on a product register. *Contact Dermatit.* **1994**, *31*, 18–21. [CrossRef]
14. Martinez-Mera, C.; González, M.A.; Hospital, M.; Turrión-Merino, L. Isothiazolinones in paint as a cause of airborne contact dermatitis in a patient with psoriasis receiving anti-interleukin-17 therapy. *Contact Dermatit.* **2019**, *80*, 328–329. [CrossRef] [PubMed]
15. Vauhkala, A.R.; Pesonen, M.; Suomela, S.; Kuuliala, O.; Suuronen, K.; Aalto-Korte, K. Occupational contact allergy to methylchloroisothiazolinone/methylisothiazolinone and methylisothiazolinone. *Contact Dermatit.* **2015**, *73*, 150–156. [CrossRef] [PubMed]
16. Todberg, T.; Opstrup, M.S.; Johansen, J.D.; Hald, M. Occupational facial contact dermatitis caused by methylchloroisothiazolinone/methylisothiazolinone in a stainless steel aerosol spray. *Contact Dermatit.* **2017**, *77*, 171–172. [CrossRef]
17. Urwin, R.; Warburton, K.; Carder, M.; Turner, S.; Agius, R.; Wilkinson, S.M. Methylchloroisothiazolinone and methylisothiazolinone contact allergy: An occupational perspective. *Contact Dermatit.* **2015**, *72*, 381–386. [CrossRef]
18. Castanedo-Tardana, M.P.; Zug, K.A. Methylisothiazolinone. *Dermatitis* **2013**, *24*, 2–6. [CrossRef]

19. Commission Regulation (EU) 2016/1198 of 22 July 2016 Amending Annex to Regulation (EU) No. 1223/2009 of the European Parliament and of the Council on Cosmetic Products (Text with EEA Relevance). Off. J. L32016R1198, 2016. Available online: https://eur-lex.europa.eu/legal-content/EN/TXT/?uri=CELEX%3A32016R1198 (accessed on 13 December 2022).
20. Commission Regulation (EU) 2017/1224 of 6 July 2017 Amending Annex V to Regulation (EC) No 1223/2009 of the European Parliament and of the Council on Cosmetic Products (Text with EEA Relevance). Off. J. 32017R1224, 2017. Available online: https://eur-lex.europa.eu/legal-content/EN/TXT/?toc=OJ:L:2017:174:TOC&uri=uriserv:OJ.L_.2017.174.01.0016.01.ENG (accessed on 13 December 2022).
21. Baranowska, I.; Wojciechowska, I. The determination of preservatives in cosmetics and environmental waters by HPLC. *Pol. J. Environ. Stud.* **2013**, *22*, 1609–1625.
22. Abad-Gil, L.; Lucas-Sánchez, S.; Gismera, M.J.; Sevilla, M.T.; Procopio, J.R. Determination of paraben-, isothiazolinone- and alcohol-type preservatives in personal care products by HPLC with dual (diode-array and fluorescence) detection. *Microchem. J.* **2021**, *160*, 105613. [CrossRef]
23. Rastogi, S.C. Analytical control of preservative labelling on skin creams. *Contact Dermatit.* **2000**, *43*, 339–343. [CrossRef] [PubMed]
24. Kawakami, T.; Isama, K.; Nishimura, T. Analysis of Isothiazolinones and Other Preservatives in Gel-Products Used for Cooling in Japan. *J. Environ. Chem.* **2012**, *22*, 205–211. [CrossRef]
25. Heo, J.J.; Kim, U.J.; Oh, J.E. Simultaneous quantitative analysis of four isothiazolinones and 3-iodo-2-propynyl butyl carbamate in hygienic consumer products. *Environ. Eng. Res.* **2019**, *24*, 137–143. [CrossRef]
26. Bester, K.; Lamani, X. Determination of biocides as well as some biocide metabolites from facade run-off waters by solid phase extraction and high performance liquid chromatographic separation and tandem mass spectrometry detection. *J. Chromatogr. A* **2010**, *1217*, 5204–5214. [CrossRef]
27. Speksnijder, P.; van Ravestijn, J.; de Voogt, P. Trace analysis of isothiazolinones in water samples by large-volume direct injection liquid chromatography tandem mass spectrometry. *J. Chromatogr. A* **2010**, *1217*, 5184–5189. [CrossRef]
28. Rafoth, A.; Gabriel, S.; Sacher, F.; Brauch, H.J. Analysis of isothiazolinones in environmental waters by gas chromatography-mass spectrometry. *J. Chromatogr. A* **2007**, *1164*, 74–81. [CrossRef] [PubMed]
29. Baranowska, I.; Wojciechowska, I.; Solarz, N.; Krutysza, E. Determination of preservatives in cosmetics, cleaning agents and pharmaceuticals using fast liquid chromatography. *J. Chromatogr. Sci.* **2014**, *52*, 88–94. [CrossRef] [PubMed]
30. Wu, T.; Wang, C.; Wang, X.; Ma, Q. Simultaneous determination of 21 preservatives in cosmetics by ultra performance liquid chromatography. *Int. J. Cosmet. Sci.* **2008**, *30*, 367–372. [CrossRef] [PubMed]
31. Kim, H.Y.; Lee, Y.J.; Kim, K.B. Analytical method development of methylisothiazolinone, a preservative, in rat plasma using LC-MS/MS. *J. Chromatogr. B Anal. Technol. Biomed. Life Sci.* **2018**, *1100–1101*, 27–32. [CrossRef]
32. Hoa, L.T.H.; Hung, V.T.N.; Trang, D.T.; Thu, T.N.H.; Le, D.C. Development and Validation of an HPLC Method for Simultaneous Assay of MCI and MI in Shampoos Containing Plant Extracts. *Int. J. Anal. Chem.* **2019**, *2019*, 1851796. [CrossRef]
33. Abad-Gil, L.; Lucas-Sánchez, S.; Jesús Gismera, M.; Teresa Sevilla, M.; Procopio, J.R. HPLC method with electrochemical detection on gold electrode for simultaneous determination of different antimicrobial agents in cosmetics. *Microchem. J.* **2022**, *182*, 107881. [CrossRef]
34. Abad-Gil, L.; Gismera, M.J.; Sevilla, M.T.; Procopio, J.R. Determination of methylisothiazolinone in waters. Comprehensive study about electrochemical behaviour on gold electrode and optimization of square-wave voltammetric methods. *J. Electroanal. Chem.* **2021**, *880*, 114831. [CrossRef]
35. Abad-Gil, L.; Gismera, M.J.; Sevilla, M.T.; Procopio, J.R. Methylisothiazolinone response on disposable electrochemical platforms modified with carbon, nickel or gold-based nanomaterials. *Microchim. Acta* **2020**, *187*, 1–10. [CrossRef] [PubMed]
36. Wang, L.; Fang, F.-G.; Tu, W.-J. Determination of isothiazolinones in cosmetics using differential pulse voltammetry on an ultramicroelectrode. *Int. J. Pharm. Ther.* **2014**, *5*, 207–212.
37. Ruiz Montoya, M.; Giráldez, I.; Morales, E.; Estévez Brito, R.; Rodríguez Mellado, J.M. Electrochemical oxidation of isothiazolinone biocides and their interaction with cysteine. *Electrochim. Acta* **2020**, *337*, 135770. [CrossRef]
38. Han, W.; Chen, Y.; Wang, L.; Sun, X.; Li, J. Mechanism and kinetics of electrochemical degradation of isothiazolin-ones using Ti/SnO_2-Sb/PbO_2 anode. *Desalination* **2011**, *276*, 82–88. [CrossRef]
39. Wang, Y.; Chen, M.; Wang, C.; Meng, X.; Zhang, W.; Chen, Z.; Crittenden, J. Electrochemical degradation of methylisothiazolinone by using Ti/SnO_2-Sb_2O_3/A, B-PbO_2 electrode: Kinetics, energy efficiency, oxidation mechanism and degradation pathway. *Chem. Eng. J.* **2019**, *374*, 626–636. [CrossRef]
40. Kandavelu, V.; Yoshihara, S.; Kumaravel, M.; Murugananthan, M. Anodic oxidation of isothiazolin-3-ones in aqueous medium by using boron-doped diamond electrode. *Diam. Relat. Mater.* **2016**, *69*, 152–159. [CrossRef]
41. Peng, L.; Liu, H.; Wang, W.L.; Xu, Z.B.; Ni, X.Y.; Wu, Y.H.; Wu, Q.Y.; Hu, H.Y. Degradation of methylisothiazolinone biocide using a carbon fiber felt-based flow-through electrode system (FES) via anodic oxidation. *Chem. Eng. J.* **2020**, *384*, 123239. [CrossRef]
42. Scholz, F. *Electroanalytical Methods. Guide to Experiments and Applications. Second Revised and Extended Edition*; Springer: Berlin/Heidelberg, Germany, 2010.
43. Gosser, D.K. *Cyclic Voltammetry; Simulation and Analysis of Reaction Mechanisms*; VCH Publishers, Inc.: New York, NY, USA, 1994.

44. Michalkiewicz, S.; Skorupa, A.; Jakubczyk, M. Carbon materials in electroanalysis of preservatives: A review. *Materials* **2021**, *14*, 7630. [CrossRef]
45. Oliveira, S.C.B.; Oliveira-Brett, A.M. Boron doped diamond electrode pre-treatments effect on the electrochemical oxidation of dsDNA, DNA bases, nucleotides, homopolynucleotides and biomarker 8-oxoguanine. *J. Electroanal. Chem.* **2010**, *648*, 60–66. [CrossRef]

molecules

MDPI

Communication

Rhombohedral Boron Monosulfide as a p-Type Semiconductor

Norinobu Watanabe [1], Keisuke Miyazaki [2], Masayuki Toyoda [3], Kotaro Takeyasu [4,5], Naohito Tsujii [6], Haruki Kusaka [1], Akiyasu Yamamoto [7], Susumu Saito [3], Masashi Miyakawa [8], Takashi Taniguchi [6], Takashi Aizawa [6], Takao Mori [6], Masahiro Miyauchi [2,*] and Takahiro Kondo [4,5,9,*]

1 Graduate School of Pure and Applied Sciences, University of Tsukuba, Tsukuba 305-8573, Japan
2 Department of Materials Science and Engineering, School of Materials and Chemical Technology, Tokyo Institute of Technology, Meguro-ku, Tokyo 152-8552, Japan
3 Department of Physics, Tokyo Institute of Technology, Meguro-ku, Tokyo 152-8551, Japan
4 Tsukuba Research Center for Energy Materials Science, Department of Materials Science, Faculty of Pure and Applied Sciences, University of Tsukuba, Tsukuba 305-8573, Japan
5 R&D Center for Zero CO_2 Emission with Functional Materials, University of Tsukuba, Tsukuba 305-8573, Japan
6 International Center for Materials Nanoarchitectonics (WPI-MANA), National Institute for Materials Science (NIMS), 1-1 Namiki, Tsukuba, Ibaraki 305-0044, Japan
7 Institute of Engineering, Tokyo University of Agriculture and Technology, Tokyo 183-8538, Japan
8 Research Center for Functional Materials, National Institute for Materials Science, Tsukuba 305-0044, Japan
9 Advanced Research Center for Quantum Physics and Nanoscience, Tokyo Institute of Technology, Meguro-ku, Tokyo 152-8551, Japan
* Correspondence: mmiyauchi@ceram.titech.ac.jp (M.M.); takahiro@ims.tsukuba.ac.jp (T.K.)

Abstract: Two-dimensional materials have wide ranging applications in electronic devices and catalysts owing to their unique properties. Boron-based compounds, which exhibit a polymorphic nature, are an attractive choice for developing boron-based two-dimensional materials. Among them, rhombohedral boron monosulfide (r-BS) has recently attracted considerable attention owing to its unique layered structure similar to that of transition metal dichalcogenides and a layer-dependent bandgap. However, experimental evidence that clarifies the charge carrier type in the r-BS semiconductor is lacking. In this study, we synthesized r-BS and evaluated its performance as a semiconductor by measuring the Seebeck coefficient and photo-electrochemical responses. The properties unique to p-type semiconductors were observed in both measurements, indicating that the synthesized r-BS is a p-type semiconductor. Moreover, a distinct Fano resonance was observed in Fourier transform infrared absorption spectroscopy, which was ascribed to the Fano resonance between the E(2) (TO) phonon mode and electrons in the band structures of r-BS, indicating that the p-type carrier was intrinsically doped in the synthesized r-BS. These results demonstrate the potential future application prospects of r-BS.

Keywords: rhombohedral boron monosulfide; two-dimensional materials; seebeck coefficient

Citation: Watanabe, N.; Miyazaki, K.; Toyoda, M.; Takeyasu, K.; Tsujii, N.; Kusaka, H.; Yamamoto, A.; Saito, S.; Miyakawa, M.; Taniguchi, T.; et al. Rhombohedral Boron Monosulfide as a p-Type Semiconductor. *Molecules* **2023**, *28*, 1896. https://doi.org/10.3390/molecules28041896

Academic Editor: Minghao Yu

Received: 12 January 2023
Revised: 9 February 2023
Accepted: 14 February 2023
Published: 16 February 2023

1. Introduction

Two-dimensional (2D) materials exhibit unique properties such as a large surface area and unique electronic states [1–3]. Hence, they have potential applications in the development of superior electronic devices and catalysts. Among them, boron has a number of stable structures owing to its polymorphism [4–6]. Theoretical predictions indicate that borophene [6–9], boron hydride [10], boron sulfide [11], boron oxide [12], and boron phosphide [13] form stable 2D phases; hence, these structures have attracted much attention as targets for developing new 2D materials [14]. Among them, rhombohedral boron monosulfide (r-BS) has been successfully synthesized experimentally [15] and is theoretically predicted to exhibit excellent thermal conductivity [16] and high hydrogen storage performance via alkali modification [17]. Moreover, experimental reports have indicated that r-BS can be easily exfoliated from bulk to nanosheets physically in the air [18]. Furthermore, experimental results as well as density functional theory (DFT) calculations have shown that the bandgap of r-BS varies with the number of layers [18]. Hence, r-BS has the potential for a wide range of applications. By contrast, the bandgap of r-BS is always indirect and independent of the number of layers, according to the theoretical calculations [18], which is in sharp contrast to the case of intriguing 2D materials like MoS_2, which shows a transition from indirect to direct bandgap by changing the number of layers from bulk to monolayer [1]. As for the effective mass of r-BS, the average effective electron and hole masses are reported to change from 0.41 to 0.29 and 0.57 to 6.95 by changing the number of layers from bulk to monolayer, respectively, based on the theoretical calculations [18]. This means that the mobility of the carrier in r-BS is quite different depending on the type of carrier. However, the properties of r-BS as a semiconductor have not been experimentally investigated. Semiconducting borides have predominantly exhibited p-type characteristics. In 3D-structured borides, this is due to the particular bonding of the boron clusters [19] and electron deficiency. The 2D borides are much less well known, although n-type behavior has generally been proven to be much more difficult to induce in h-BN [20]. Therefore, it is of high interest to investigate r-BS.

In this study, r-BS was synthesized and its structure and optical properties were evaluated via X-ray diffraction (XRD), Fourier transform infrared absorption spectroscopy (FT-IR), Raman scattering, and UV-visible spectroscopy. The Seebeck coefficient and photo-electrochemical responses were measured as well. These measurements revealed that the synthesized r-BS shows the properties of a p-type semiconductor driven by defects.

2. Results and Discussion

2.1. Structure of Synthesized r-BS

r-BS was synthesized based on the methods described in previous studies [15,18,21,22]. Powdered sulfur and amorphous boron—used as starting materials—were mixed at an atomic ratio of 1:1, and pellets of the samples were formed. The pellets were heated to 1873 K at a pressure of 5.5 GPa and then quenched to room temperature (approximately 300 K) to obtain the r-BS samples. The resulting r-BS sample was a pink pellet that could be easily powdered. The synthesis procedure is detailed in the Materials and Methods section.

As shown in Figure 1a, the structure of r-BS has an R-3m symmetry. r-BS consists of boron and sulfur layers periodically stacked in an A-B-C stacking manner by van der Waals forces. This unique structure is similar to those of the transition metal dichalcogenides such as MoS_2: the structure of MoS_2 is an arrangement wherein one Mo atom in the MoS_2 structure is replaced by a pair of B atoms. However, according to a previous study [18], the nature of bonding with sulfur is rather different. B and S atoms are covalently bonded in r-BS, whereas bonds between S and transition metals (such as Mo) are ionic (the transition metal is positively charged and sulfur is negatively charged) [18]. The results of XRD measurements of the r-BS synthesized in this study are shown in Figure 1b. The structure of r-BS synthesized under the conditions of this experiment is consistent with the previously reported structure [15,18,21,22].

The size of the r-BS crystallite was roughly estimated to be 30 ± 10 nm independent of the crystal orientation, using Scherrer's formula:

$$B_{hkl} = \frac{0.94\lambda}{L_{hkl}\cos\theta}, \quad (1)$$

where B_{hkl} is the peak half-width (rad), λ is the wavelength of the X-rays used for irradiation (CuKα: 1.5406 Å), L_{hkl} is the crystallite size, and θ is the diffraction angle.

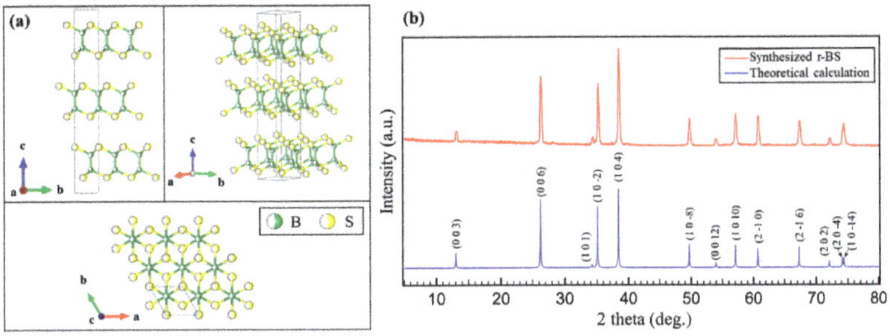

Figure 1. (**a**) The structure of r-BS [23] and (**b**) X-ray diffraction patterns (CuKα: λ = 1.5418 Å) of r-BS obtained via a high pressure synthesis method, and those of the theoretically calculated structure [18].

2.2. Optical Properties of r-BS

The optical properties examined in this work are shown in Figure 2. Figure 2a shows the results of FT-IR measurements of the synthesized powder, r-BS. A relatively large mountain-valley shape was observed at 673 cm^{-1}, and a smaller peak was observed at the bottom of the valley at 701 cm^{-1}. The smaller peak at 701 cm^{-1} was assigned to the $A_1(2)$ (TO) phonon mode, based on a previous study [22] and theoretical predictions [24]. Our DFT calculations also predicted the same peak, as shown by a purple curve in Figure 2a (the calculated phonon dispersion and phonon density of states are shown in Figure 2d,e, respectively). The significant smaller peak intensity is attributed to the plane-perpendicular and smaller dynamic dipole moment. The mountain-valley shape at around 673 cm^{-1} was fitted with a Fano resonance cross section σ as a function of energy E given by

$$\sigma = \frac{\left(\frac{q\Gamma_{res}}{2} + E - E_{res}\right)^2}{\left(\frac{\Gamma_{res}}{2}\right)^2 + (E - E_{res})^2} \tag{2}$$

where E_{res} is the resonance energy, Γ_{res} is the resonance width, and q is the Fano parameter determined by the coupling between a vibrational mode and conductive electrons [25]. As shown in Figure 2a, the mountain-valley shape with the background subtraction was fitted by E_{res} of 82.9 meV (669 cm^{-1}), Γ_{res} of 9.2 meV (74 cm^{-1}), and q of -1.19. This means that the mountain-valley shape is derived from the Fano resonance between the E(2) (TO) phonon mode [20] and electrons in the band structures. Fano resonance has also been observed between the E_u phonon mode and conductive electrons in bilayer graphene [26–28]. The appearance of Fano resonance requires doping of carriers in semiconducting materials [29,30]. Therefore, herein, the Fano resonance proves that the carrier was intrinsically doped in the synthesized r-BS. In addition, a reported Fano parameter for the bilayer graphene was approximately -0.8 [27]. The slightly larger absolute value of Fano parameter for the r-BS means that the coupling of the phonon mode and electrons is more than comparable to that of bilayer graphene.

As shown in Figure 2b, the Raman scattering spectrum of the synthesized r-BS powder shows three distinct peaks at 319, 686, and 1041 cm^{-1}, which are ascribed to the A1(3), E(4), and A1(4) modes of r-BS, respectively [18,22]. The absence of other peaks corresponding to boron or sulfur compounds in the Raman spectra indicates that r-BS was synthesized as a single phase.

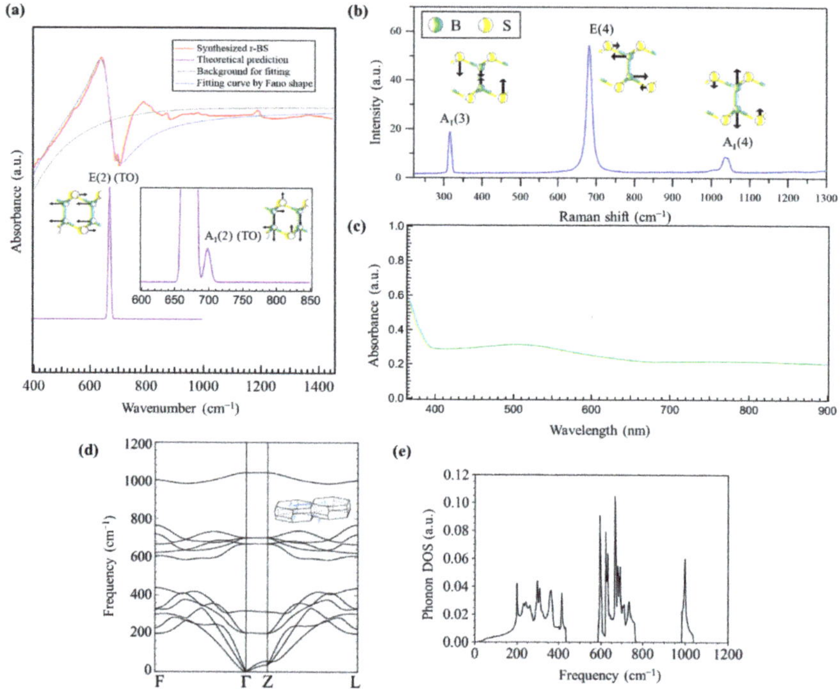

Figure 2. Optical properties of the synthesized r-BS powder. (**a**) Fourier transform infrared absorption spectra with fitting by Fano shapes and theoretical predictions of the E(2) (TO) and A1(2) (TO) phonon modes. (**b**) Raman scattering and (**c**) Ultraviolet-Visible absorption spectra. r-BS calculated phonon dispersion (**d**) and density of states (DOS) (**e**) of r-BS. The *k*-point path in the rhombohedral Brillouin zone is shown in the inset in panel d.

Figure 2c shows the ultraviolet-visible adsorption spectroscopy (UV-vis) results of the synthesized r-BS powder. Two types of absorption were observed: a sharp increase at 400 nm and a gradual increase from 680 nm to 510 nm. A wavelength of 400 nm corresponds to an energy of 3.1 eV. The theoretical bandgap of bulk r-BS is approximately 2.8 eV [18]. The theoretical bandgap is usually underestimated below the experimentally determined bandgap [31,32]. Therefore, the present bandgap of r-BS is consistent with the previously reported theoretical value. The gradual increase in the absorption at 680 nm (corresponding energy is 1.8 eV) is considered to be the transition from the valence band to the localized state in the bandgap, possibly due to the presence of defects because of the p-type nature of the r-BS carrier, as discussed below. The UV-Vis spectrum of r-BS was reported by Sasaki et al., who also observed two steps of optical absorption [15]. The strong absorption in the UV region was assigned to bandgap excitation, while the weak absorption in the visible light range was ascribed to defects in r-BS. The presence of defects in r-BS is indicated by the electron spin resonance (Figure S1 from Supplementary Materials), i.e., the observed distinct peaks can be ascribed to the unpaired electrons due to the presence of defects in r-BS, such as sulfur and/or boron vacancies and/or impurities. Previously reported ESR signals with g = 2.0037 for r-BS were consistently attributed to r-BS electric carriers originating from deficiency [15].

2.3. *p-Type Property of Synthesized r-BS*

To identify the semiconductor type of the synthesized r-BS, the Seebeck coefficient was measured. First, an r-BS pellet was placed with flattened edges between the electrodes and chromel-alumel thermocouples on its side. A temperature difference was applied across the sample using a heater while passing a current through it, and the voltage was monitored. Figure 3a shows the Seebeck coefficient of r-BS obtained in this measurement (the measurement method is detailed in the Materials and Methods section). The measured Seebeck coefficient was in the range of 500–520 μVK^{-1}. This indicates that holes are the charge carriers in the synthesized r-BS, and hence it is a p-type semiconductor. The sample has a high resistance (several MΩ) and the contact between the sample pellet and electrode may cause an error of approximately 20%. Hence, the Seebeck coefficient measurement result is merely a qualitative indication that the synthesized r-BS is a p-type semiconductor. The decrease in the Seebeck coefficient with an increase in temperature is attributed to the increase in carrier concentration. XRD measurements and FT-IR spectroscopy of the r-BS were conducted thereafter (Figures S2 and S3 from Supplementary Materials). The results show that there were no significant changes in the structure of r-BS caused by the heating during the Seebeck coefficient measurements. The sample was heated twice during the Seebeck coefficient measurements: the first round of heating included the desorption of adsorbed water (Figures S4–S6 from Supplementary Materials); the results shown in Figure 3a correspond to the Seebeck coefficient measured during the second round of heating.

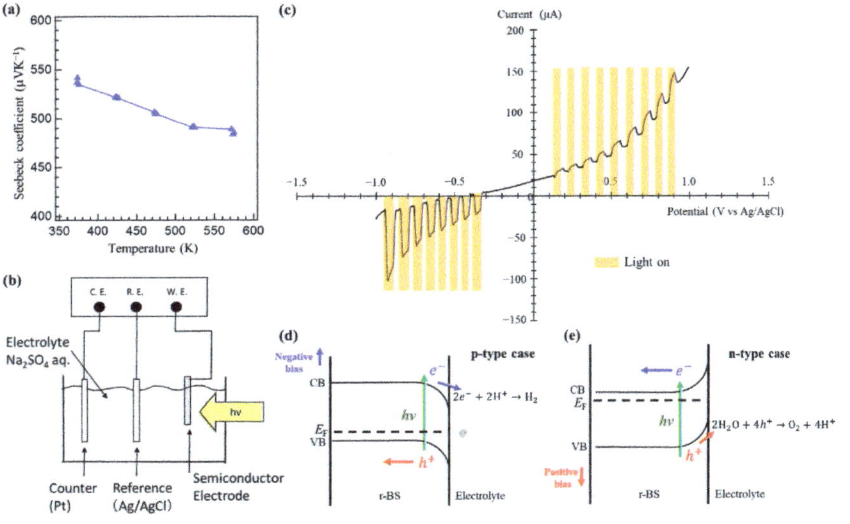

Figure 3. p-type property of the synthesized r-BS. (**a**) Temperature dependence of the Seebeck coefficient of the synthesized r-BS. (**b**) The experimental system used for the photoelectrochemical measurements. (**c**) Photoelectrochemical measurements of r-BS. Linear sweep voltammetry (LSV) was conducted with and without light irradiation. Schematic expected energy diagrams and photoelectrochemical reactions in the case that (**d**) r-BS is a p-type semiconductor and the case that (**e**) r-BS is an n-type semiconductor.

Photo-electrochemical measurements were conducted to further examine the semiconductor property of r-BS. A schematic of the setup used for the measurement is shown in Figure 3b. The electrolyte was 0.5 M Na_2SO_4 (pH $\approx$ 7). The r-BS film coated on a fluorine-doped tin oxide (FTO) glass was used as a working electrode. Pt was used as the counter electrode, and Ag/AgCl was used as the reference. As shown in Figure 3c, in the dark condition, a p-type rectification profile was observed in the linear sweep voltammetry (LSV). The current was larger for the positive bias condition because the holes contributed to the current, while the photo-electrochemical current was more pronounced on the cathodic (negative bias) side. Figure 3d,e show the schematic expected energy diagrams and photo-electrochemical reactions for p-type and n-type semiconductors, respectively. For the p-type, the photo-electrochemical current is dominant when a negative bias is applied because of the hydrogen production driven by photogenerated electrons in the conduction band [33] (Figure 3d). On the other hand, an n-type semiconductor shows photocurrent at positive bias because of a water oxidation reaction, similar to the TiO_2 and/or WO_3 cases [34] (Figure 3e). Thus, the result in Figure 3c clearly indicates that the synthesized r-BS has a p-type nature, with holes being the majority carriers under dark conditions.

Electron spin resonance (ESR) of the r-BS before and after heating at 573 K (Figure S1 from Supplementary Materials) was performed to examine the defect state of the sample. The results indicate that finite defects exist intrinsically in the synthesized r-BS, and the amount of defect does not considerably change after heating at 573 K. These intrinsic defects may be the origin of the p-type carrier in the synthesized r-BS. Further studies are required to clarify the origin of the p-type carrier of r-BS.

3. Materials and Methods

3.1. Starting Material

Amorphous boron (>99.5%) was prepared by the decomposition of B_2H_6 (Primary Metal Chemical Japan, Kanagawa, Japan). Sulfur (99%) was purchased from Wako Pure Chemical Industries Ltd., Osaka, Japan.

3.2. Synthesis of r-BS

Boron and sulfur were mixed at an atomic ratio of 1:1 in a mortar and then pressed at 200 $kgfcm^{-2}$ to form pellets. The pellets were packed in h-BN capsules sandwiched between NaCl discs; they were then sandwiched between graphite discs to prepare the cell. The cells were heated at 1873 K for 40 min at a pressure of 5.5 GPa using a belt-type high-pressure apparatus with a cylinder bore diameter of approximately 32 mm [35]. The cell was then quenched and removed; the cell was crushed and separated from the r-BS sample. The sample was pink in color and could be easily crushed to obtain the powdered form.

3.3. X-ray Diffraction (XRD)

XRD patterns were acquired at room temperature (approximately 300 K) using a MiniFlex600 (Rigaku, Tokyo, Japan) with CuKα radiation. X-rays were generated using the line-focus principle. A reflection-free Si plate was used as the sample stage. Diffraction patterns were recorded using a D/teX Ultra silicon strip detector (Rigaku) at 0.1° s^{-1} up to a 2θ value of 80°.

3.4. Fourire Transform Infrared Absorption Spectroscopy (FT-IR)

FT-IR measurements were performed at room temperature using a benchtop IR equipment (ALPHAII, Bruker, Billerica, MA, USA).

3.5. Ultraviolet-Visible Adsorption Spectroscopy (UV-vis)

The UV-Vis absorption spectra of the r-BS powders were measured using spectrophotometer (V-750, Jasco, Tokyo, Japan) with an integration sphere unit at room temperature.

3.6. Seebeck Coefficient Measurement

The Seebeck coefficient values of r-BS were obtained using a ZEM-2 (ADVANCE RIKO, Yokohama, Japan). The edges of the r-BS pellet were flattened via sanding. Since the r-BS pellets alone could easily be broken up by a slight shock, pyrophyllite, which is not conductive, was left around the pellet. A carbon sheet was laid on the electrodes, with the pellets sandwiched between them. Chromel-alumel thermocouples were applied from the side. The Seebeck coefficient was measured at 373 K, 423 K, 473 K, 523 K, and 573 K. The temperature difference ΔT was applied to the sample in each temperature condition. The measurements were performed three times with different ΔT (1, 2, and 3 K) at each temperature condition.

3.7. *Photoelectrochemical Measurements*

The r-BS electrode was prepared on a fluorine-doped tin oxide (FTO)-coated glass by a drop casting method using r-BS powder dispersed in ethanol with a 5% NafionTM solution (DE520 CS type, Wako Pure Chemical Industries Ltd., Osaka, Japan). The photocurrent of r-BS was evaluated in an aqueous solution with 0.5 M sodium sulfate (Wako Pure Chemical Industries Ltd., Osaka, Japan) at a pH of 7.0. The working, counter, and reference electrodes were r-BS, Pt plates and Ag/AgCl, respectively. The chopping light was irradiated on the r-BS electrode using a 150 W xenon lamp, and its photocurrent was recorded by a potentiostat (Hokuto Denko Corp., HZ-7000, Tokyo, Japan) during the linear sweep voltammogram (LSV).

3.8. *Thermogravimetry Analysis (TGA)*

The experiment was performed using an STA 2500 Regulus (Netzsch Japan, Japan). The sample was placed on an Al_2O_3 holder, and the experiments were conducted by heating at 10 Ks^{-1} under an Ar flow.

3.9. *ESR*

ESR measurements were performed using an EMX nano (Bruker, USA). The r-BS powder was placed in a quartz tube and cooled by liquid nitrogen.

3.10. *DFT Calculation and Simulation of Vibrational Spectrum*

Density-functional theory (DFT) calculations were performed using the Quantum ESPRESSO program package [36,37]. The local density approximation was used for the exchange-correlation energy function. The norm-conserving (for electronic structures) and ultrasoft (for phonon dispersion) pseudopotentials [38,39] were used for the core electrons. The valence (B-2s, B-2p, S-3s, and S-3p) wavefunctions were expanded on a plane-wave basis with a kinetic cutoff energy of 120 Ry. The structural optimization was performed until the remaining atomic forces fell below 10^{-4} a.u. In order to calculate the vibrational properties, the density-functional perturbation theory calculations were performed using the Phonon package of Quantum ESPRESSO. The dynamical matrix in the harmonic approximation was calculated at Γpoint using a linear response approach. The calculated IR spectrum was broadened by a Gaussian function with a FWHM of 10 cm^{-1}.

4. Conclusions

Based on the Seebeck coefficient measurements and photo-electrochemical responses, we identified that the r-BS synthesized herein is a p-type semiconductor. Moreover, the distinct Fano resonance observed in the FT-IR results, which was ascribed to the Fano resonance between the E(2) (TO) phonon mode and electrons in the band structures of r-BS, indicates that the p-type carrier was intrinsically doped in the synthesized r-BS.

Supplementary Materials: The following supporting information can be downloaded at: https://www.mdpi.com/article/10.3390/molecules28041896/s1, Figure S1: ESR measurement results of the r-BS sample before and after heating to 573 K; Figure S2: XRD pattern of the r-BS sample after Seebeck coefficient measurements (heating at 573 K); Figure S3: FT-IR spectra before and after Seebeck coefficient measurements (heating at 573 K); Figure S4: All Seebeck coefficient measurements for the r-BS sample; Figure S5: Thermogravimetric analysis (TGA) results of r-BS; Figure S6: Water desorption from r-BS during heating.

Author Contributions: A.Y. performed the amorphous boron synthesis. N.W., K.M., H.K., M.M. (Masashi Miyakawa) and T.T. conducted the r-BS synthesis. N.W., H.K. and T.K. conducted the XRD analysis. N.W., K.T., M.T., S.S. and T.K. conducted the IR analysis. M.T. and S.S. conducted the DFT IR analysis. N.W., K.M., M.M. (Masahiro Miyauchi) and T.K. conducted the UV-Vis analysis. N.T., T.A. and T.M. conducted the Seebeck coefficient analysis. K.M. and M.M. (Masahiro Miyauchi) conducted the LSV analysis. N.W. and T.K. conducted the TGA and gas analysis. N.W., K.M., M.M. (Masahiro Miyauchi) and T.K. conducted the ESR analysis. N.W., K.M., M.T., K.T., N.T., H.K., A.Y., S.S., M.M.

(Masashi Miyakawa), T.T., T.A., T.M., M.M. (Masahiro Miyauchi) and T.K. discussed the results and edited the manuscript. N.W. and T.K. wrote the draft of the manuscript. All authors have read and agreed to the published version of the manuscript.

Funding: This research was funded by the MEXT Element Strategy Initiative: To Form Core Research Centers (JPMXP0112101001), JSPS KAKENHI (Grant Nos. JP19H02551, JP21H05012, JP22K18964, and Hydrogenomics [JP19H05046:A01 and JP21H00015:B01]), JST A-STEP (Grant No. JPMJTR22T4), JST Mirai (Grant No. JPMJMI19A1), and MHI Innovation Accelerator LLC.

Institutional Review Board Statement: Not applicable.

Informed Consent Statement: Not applicable.

Data Availability Statement: Data is available on request from the corresponding author.

Conflicts of Interest: The authors declare no conflict of interest.

Sample Availability: Samples of the compounds are available from the authors.

References

1. Tan, C.; Cao, X.; Wu, X.J.; He, Q.; Yang, J.; Zhang, X.; Chen, J.; Zhao, W.; Han, S.; Nam, G.H.; et al. Recent advances in ultrathin two-dimensional nanomaterials. *Chem. Rev.* **2017**, *117*, 6225–6331. [CrossRef] [PubMed]
2. Deng, D.; Novoselov, K.S.; Fu, Q.; Zheng, N.; Tian, Z.; Bao, X. Catalysis with two-dimensional materials and their heterostructures. *Nat. Nanotechnol.* **2016**, *11*, 218–230. [CrossRef] [PubMed]
3. Osada, M.; Sasaki, T. Two-dimensional dielectric nanosheets: Novel nanoelectronics from nanocrystal building blocks. *Adv. Mater.* **2012**, *24*, 210–228. [CrossRef] [PubMed]
4. Kondo, T.; Matsuda, I. Chemically modified borophene. In *2D Boron: Boraphene, Borophene, Boronene*; Matsuda, I., Wu, K., Eds.; Springer International Publishing: Cham, Switzerland, 2021; pp. 89–119. [CrossRef]
5. Kondo, T. Recent progress in boron nanomaterials. *Sci. Technol. Adv. Mater.* **2017**, *18*, 780–804. [CrossRef]
6. Zhang, Z.; Penev, E.S.; Yakobson, B.I.; Boron, T.-D. Two-dimensional boron: Structures, properties and applications. *Chem. Soc. Rev.* **2017**, *46*, 6746–6763. [CrossRef]
7. Penev, E.S.; Bhowmick, S.; Sadrzadeh, A.; Yakobson, B.I. Polymorphism of two-dimensional boron. *Nano Lett.* **2012**, *12*, 2441–2445. [CrossRef]
8. Wu, X.; Dai, J.; Zhao, Y.; Zhuo, Z.; Yang, J.; Zeng, X.C.; Boron, T.-D. Two-dimensional boron Monolayer sheets. *ACS Nano* **2012**, *6*, 7443–7453. [CrossRef]
9. Boustani, I. New quasi-planar surfaces of bare boron. *Surf. Sci.* **1997**, *370*, 355–363. [CrossRef]
10. Jiao, Y.; Ma, F.; Bell, J.; Bilic, A.; Du, A.; Boron, T.-D. Two-Dimensional Boron Hydride sheets: High stability, massless dirac fermions, and excellent mechanical properties. *Angew. Chem.* **2016**, *128*, 10448–10451. [CrossRef]
11. Fan, D.; Lu, S.; Chen, C.; Jiang, M.; Li, X.; Hu, X. Versatile two-dimensional boron monosulfide polymorphs with tunable bandgaps and superconducting properties. *Appl. Phys. Lett.* **2020**, *117*, 013103. [CrossRef]
12. Arnold, F.M.; Seifert, G.; Kunstmann, J. Thermodynamic stability of borophene, B_2O_3 and other $B_{1-x}O_x$ sheets thermodynamic stability of borophene, B_2O_3 and other $B_{1-x}O_x$ sheets. *J. Phys. Commun.* **2020**, *4*, 031001. [CrossRef]
13. Şahin, H.; Cahangirov, S.; Topsakal, M.; Bekaroglu, E.; Akturk, E.; Senger, R.T.; Ciraci, S. Monolayer honeycomb structures of Group-IV elements and III-V binary compounds: First-principles calculations. *Phys. Rev. B* **2009**, *80*, 155453. [CrossRef]
14. Oganov, A.R.; Solozhenko, V.L. Boron: A hunt for superhard polymorphs. *J. Superhard Mater.* **2009**, *31*, 285–291. [CrossRef]
15. Sasaki, T.; Takizawa, H.; Uheda, K.; Endo, T. High pressure synthesis of binary B-S compounds. *Phys. Status Solidi (B)* **2001**, *223*, 29–33. [CrossRef]
16. Mishra, P.; Singh, D.; Sonvane, Y.; Ahuja, R.; Boron Monochalcogenide, T.-D. Monolayer for thermoelectric material. *Sustain. Energy Fuels.* **2020**, *4*, 2363–2369. [CrossRef]
17. Mishra, P.; Singh, D.; Sonvane, Y.; Ahuja, R. Metal-functionalized 2D boron sulfide monolayer material enhancing hydrogen storage capacities. *J. Appl. Phys.* **2020**, *127*, 184305. [CrossRef]
18. Kusaka, H.; Ishibiki, R.; Toyoda, M.; Fujita, T.; Tokunaga, T.; Yamamoto, A.; Miyakawa, M.; Matsushita, K.; Miyazaki, K.; Li, L.; et al. Crystalline boron monosulfide nanosheets with tunable bandgaps. *J. Mater. Chem. A* **2021**, *9*, 24631–24640. [CrossRef]
19. Mori, T. Thermoelectric and magnetic properties of rare earth borides: Boron cluster and layered compounds. *J. Solid State Chem.* **2019**, *275*, 70–82. [CrossRef]
20. Lu, S.; Shen, P.; Zhang, H.; Liu, G.; Guo, B.; Cai, Y.; Chen, H.; Xu, F.; Zheng, T.; Xu, F.; et al. Towards n-type conductivity in hexagonal boron nitride. *Nat. Commun.* **2022**, *13*, 3109. [CrossRef]
21. Cherednichenko, K.A.; Kruglov, I.A.; Oganov, A.R.; Le Godec, Y.; Mezouar, M.; Solozhenko, V.L. Boron monosulfide: Equation of state and pressure-induced phase transition. *J. Appl. Phys.* **2018**, *123*, 135903. [CrossRef]
22. Cherednichenko, K.A.; Sokolov, P.S.; Kalinko, A.; Le Godec, Y.; Polian, A.; Itié, J.P.; Solozhenko, V.L. Optical phonon modes in rhombohedral boron monosulfide under high pressure. *J. Appl. Phys.* **2015**, *117*, 185904. [CrossRef]

23. Momma, K.; Izumi, F. VESTA 3 for three-dimensional visualization of crystal, volumetric and morphology data. *J. Appl. Crystallogr.* **2011**, *44*, 1272–1276. [CrossRef]
24. Zhang, Y.; Zhou, M.; Yang, M.; Yu, J.; Li, W.; Li, X.; Feng, S. Experimental realization and computational investigations of B_2S_2as a new 2D material with potential applications. *ACS Appl. Mater. Interfaces.* **2022**, *14*, 32330–32340. [CrossRef] [PubMed]
25. Fano, U. Effects of configuration interaction on intensities and phase shifts. *Phys. Rev.* **1961**, *124*, 1866–1878. [CrossRef]
26. Kuzmenko, A.B.; Benfatto, L.; Cappelluti, E.; Crassee, I.; van der Marel, D.; Blake, P.; Novoselov, K.S.; Geim, A.K. Gate tunable infrared phonon anomalies in bilayer graphene. *Phys. Rev. Lett.* **2009**, *103*, 116804. [CrossRef]
27. Cappelluti, E.; Benfatto, L.; Kuzmenko, A.B. Phonon switching and combined Fano-Rice effect in optical spectra of bilayer graphene. *Phys. Rev. B* **2010**, *82*, 041402(R). [CrossRef]
28. Cappelluti, E.; Benfatto, L.; Manzardo, M.; Kuzmenko, A.B. Charged-phonon theory and Fano effect in the optical spectroscopy of bilayer graphene. *Phys. Rev. B* **2012**, *86*, 115439. [CrossRef]
29. Humlíček, J. Ellipsometric study of Fano resonance in heavily doped p-type Si and SiGe alloys. *Thin Solid Films* **1998**, *313–314*, 656–660. [CrossRef]
30. Simonian, A.W.; Sproul, A.B.; Shi, Z.; Gauja, E. Observation of Fano resonance in heavily doped p-type silicon at room temperature. *Phys. Rev. B* **1995**, *52*, 5672–5674. [CrossRef]
31. Perdew, J.P. Density functional theory and the band gap problem. *Int. J. Quant. Chem.* **1985**, *28*, 497–523. [CrossRef]
32. Perdew, J.P.; Yang, W.; Burke, K.; Yang, Z.; Gross, E.K.; Scheffler, M.; Scuseria, G.E.; Henderson, T.M.; Zhang, I.Y.; Ruzsinszky, A.; et al. Understanding band gaps of solids in generalized Kohn–Sham theory. *Proc. Natl. Acad. Sci. USA* **2017**, *114*, 2801–2806. [CrossRef] [PubMed]
33. Miyamoto, H.; Matsuzaki, K.; Yamaguchi, A.; Miyauchi, M. Visible-Light-Active Photoelectrochemical Z-Scheme System Based on Top 5 Clarke-Number Elements. *ACS Appl. Energy Mater.* **2018**, *1*, 5954. [CrossRef]
34. Nukui, Y.; Srinivasan, N.; Shoji, S.; Atarashi, D.; Sakai, E.; Miyauchi, M. Vertically aligned hexagonal WO_3 nanotree electrode for photoelectrochemical water oxidation. *Chem. Phys. Lett.* **2015**, *635*, 306–311. [CrossRef]
35. Yamaoka, S.; Akaishi, M.; Kanda, H.; Osawa, T.; Taniguchi, T.; Sei, H.; Fukunaga, O. Development of belt type high pressure apparatus for material synthesis at 8 GPa. *J. High Press. Inst. Jpn.* **1992**, *30*, 249–258.
36. Giannozzi, P.; Baroni, S.; Bonini, N.; Calandra, M.; Car, R.; Cavazzoni, C.; Ceresoli, D.; Chiarotti, G.L.; Cococcioni, M.; Dabo, I.; et al. QUANTUM ESPRESSO: A modular and open-source software project for quantum simulations of materials. *J. Phys. Condens. Matter* **2009**, *21*, 395502. [CrossRef]
37. Giannozzi, P.; Andreussi, O.; Brumme, T.; Bunau, O.; Nardelli, M.B.; Calandra, M.; Car, R.; Cavazzoni, C.; Ceresoli, D.; Cococcioni, M.; et al. Advanced capabilities for materials modelling with Quantum ESPRESSO. *J. Phys. Condens. Matter* **2017**, *29*, 465901. [CrossRef]
38. Hartwigsen, C.; Goedecker, S.; Hutter, J. Relativistic separable dual-space Gaussian pseudopotentials from H to Rn. *Phys. Rev. B* **1998**, *58*, 3641–3662. [CrossRef]
39. Rappe, A.M.; Rabe, K.M.; Kaxiras, E.; Joannopoulos, J.D. Optimized pseudopotentials. *Phys. Rev. B* **1990**, *41*, 1227. [CrossRef]

molecules

Article
Hall Effect Anisotropy in the Paramagnetic Phase of $Ho_{0.8}Lu_{0.2}B_{12}$ Induced by Dynamic Charge Stripes

Artem L. Khoroshilov [1,*], Kirill M. Krasikov [1,*], Andrey N. Azarevich [1,2], Alexey V. Bogach [1], Vladimir V. Glushkov [1], Vladimir N. Krasnorussky [1,3], Valery V. Voronov [1], Natalya Y. Shitsevalova [4], Volodymyr B. Filipov [4], Slavomir Gabáni [5], Karol Flachbart [5] and Nikolay E. Sluchanko [1,*]

1 Prokhorov General Physics Institute of the Russian Academy of Sciences, 38, Vavilov Str., 119991 Moscow, Russia
2 Moscow Institute of Physics and Technology, State University, 141700 Moscow, Russia
3 Vereshchagin Institute for High Pressure Physics of RAS, 14 Kaluzhskoe Shosse, 142190 Troitsk, Russia
4 Institute for Problems of Materials Science, NASU, Krzhizhanovsky Str., 3, 03142 Kyiv, Ukraine
5 Institute of Experimental Physics SAS, 47, Watsonova, 04001 Košice, Slovakia
* Correspondence: krasikovkm@lt.gpi.ru (K.M.K.); poligon-5l@yandex.ru (A.L.K.); nes@lt.gpi.ru (N.E.S.)

Abstract: A detailed study of charge transport in the paramagnetic phase of the cage-cluster dodecaboride $Ho_{0.8}Lu_{0.2}B_{12}$ with an instability both of the *fcc* lattice (cooperative Jahn–Teller effect) and the electronic structure (dynamic charge stripes) was carried out at temperatures 1.9–300 K in magnetic fields up to 80 kOe. Four mono-domain single crystals of $Ho_{0.8}Lu_{0.2}B_{12}$ samples with different crystal axis orientation were investigated in order to establish the singularities of Hall effect, which develop due to (i) the electronic phase separation (stripes) and (ii) formation of the disordered cage-glass state below T*~60 K. It was demonstrated that a considerable intrinsic anisotropic positive component $\rho^{an}{}_{xy}$ appears at low temperatures in addition to the ordinary negative Hall resistivity contribution in magnetic fields above 40 kOe applied along the [001] and [110] axes. A relation between anomalous components of the resistivity tensor $\rho^{an}{}_{xy} \sim \rho^{an}{}_{xx}{}^{1.7}$ was found for **H** | | [001] below T*~60 K, and a power law $\rho^{an}{}_{xy} \sim \rho^{an}{}_{xx}{}^{0.83}$ for the orientation **H** | | [110] at temperatures $T < T_S \sim 15$ K. It is argued that below characteristic temperature $T_S \sim 15$ K the anomalous odd $\rho^{an}{}_{xy}(T)$ and even $\rho^{an}{}_{xx}(T)$ parts of the resistivity tensor may be interpreted in terms of formation of long chains in the filamentary structure of fluctuating charges (stripes). We assume that these $\rho^{an}{}_{xy}$(**H** | | [001]) and $\rho^{an}{}_{xy}$(**H** | | [110]) components represent the intrinsic (Berry phase contribution) and extrinsic (skew scattering) mechanism, respectively. Apart from them, an additional ferromagnetic contribution to both isotropic and anisotropic components in the Hall signal was registered and attributed to the effect of magnetic polarization of 5*d* states (ferromagnetic nano-domains) in the conduction band of $Ho_{0.8}Lu_{0.2}B_{12}$.

Keywords: dynamic charge stripes; anomalous Hall effect; Jahn-Teller instability

PACS: 72.15.Gd; 72.20.My

Citation: Khoroshilov, A.L.; Krasikov, K.M.; Azarevich, A.N.; Bogach, A.V.; Glushkov, V.V.; Krasnorussky, V.N.; Voronov, V.V.; Shitsevalova, N.Y.; Filipov, V.B.; Gabáni, S.; et al. Hall Effect Anisotropy in the Paramagnetic Phase of $Ho_{0.8}Lu_{0.2}B_{12}$ Induced by Dynamic Charge Stripes. *Molecules* **2023**, *28*, 676. https://doi.org/10.3390/molecules28020676

Academic Editors: Iwao Matsuda, Takahiro Kondo and Josep M. Oliva-Enrich

Received: 7 November 2022
Revised: 22 December 2022
Accepted: 31 December 2022
Published: 9 January 2023

1. Introduction

Numerous fundamental studies on strongly correlated electron systems (SCES) such as manganites [1–4], high-temperature superconducting (HTSC) cuprates [5–8], iron-based superconductors [9–13], chalcogenides [14], etc., have allowed for the discovery of a diversity of physical phenomena universal to SCES. Indeed, all these systems are characterized by a complexity of phase diagrams induced by strong phase separation due to structural or electronic instability [15]. The spatial electronic/magnetic inhomogeneity turns out to be directly related to simultaneously active spin, charge, orbital, and lattice degrees of freedom, which are considered as factors responsible for the appearance of high-temperature superconductivity in cuprates, as well as for the emergence of colossal magnetoresistance in manganites [5,16–18]. In particular, there are two possible mechanisms of the formation of spatially inhomogeneous ground states in SCES [19]: (i) disorder resulting from phase separation near a first-order metal–insulator transition caused by an external factor [19,20] and (ii) frozen disorder in the glass phase with short-range order formed by nanoscale clusters [21–23]. In the second case, one more and among the most significant mechanisms leading to an inhomogeneous glass state in HTSC oxides is the formation of static and dynamic charge stripes [24]. Such structures have been repeatedly observed in HTSC cuprates and nickelates by both direct and indirect techniques [25–29].

Studying the effect of spatial charge inhomogeneity on the scattering of charge carriers in HTSC cuprates, manganites, and other SCES is rather difficult due to their complex composition, low symmetry crystal structure, and high sensitivity to external conditions (pressure, magnetic field, etc., see, e.g., [4]). In this respect, it looks promising to use another model SCES—rare-earth dodecaborides (RB_{12}). The RB_{12} (R—Tb, Dy, Ho, Er, Tm, Yb and Lu) attract considerable attention due to the unique combination of their physical properties, such as high melting point, microhardness, and high chemical resistance, which create prospects for practical applications [30–33]. These materials are also extremely interesting for fundamental studies. Indeed, both electronic (dynamic charge stripes) and structural (cooperative dynamic Jahn–Teller (JT) effect of the boron sub-lattice) instabilities take place in these high-boron borides with a simple *fcc* lattice (space group $Fm\bar{3}m$, see Figure 1a), in which stoichiometry can be reliably controlled during crystal growth [34].

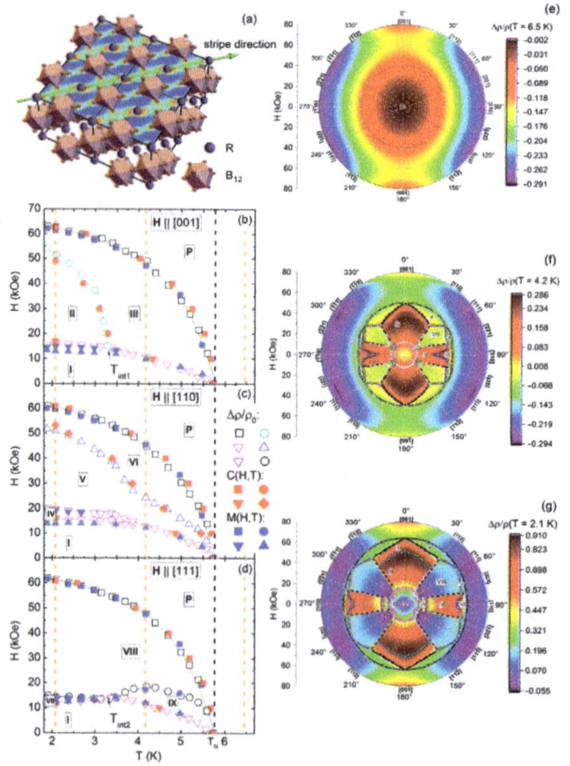

Figure 1. (**a**) Sketch of charge stripes arrangement (green lines) in the RB_{12} crystal structure, (**b–d**) H-T magnetic phase diagrams of the AF state of $Ho_{0.8}Lu_{0.2}B_{12}$ in different field directions, (**e**) polar plot of the field-angular magnetoresistance dependence in the paramagnetic state and (**f,g**) angular H-φ magnetic phase diagrams (color shows the magnetoresistance amplitude) of the AF state of $Ho_{0.8}Lu_{0.2}B_{12}$ at low temperatures (reproduced from [35]).

Let us name the main factors that determine the appearance of spatial inhomogeneity, leading to symmetry lowering in *fcc* rare earth (RE) dodecaborides. Firstly, the cooperative dynamic Jahn–Teller effect in the rigid boron sub-lattice with covalent bonds that leads to lifting of degeneracy of the highest occupied molecular orbitals (HOMO) in B_{12} octahedrons and produces static structural distortions with related splitting E_{JT}~500–1500 K (~50–150 meV) of HOMO [36]. Secondly, reaching the Ioffe–Regel limit near T_E~150 K (~15 meV) causes a development of vibrational instability, which leads to an increase in the density of phonon states at T~T_E [37]. Thirdly, order–disorder transition to the cage-glass state at T*~60 K (~5–6 meV) causes random displacements of RE ions from central symmetric positions in B_{24} cuboctahedra, which form a rigid covalent boron framework [37]. Fourthly, well below T* large amplitude vibrations of neighbored RE ions in trigonal planes (transverse to the axis of the ferrodistortive JT effect [36]) produce periodic changes of hybridization between the 5*d*(RE) and 2*p*(B) states in the conduction band. This leads to the emergence of high-frequency charge density fluctuations with frequencies ν_S~240 GHz [38] (denoted also as dynamic charge stripes, see Figure 1a) along one of the <110> direction in the *fcc* structure [39–42]. Stripe patterns are formed at characteristic temperature T_S~$h\nu_S/k_B$~15 K (~1 meV). These structural and electronic instabilities initiate nanoscale phase separation and inevitably cause strong charge transport anisotropy in external magnetic field both in the nonmagnetic reference LuB_{12} [37] and in magnetic RB_{12} [41–46]. In particular, the above features of the crystal and electronic structure of RB_{12} have a decisive effect on the characteristics of charge transport when Lu ions (with a filled *f*-shell, $4f^{14}$ configuration) are partially replaced by magnetic Ho ($4f^{10}$) ions in $Ho_xLu_{1-x}B_{12}$ compounds.

The spatial inhomogeneity of fluctuating electron density is the origin for the strong anisotropy of magnetic phase diagrams in these systems (see, e.g., Figure 1b–d for the best quality single-domain crystals of $Ho_{0.8}Lu_{0.2}B_{12}$ and [35,43–46]). Indeed, strong magnetic anisotropy is observed, for instance, in $Ho_xLu_{1-x}B_{12}$ with a high concentration of magnetic ions both in the paramagnetic (P) state (see the color plot in Figure 1e demonstrating the anisotropy of magnetoresistance) and on the angular antiferromagnetic (AF) phase diagrams, which reveal a Maltese cross-symmetry (see Figure 1f,g, [35,43,44] and also [46] for TmB_{12}). It is worth noting that, like in the nonmagnetic reference compound LuB_{12}, strong charge transport anisotropy is observed in the paramagnetic state of $Ho_xLu_{1-x}B_{12}$, ErB_{12} [45], and TmB_{12} [46] (see, for example, Figure 1e and [47]) and attributed to interaction of electron density fluctuations (stripes) with external steady magnetic field (for recent review see [48] and references therein).

Until now, studies of electron transport in $Ho_xLu_{1-x}B_{12}$ have mainly focused on transverse magnetoresistance (see, e.g., [35,43,47]). Nevertheless, the recent study of LuB_{12} [49] and the initial short research on $Ho_{0.8}Lu_{0.2}B_{12}$ (Ref. [50]) have demonstrated a significant anisotropy of the Hall effect due to an anomalous positive anisotropic contribution that appeared below T*~60 K. Thus, it is of great interest to study in detail the effect of electronic phase separation on the off-diagonal component of the resistivity tensor in model magnetic compound $Ho_{0.8}Lu_{0.2}B_{12}$ with dynamic charge stripes. As a continuation of the short study conducted in [50], this work presents results and detailed analyses of the normal and anomalous contributions to the Hall effect in the paramagnetic phase of $Ho_{0.8}Lu_{0.2}B_{12}$. We investigated both the angular and magnetic field dependences of Hall resistivity in detail and determined the anisotropic component of the resistivity tensor for this model system with electronic phase separation (dynamic charge stripes). The observed complex angular behavior of the anisotropic Hall resistivity is attributed to interaction of the filamentary structure of fluctuating charges with the external magnetic field. The arguments presented here favor both intrinsic and extrinsic mechanisms of the anomalous Hall effect formation.

2. Experimental Results and Data Analysis

2.1. Temperature Dependences of Resistivity and Hall Resistivity

In conventional Hall effect experiments the Hall coefficient is calculated as $R_H = \rho_H/H = ((V_H(+H) - V_H(-H))/(2I))\cdot d/H$, where d is the sample thickness, I the excitation current and $V_H(\pm H)$ the voltage measured on Hall probes in two opposite directions of external magnetic field. Taking into account the complex field dependence of Hall effect in parent compound LuB_{12} [49], and the different origin of the detected anomalous contributions to Hall resistivity [51], the term "reduced Hall resistivity" for ρ_H/H is used below in the present study instead of the Hall coefficient R_H.

Figure 2 shows the temperature dependences of resistivity ρ(T) at H = 0 and 80 kOe, as well as the reduced Hall resistivity $\rho_H(T)/H$ in $Ho_{0.8}Lu_{0.2}B_{12}$ calculated from experimental results recorded for three different crystals with principal directions **H** | | **n** | | [001], **H** | | **n** | | [110], and **H** | | **n** | | [111], and identical DC current direction **I** | | [$\overline{110}$]. Vertical solid lines point to the transition to the cage-glass state at T*~60 K [37] and to the formation of stripes at T_S~15 K (see also discussion below). In the zero-magnetic field, the ρ(T) curves measured for all three $Ho_{0.8}Lu_{0.2}B_{12}$ samples correspond to metallic conductivity with the same RRR value ρ(300 K)/ρ(4.2 K) = 13.47 (Figure 2a). The data for H = 0 kOe and 80 kOe are clearly separated below T*~ 60 K, indicating a pronounced sign-alternating magnetoresistance. Note that the ρ(T, H = 80 kOe) curves for crystals with **n** | | [110] and **n** | | [111] match together above the characteristic temperature T_S~15 K and differ noticeably at lower temperatures. On the contrary, at T < T*~60 K, the ρ(T, H = 80 kOe) dependence for field direction **H** | | **n** | | [001] lies well above those for **H** | | **n** | | [110] and **H** | | **n** | | [111], so the MR anisotropy reaches values ρ(**n** | | [001])/ρ(**n** | | [111]) ≈ 1.8 (MR~80%) at T = 2.1 K. Open symbols in Figure 2b show the results of ρ_H/H(T) measurements in the scheme with two opposite orientations of **H**. Significant differences between the ρ_H/H(T) dependences for different field directions appear below T*~60 K, while the curves for samples with **H** | | **n** | | [110] and **H** | | **n** | | [111] start to diverge below 15 K. In this case, the lowest negative values of ρ_H/H(T) are detected for the **n** | | [001] sample, while the highest values are observed for **H** | | **n** | | [111]. The maximal anisotropy of the reduced Hall resistivity at T = 2.1 K and H = 80 kOe equals to ρ_H/H(**n** | | [111])/ρ_H/H(**n** | | [001]) ≈ 1.8 (~80%), which is very similar to resistivity anisotropy. Thus, the temperature dependences of ρ_H/H(T) allow to identify some anisotropic positive component of the Hall signal, which appears in $Ho_{0.8}Lu_{0.2}B_{12}$ in strong magnetic fields. It is worth noting that temperature-lowering results in the increase in anisotropy for both Hall resistivity and MR components (Figure 2).

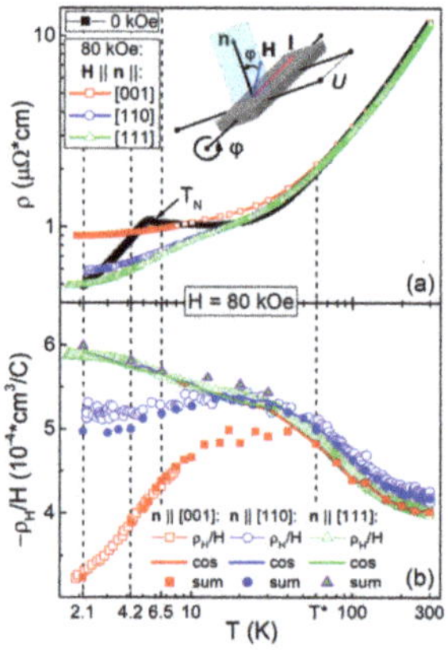

Figure 2. Temperature dependences (**a**) of resistivity ρ(T) at H = 0 and H = 80 kOe, as well as (**b**) the absolute value of reduced Hall resistivity $-\rho_H(T)/H$ in $Ho_{0.8}Lu_{0.2}B_{12}$ for samples with **H** | | **n** | | [001],

H | | **n** | | [110], and **H** | | **n** | | [111] (see inset). On panel (**b**) open and closed symbols show the experimental data for $-\rho_H/H$ and the sum of isotropic and anisotropic contributions to Hall effect sum = $\rho_{H0}/H + \rho_H^{an}/H$, correspondingly. Thick solid lines show the reduced amplitudes of the isotropic component cos = ρ_{H0}/H (see Section 2.3 for details). Vertical dashed lines point to the transition to the cage-glass state at T*~60 K [37] and to the formation of stripes at T_S~15 K (see discussion below), and denote the temperatures 2.1 K, 4.2 K, and 6.5 K at which Hall resistivity was studied in more detail.

2.2. Field Dependences of Hall Resistivity and Magnetization

Figure 3a–c show the reduced Hall resistivity $\rho_H/B(B)$ vs magnetic induction **B** at temperatures of 2.1, 4.2, and 6.5 K measured in the conventional scheme on three different samples with magnetic field **H** applied along their normal vectors- **n** | | [001], **n** | | [110], and **n** | | [111], correspondingly. The related magnetic susceptibility M/B(B) is shown in Figures 3d–f and 4 shows the temperature dependence of magnetic susceptibility M/B(T) ≡ χ(T) measured in magnetic field H = 100 Oe. The data were corrected by demagnetizing fields. It is seen from Figure 3d–f, that M/B(B) decreases with increasing both field and temperature in the paramagnetic phase, indicating a trend towards saturation of magnetization in strong magnetic fields. It can be discerned from Figure 3e,f that in the paramagnetic state, these M/B(B) curves are very similar, and the magnetic anisotropy at H~70 kOe does not exceed 1.4% even at lowest available temperature 2.1 K. Therefore, below we analyze the Hall effect using the same dependence M/B(**n** | | [001]) for all three orientations of applied magnetic field.

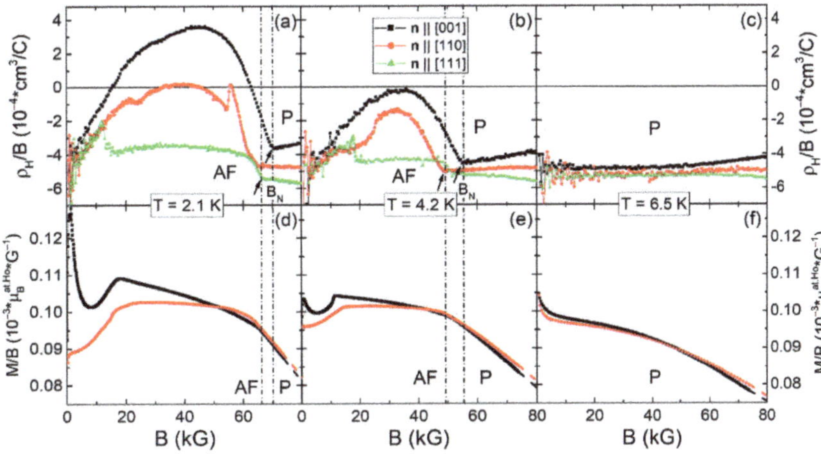

Figure 3. (**a–c**) Field dependences of the reduced Hall resistivity ρ_H/B vs magnetic induction **B** at T = 2.1, 4.2, and 6.5 K for samples with **H** | | **n** | | [001], **n** | | [110], and **n** | | [111], respectively. Arrows at $\mathbf{B_N}$ indicate the AF-P transitions. (**d–f**) Corresponding curves of magnetic susceptibility M/B(B). Dashed lines show the approximation in the interval 7–8 T (see text).

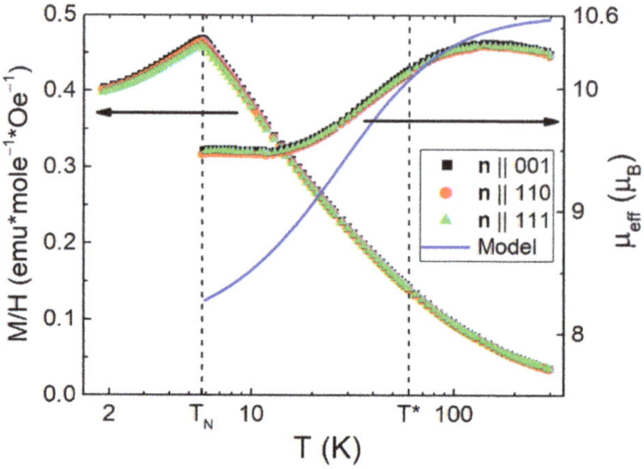

Figure 4. Temperature dependences of magnetic susceptibility (left scale) and of the effective moment (right scale) at H = 100 Oe for different samples with **H** | | **n** | | [001], **n** | | [110], and **n** | | [111] (see text). Solid line indicates the changes of the magnetic moment of Ho $^5I_8$-multiplet splitting by CEF in HoB_{12} (see [52]).

The AF-P phase transition at T_N = 5.75 K can be clearly recognized on the temperature dependence of magnetic susceptibility measured at H = 100 Oe (see Figure 4). Above T_N, the low field magnetic susceptibility χ(*T*) may be described approximately by a Curie–Weiss type dependence

$$\chi = M/H = N_{Ho}\, {\mu_{eff}}^2/(3kB\,(T - \theta_p)) + \chi_0 \quad (1)$$

where $N_{Ho} = 0.95{\cdot}x(\mathrm{Ho}){\cdot}10^{22}$ cm^{-3}, and μ_{eff}~10 μ_B are the concentration and the effective magnetic moment of Ho-ions, correspondingly (μ_B and k_B denote Bohr magneton and Boltzmann constant), $\theta_p \approx -14$ K is the paramagnetic Curie temperature corresponding to AF exchange between magnetic dipoles. $\chi_0 \approx -1.78{\cdot}10^{-3}$ μ_B/mole/Oe is the temperature-independent additive combination of (i) diamagnetic susceptibility of the boron cage and (ii) Pauli paramagnetism and Landau diamagnetism of conduction electrons.

Fitting of χ(T) by Equation (1) with temperature-dependent μ_{eff}(T) indicates that within experimental accuracy the susceptibility follows the Curie–Weiss dependence with magnetic moment, which is only slightly below the total moment $\mu_{eff} \approx 10.6$ μ_B of Ho^{3+} 4*f*-shell in the range 80–300 K. As the population of excited magnetic states of the Ho^{3+} $^5I_8$ multiplet (that is split by crystalline electric field (CEF) [52]) declines significantly in the range 8–80 K, μ_{eff} decreases moderately (to 9.5 μ_B; see Figure 4, right scale). Thus, even at T_N, the value of μ_{eff} noticeably exceeds the magnetic moment of the Γ_{51} ground state triplet μ_{eff} (Γ_{51}) ≈ 8 μ_B (solid line in Figure 4, right scale). The difference ($\Delta\mu_{eff}$~1.5 μ_B, Figure 4) may be related to ferromagnetic correlations, which develop in this SCES below *T**~60 K. Note that below 25 K, various short-range ordering features including ferromagnetic components were previously observed in magnetic RB_{12} [53–55].

As can be seen from Figure 3a–c, the behavior of reduced Hall resistivity ρ_H/B(B) differs significantly depending on **B** direction. Indeed, in the paramagnetic region for **B** | | **n** | | [001] the value of ρ_H/B(B) turns out to decrease, for **B** | | **n** | | [110] the curve is practically field independent, and for **B** | | **n** | | [111], an increase of negative ρ_H/B(B) values is observed. These trends persist in temperature range 2.1–6.5 K in the paramagnetic phase (above Neel field, B > B_N in Figure 3a–c), and the anisotropy of ρ_H/B(**n** | | [001])/ρ_H/B(**n** | | [111]) reaches values of ~80% at 2.1 K for B = 80 kG in accordance with the results of Figure 2b. Such strong anisotropy is very unusual for the paramagnetic state of *fcc* metals (as $Ho_xLu_{1-x}B_{12}$) with intense charge carrier scattering in the disordered cage-glass phase.

2.3. Angular Dependences of Hall Resistivity in the Paramagnetic State of $Ho_{0.8}Lu_{0.2}B_{12}$

To reveal the nature of the strong anisotropy of ρ_H/H Hall resistivity (Figures 2b and 3a–c) as well as to separate different contributions to Hall effect, it is of interest to study the angular dependencies of Hall resistivity $\rho_H(\varphi)$ in $Ho_{0.8}Lu_{0.2}B_{12}$ for different configurations of magnetic field with respect to principal crystallographic directions. Here we present precision measurements of Hall resistivity $\rho_H(\varphi, T_0, H_0, \mathbf{n})$ angular dependencies performed at 2.1–300 K in magnetic field up to 80 kOe for four crystals of $Ho_{0.8}Lu_{0.2}B_{12}$ with different orientations of normal vector **n** to sample surface: **n** | | [001], **n** | | [110], **n** | | [111], and **n** | | [112] (see inset in Figure 2a). In these cases, each sample was rotated around current axis **I** | | [1–10]. Thus, both fixed vector **H** and rotating vector **n** were lying in the same plane (1–10). For clarity, Figure S1 in Supplementary Materials demonstrates a direct correlation between the results of two different measurements of the Hall effect: (i) in the conventional field-sweep scheme with two opposite directions of ±**H** | | **n** and (ii) in the step-by-step rotation of the sample around **I** | | [1–10] with a fixed **H** vector in the plane perpendicular to the rotation axis (see the inset in Figure 2a).

Figure 5 shows the results of angular $\rho_H(\varphi)$ measurements at H = 80 kOe for samples with normal directions **n** | | [001] and **n** | | [110] in temperature ranges 40–300 K (Figure 5a,c) and 2.1–25 K (Figure 5b,d). The experimental results were fitted by formula.

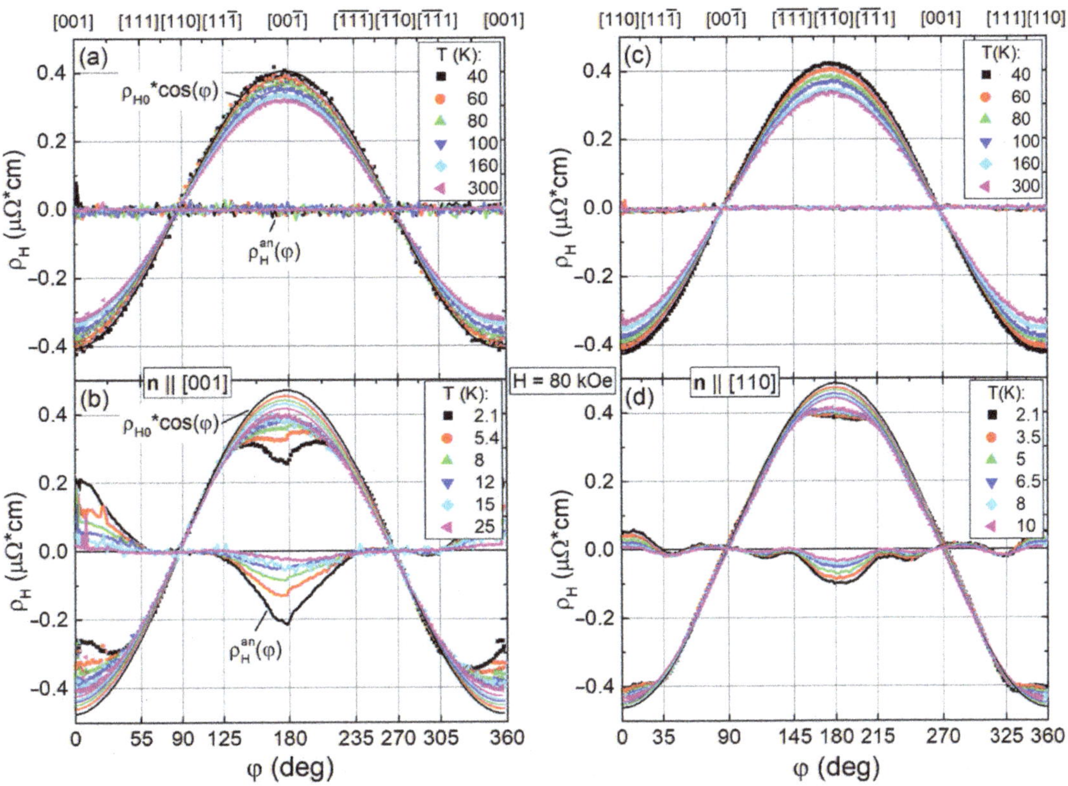

Figure 5. (**a–d**) Angular dependencies of Hall resistivity $\rho_H(\varphi)$ measured in H = 80 kOe for samples with **n** | | [001] and **n** | | [110] in the temperature range 2–300 K. Symbols show the experimental $\rho_H(\varphi)$ data, thin and thick curves demonstrate the isotropic $f_{cos}(\varphi) \approx \rho_{H0} \cdot \cos(\varphi)$ and anisotropic $\rho_H^{an}(\varphi) = \rho_H(\varphi) - f_{cos}(\varphi)$ contributions, correspondingly.

$$\rho_H(\varphi) = \rho_H^{const} + \rho_{H0}\cdot\cos(\varphi + \varphi_{sh}) + \rho_H^{an}(\varphi) \quad (2)$$

where ρ_H^{const} is an angle independent component, ρ_{H0} is the amplitude of the isotropic cosine-like contribution to Hall resistivity $f_{cos}(\varphi) = \rho_{H0}\cdot\cos(\varphi + \varphi_{sh})$, φ_{sh} is the phase shift, and $\rho_H^{an}(\varphi) = \rho_{H0}^{an}\cdot g(\varphi)$ the anisotropic contribution to Hall resistivity (see Figure 5). The approximation of $\rho_H(\varphi)$ within the framework of Equation (2) for two crystals with normal directions **n**| |[001] and **n**| |[110] was carried out in two intervals $\Delta\varphi = 90 \pm 35°$ and $\Delta\varphi = 270 \pm 35°$ where the cosine-type behavior is almost perfect. By analogy, $\rho_H(\varphi)$ curves for samples with **n**| |[111] and **n**| |[112] were approximated in same intervals $\Delta\varphi = 90 \pm 35°$ and $\Delta\varphi = 270 \pm 35°$ (near the zeros of angular dependencies), but without reference to certain crystallographic directions (see Figure S2 in Supplementary Materials). As a result, the ρ_{H0}(T, H, **n**) and φ_{sh}(T, H, **n**) parameters of the isotropic contribution $f_{cos}(\varphi)$ in (2) were found directly from this approximation. The anisotropic contribution ρ_H^{an}(T, H, **n**) at fixed direction **H**| |**n** was determined as an average of the sum of absolute $\rho_H^{an}(\varphi)$ values found for **n** at $\varphi = 0°$, $\varphi = 180°$, and $\varphi = 360°$ in the rotation experiment. As can be seen from the analysis of angular $\rho_H^{an}(\varphi)$ dependencies undertaken below, the proposed approach reveals significant limitations and inaccuracies inherent in Hall effect measurements according to the conventional field-sweep scheme. Taking into account that ρ_H^{const} and $\varphi_{sh} \approx 3$–$5°$ lead only to small corrections in determining the $\rho_{H0} = \rho_{H0}$(T, H, **n**) and $\rho_H^{an} = \rho_H^{an}$(T, H, **n**) amplitudes in Equation (2), the experimentally measured Hall resistivity is discussed below as a sum of isotropic and anisotropic contributions $\rho_H(\varphi) \approx \rho_{H0}\cdot\cos(\varphi) + \rho_H^{an}\cos(\varphi)\cdot g(\varphi)$.

In the range 40–300 K at H = 80 kOe the experimental data for **n**| |[001] and **n**| |[110] samples (Figure 5a,c) are well fitted by a cosine dependence, indicating the absence of anisotropic contribution—$\rho_H^{an}(\varphi)$~0. On the contrary, below 40 K, the $\rho_H^{an}(\varphi)$ curves for **n**| |[001] exhibit a broad feature in a wide range of angles around <001> (between nearest <111> axes (see Figure 5b)) with a step-like singularity just at <001>. Several peaks of relatively small amplitude may be identified on $\rho_H^{an}(\varphi)$ dependence for the **n**| |[110] sample (Figure 5d). The $\rho_H^{an}(\varphi)$ curves for samples with **n**| |[111] and **n**| |[112] in the range 2.1–30 K and at H = 80 kOe are presented in Supplementary Materials (see Figure S2). Note that the $\rho_H(\varphi)$ dependences for **n**| |[111] and **n**| |[112] being similar to each other differ from curves recorded for **n**| |[001] and **n**| |[110] samples, and deviate significantly from cosine dependence in a wide range of angles. The anisotropic contribution of $\rho_H^{an}(\varphi)$ extracted for **n**| |[111] and **n**| |[112] samples is close to zero near their normal directions **n** (for more details see Figure S2 in Supplementary Materials).

Figure 6 shows the result of approximation by Equation (2) of the measured Hall resistivity $\rho_H(\varphi)$ at T = 6.5 K in fixed fields up to 80 kOe for the studied four crystals. It is seen that the anisotropic contribution $\rho_H^{an}(\varphi)$ appears just above 40 kOe having the largest amplitude for sample **n**| |[001], it decreases by a factor of 2 for **n**| |[110] and goes to zero for **n**| |[111] and **n**| |[112] samples (Figure 6). Indeed, below 40 kOe the experimental data (shown by symbols) and the cosine fits (thin solid lines) coincide with a good accuracy indicating the absence of an anisotropic component $\rho_H^{an}(\varphi)$ in the low field region (see also Figure S3 in Supplementary Materials for the **n**| |[111] at T = 20 K).

It is worth noting that in the range T > T*~60 K, the temperature dependences of reduced Hall resistivity ρ_H/H(T) at H = 80 kOe for samples with **n**| |[001], **n**| |[110], and **n**| |[111] coincide within the experimental accuracy (Figure 2b). Angular $\rho_H(\varphi)$ curves can be fitted well by cosine (Figure 5a,c), and they are close to each other. Note also that for **n**| |[110] and **n**| |[111] samples, the amplitudes ρ_{H0} of $\rho_H(\varphi)$ coincide in an even wider temperature range of 15–60 K (Figure 2b). Below T_S~15 K and in the range H > 40 kOe, a significant deviation of the ρ_H/H(T) curves from cosine-type behavior is observed for crystals with **n**| |[110], **n**| |[111], and **n**| |[112] in intervals $\Delta\varphi = 90 \pm 35°$ and $\Delta\varphi = 270 \pm 35°$ (see, e.g., Figure 6). This can lead to large errors in determining the amplitude ρ_{H0} of the main contribution from conventional ±**H** field-sweep measurements. In our opinion, this finding allows us to explain the different behavior of reduced Hall resistivity ρ_H(T)/H for various field directions (see Figure 2) and shed light on the shortcoming of conventional $\pm\mathbf{H}$ field-sweep scheme commonly used for the Hall effect studies. Thus, avoiding incorrect conclusions, at low temperatures and in magnetic fields H > 40 kOe, an isotropic ordinary contribution to the Hall effect is assumed, and one common ρ_{H0} value found from the analysis of Equation (2) for **n** | | [001], which was used for these four studied crystals with different **n** directions. At the same time, in fields H $\leq$ 40 kOe at T < T_S~15 K, the $\rho_H/H(T)$ curves in intervals $\varphi = 90 \pm 35°$ and $\varphi = 270 \pm 35°$ differ only slightly from cosine curves (Figure 6); therefore, approximation by Equation (2) was carried out with individual parameters of the harmonic contribution for each of the four samples.

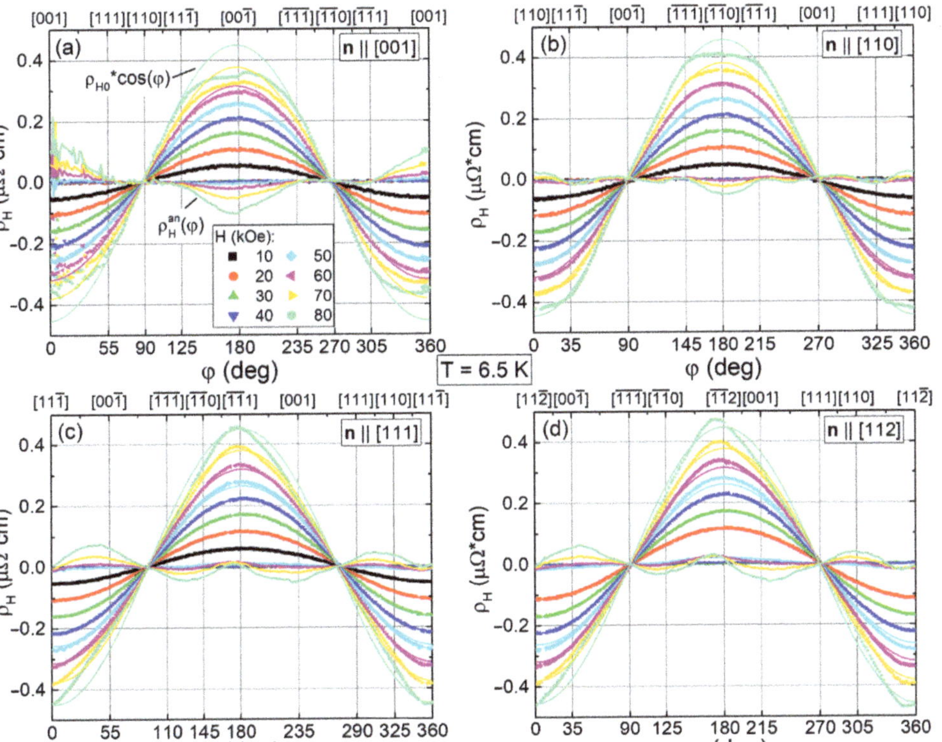

Figure 6. (**a–d**) Angular dependences of Hall resistivity $\rho_H(\varphi)$ at T = 6.5 K in fixed magnetic field up to 80 kOe for samples with **n** | | [001], **n** | | [110], **n** | | [111], and **n** | | [112]. Symbols show the experimental $\rho_H(\varphi)$ curves, thin and thick lines indicate isotropic $f_{cos}(\varphi) \approx \rho_{H0}\cdot\cos(\varphi)$ and anisotropic $\rho_H^{an}(\varphi) = \rho_H(\varphi) - f_{cos}(\varphi)$ contributions, correspondingly.

2.4. Analysis of Contributions to Hall Resistivity

Figure 7 shows the fitting parameters obtained from the approximation by Equation (2) (Figures 5 and 6 and Supplementary Materials) of $\rho_H(\varphi)$ dependences in the paramagnetic (P) phase of $Ho_{0.8}Lu_{0.2}B_{12}$. Different symbols correspond to isotropic $\rho_{H0}/H(T_0,H)$ and anisotropic $\rho_H^{an}/H(T_0,H)$ components estimated at T_0 = 2.1, 4.2, and 6.5 K (vertical dashed lines in Figure 2 denote the T_0 values). The data for different samples with **n** | | [001], **n** | | [110], **n** | | [111], and **n** | | [112] in Figure 7 are indicated by different colors. It is seen in Figure 7a that at T_0 = 6.5 K for the sample with **n** | | [001], the value of ρ_{H0}/H is about field independent below 40 kOe, while in the range H > 40 kOe, the negative values of $\rho_{H0}/H(T_0,H)$ increase linearly. For **n** | | [110], **n** | | [111], and **n** | | [112] samples, the negative values of $\rho_{H0}/H(T_0,H)$ decrease moderately with increasing magnetic field below

40 kOe. Note that in P-phase, the variation of the isotropic $\rho_{H0}/H(T_0,H)$ component may be attributed to significant (~14%) and non-monotonous change of the concentration of conduction electrons if we assume one type of charge carrier with relation $\rho_{H0}/H = R_H(T) \sim 1/n_e e$ (e is the electron charge, and n_e the concentration of charge carriers). For convenience, the reduced Hall concentration $n_e/n_R = (H/\rho_{H0})/en_R$ is shown on right axis of Figure 7a, where $n_R = 0.95 * 10^{22}$ cm^{-3} is the concentration of Ho and Lu atoms. Note also that the amplitude of anisotropic contribution $\rho_H{}^{an}/H(T_0,H)$ turns out to be almost zero below 40 kOe. In a stronger magnetic field H > 40 kOe, the anisotropic component increases for samples with **n** | | [001] and **n** | | [110] (Figure 7b), with the amplitude $\rho_H{}^{an}/H(H)$ for **n** | | [001] being more than two times higher than the corresponding values for **n** | | [110]. For **n** | | [111] and **n** | | [112] directions, the small negative values turn out to be close to zero (see also Figures S2 and S5 in Supplementary Materials).

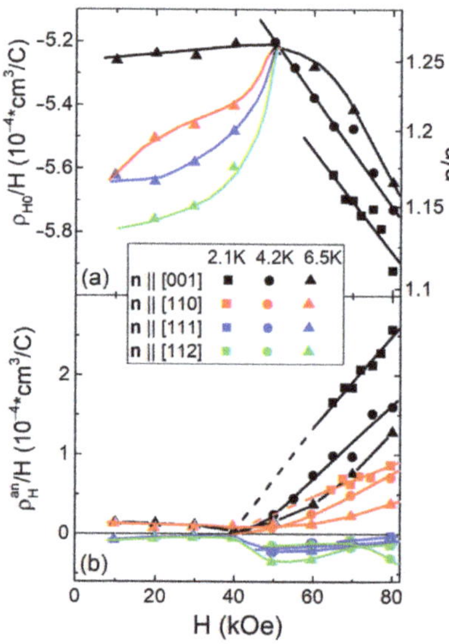

Figure 7. Reduced amplitudes of (**a**) isotropic ρ_{H0}/H and (**b**) anisotropic $\rho_H{}^{an}/H$ contributions vs external magnetic field H at temperatures of 2.1, 4.2, and 6.5 K for samples with **n** | | [001], **n** | | [110], **n** | | [111], and **n** | | [112]. Different temperatures are indicated by different shapes of symbols, while samples with different **n** directions are indicated by different colors of the symbols. Right axis on panel (**a**) shows the reduced Hall concentration n/n_R for comparison (see text).

The experimental temperature dependences $\rho_H/H(T,H_0 = 80$ kOe) obtained in the conventional, commonly used the field-sweep technique with two opposite orientations of applied magnetic field ±**H** | | **n** from one side, and the isotropic component deduced from the angular dependences of Hall resistivity $\rho_{H0}/H(T,H_0)$ from the other, are compared in Figure 2b. It can be seen that for the sample with **n** | | [111], the $\rho_{H0}/H(T,H_0)$ data coincide with good accuracy with the ρ_H/H values detected by conventional field-sweep measurements in a wide range of temperatures 1.9–300 K. For the sample with **n** | | [110], ρ_{H0}/H starts to deviate from $\rho_H/H(T,H_0)$ at T < T_S~15 K, and for **n** | | [001], noticeable differences arise already upon the transition to the cage-glass state at T*~60 K (Figure 2b). This observation allows to attribute the appearance below T* of the strong anisotropy of the Hall effect in $Ho_{0.8}Lu_{0.2}B_{12}$ to the contribution $\rho_H{}^{an}/H$, which increases additionally below T_S. Figure 2b shows a comparison of the parameter sum = $\rho_{H0}/H + \rho_H{}^{an}/H$, which corresponds to Hall effect amplitude detected from ρ_H/H angular dependences from one side, and the reduced Hall resistivity measured in the conventional ±**H** | | **n** field-sweep experiment from the other. It can be seen that the temperature behavior of the sum detected by angular measurements is in good agreement with the results of conventional $\pm$**H** | | **n** field-sweep dependences of $\rho_H/H(H)$ for all crystals under investigation (Figure 2b, **H** | | **n** | | [001], **H** | | **n** | | [110], and **H** | | **n** | | [111]) approving the validity of the proposed Hall effect separation. Then, using the simple relation $\rho_{H0}/H(T) = R_H(T) \sim 1/n_e e$, we roughly estimate from the temperature dependences presented in Figure 2b and the charge carrier concentration n_e in the conduction band. Figure 8a shows the strong field (H = 80 kOe) Arrhenius plot $lg(n_e/n_R) \sim 1/T$ for H | | **n** | | [001] and H | | **n** | | [110] directions, similar to Figure 7a (right axis), demonstrating the field dependence of the ratio n_e/n_R.

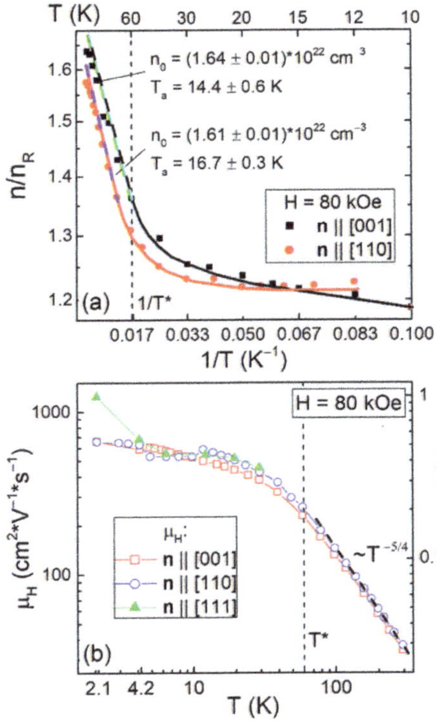

Figure 8. (**a**) Arrhenius plot *lg(n/n$_R$) = f(1/T)* of the reduced Hall concentration for **n** | | [001] and **n** | | [110] at H = 80 kOe. (**b**) Temperature dependences of the Hall mobility μ_H and the parameter $\omega_c\tau \approx \mu_H \cdot H$ for three directions **n** | | [001], **n** | | [110], and **n** | [111]. Thick dashed lines show the (**a**) activation behavior and (**b**) power law.

It can be seen in Figure 8a that the Arrhenius-type dependence $1/R_H(T) \sim e \cdot n_0 \cdot \exp(-T_a/T)$ is valid above the cage-glass transition at T*~60 K, and that the estimated activation temperatures $T_a = 14.4 \pm 0.6$ K and $T_a = 16.7 \pm 0.3$ K detected for **H** | | **n** | | [001] and **H** | | **n** | | [110], correspondingly, are close to T_S~15 K. As the detected initial concentration $n_0 = (1.61–1.64) \cdot 10^{22}$ cm^{-3} coincides within experimental accuracy for these two field directions, the reduced Hall concentration changes in the range $n_e/n_R = 1.2–1.6$ (Figure 8a) in accordance with the results of previous Hall effect measurements in RB_{12} (R = Lu, Tm, Ho, Er) [56].

Figure 8b shows the temperature dependences of Hall mobility $\mu_H(T) \approx \rho_{H0}(T)/[H \cdot \rho(T)]$ (left scale) and the related parameter $\omega_c\tau \approx \mu_H \cdot H$ (right scale, where ω_c is the cyclotron frequency and τ the carrier relaxation time) in magnetic field H = 80 kOe for samples with **n** | | [001], **n** | | [110], and **n** | | [111]. At low temperatures T < T_S~15 K the obtained $\mu_H(T)$ data tend to constant values μ_H~600–700 cm^2/(V·s) (Figure 8b), and in the range T > T*~60 K Hall mobility follows the power law $\mu_H \sim T^{-\alpha}$ with a single exponent being estimated as $\alpha \approx 5/4$ both for **H**| |**n**| |[001] and **H**| |**n**| |[110].

Similar behavior of Hall mobility was observed in the range 80–300 K previously for various LuB_{12} crystals; an $\alpha \approx 7/4$ exponent was detected for crystals with large values of RRR $\equiv \rho(300\ K)/\rho(6\ K)$ = 40–70 and $\alpha \approx 3/2$ was estimated for LuB_{12} with a small RRR~12 [49]. The $\alpha = 3/2$ exponent is typical for scattering of conduction electrons by acoustic phonons (deformation potential). The increase of α values up to 7/4 in best LuB_{12} crystals was interpreted [49] in terms of charge carriers scattering both by quasi-local vibrations of RE ions and by boron optical phonons [57] in the presence of JT distortions and rattling modes of RE ions [58–60]. In the case of $Ho_{0.8}Lu_{0.2}B_{12}$ crystals with RRR~10 the further decrease of α value from 3/2 to 5/4 could be attributed to the emergence of strong magnetic scattering in dodecaboride with Ho^{3+} magnetic ions. Note that the inequality $\omega_c\tau < 1$ (see Figure 8b), which is fulfilled in the entire temperature range 1.9–300 K in fields up to 80 kOe, corresponds to the low field regime of charge carriers in $Ho_{0.8}Lu_{0.2}B_{12}$. This indicates that the results of Hall effect measurements should be insensitive to the topology of the Fermi surface and depending mainly on the features of disorder and inhomogeneity of the crystals studied.

Obviously, the Hall effect in $Ho_{0.8}Lu_{0.2}B_{12}$ is strongly modified by the positive anisotropic contribution ρ_H^{an}/H. Figure 9a,b show the angular dependencies of $\rho_H^{an}(\varphi)/H$ at H = 80 kOe for samples with **n**| |[001] and **n**| |[110]. The same contributions for **n**| |[111] and **n**| |[112] are shown in Figure 10. As can be seen from Figures 9 and 10, the $\rho_H^{an}(\varphi)/H$ curves differ in both the amplitude and shape of angular dependence. Since very small changes of $\rho_H^{an}(\varphi)/H$ with temperature are detected for samples with **H**| |**n**| |[111] and **H**| |**n**| |[112] in the field along normal directions (φ = 180° in Figure 10), the temperature dependences of the anomalous component below are analyzed only for **n**| |[001] and **n**| |[110].

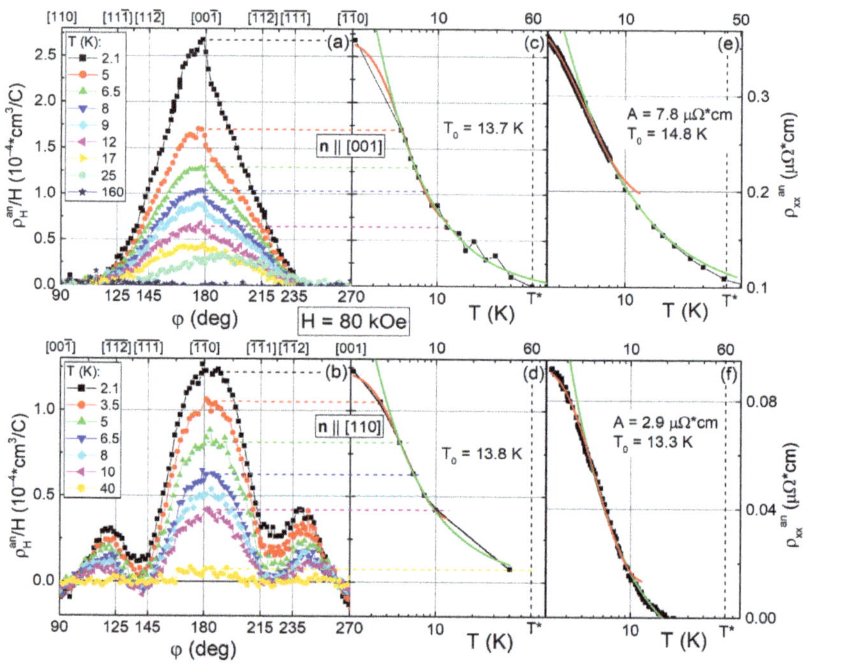

Figure 9. (**a**,**b**) Anisotropic positive contribution $\rho_H^{an}(\varphi)/H$ for samples with **n**| |[001] and **n**| |[110] at H = 80 kOe. (**c**,**d**) Temperature dependencies of ρ_H^{an}/H amplitudes for **n**| |[001] and **n**| |[110] in

the logarithmic scale. Panels (**e**,**f**) show the temperature dependencies of the anisotropic contribution to resistivity $\rho_{xx}{}^{an} = \rho(\mathbf{n}\,||\,[001], T_0, H = 80\ kOe)$-$\rho(\mathbf{n}\,||\,[111], T_0, H = 80\ kOe)$ and $\rho_{xx}{}^{an} = \rho(\mathbf{n}\,||\,[110], T_0, H = 80\ kOe) - \rho(\mathbf{n}\,||\,[111], T_0, H = 80\ kOe)$, respectively. Green and red solid lines on panels (**c**–**f**) show the approximation by Equations (3) and (4) (see text).

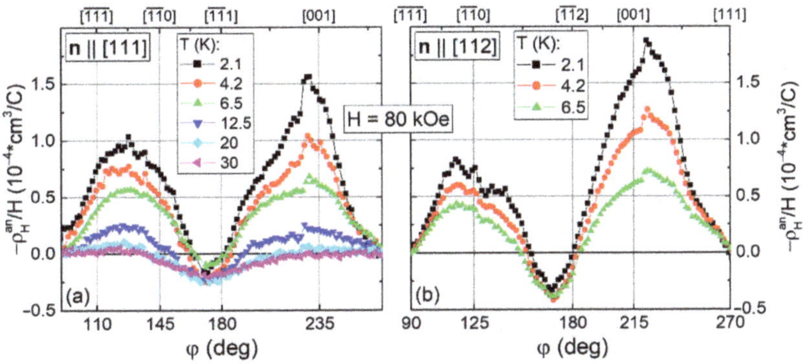

Figure 10. (**a**,**b**) Anisotropic contributions $\rho_H{}^{an}(\varphi)/H$ in magnetic field H = 80 kOe for the samples with $\mathbf{n}\,||\,[111]$ and $\mathbf{n}\,||\,[112]$, respectively.

The amplitude $\rho_H{}^{an}/H$ variation with temperature is shown in Figure 9c,d in the logarithmic plot. It is worth noting that two phenomenological relations

$$\rho_H{}^{an}/H \approx C^* \cdot (1/T - 1/T_E) \quad (3)$$

$$\rho_H{}^{an}/H \approx (\rho_H{}^{an}/H)_0 - A_H \cdot T^{-1} \cdot \exp(-T_a{}^H/T) \quad (4a)$$

were used in [49,55,61] for the Hall effect analysis in strongly correlated electron systems with a filamentary structure of conducting channels. Among these, a hyperbolic dependence (3) of Hall resistivity was observed previously in SCES $CeCu_{6-x}Au_x$ [55] and $Tm_{1-x}Yb_xB_{12}$ [61]. The authors of [55,61] argued in favor of a transverse even component of the Hall signal associated with the formation of stripes (see also [38]) on the surface and in the bulk of the crystal, similar to the chains of nanoscale stripes detected in the normal phase of HTSC [62]. Equation (4a) was applied to discuss the temperature induced destruction of the coherent state of stripes in LuB_{12} [49]. Below, we use the analysis based on Equations (3) and (4a) to highlight quantitatively the changes caused by various orientations of the external magnetic field.

Indeed, the approximation by Equation (3) in the range 5–40 K results in values $T_E \approx 132$ K and $C^* \approx 8.9 \cdot 10^{-4}$ cm^3/C for the sample with $\mathbf{n}\,||\,[001]$, while for $\mathbf{n}\,||\,[110]$, the parameter $T_E \approx 135$ K is almost the same and $C^* \approx 4.3 \cdot 10^{-4}$ cm^3/C turns out to be about half the size (see green curves in Figure 9c,d). Then, the analysis based on Equation (4a) provides very similar values $T_a{}^H \approx 13.7$–13.8 K for these two field directions (see Figure 9c,d, red curves). Note that the estimation of $T_a{}^H$ agrees both with the characteristic temperature of stripe chains formation $T_S \sim 15$ K [48] from one side, and with the activation energy $E_a/k_B = T_a \sim 14$–16 K detected above in the Arrhenius type approximation of the main contribution to the Hall effect (see Figure 8a) from the other. It can be seen from Figure 9c,d that for the case $\mathbf{n}\,||\,[001]$ and $\mathbf{n}\,||\,[110]$, Equation (4a) provides a good description of the experimental $\rho_H{}^{an}/H(T)$ curves at temperatures up to 10 K. Above 10 K, these fits (see red curves in Figure 9c,d) differ sharply from experiment, indicating the restriction of the phenomenological approach applied. The A_H coefficients in Equation (4a) differ for these two field directions by more than two times ($A_H = 73.8 \cdot 10^{-4}$ cm^3/C for $\mathbf{n}\,||\,[001]$ and $A_H = 32.3 \cdot 10^{-4}$ cm^3/C for $\mathbf{n}\,||\,[110]$), being in good agreement with the amplitude ratio for $\rho_H{}^{an}/H$ (Figure 9c,d). Furthermore, similar to the approach developed in [43] for LuB_{12}, an analysis of the anisotropic positive contribution to magnetoresistance in Figure 9e,f for $Ho_{0.8}Lu_{0.2}B_{12}$ is carried out within the relation

$$\rho_{xx}{}^{an}(\mathbf{n},T_0,H = 80\ kOe) \approx (\rho_{xx}{}^{an})_0 - A_{xx}\cdot T^{-1}\cdot \exp(-T_a{}^{\rho}/T) \quad (4b)$$

For each of the samples with **H** | | **n** | | [001] and **H** | | **n** | | [110], the anisotropic component $\rho_{xx}{}^{an}(\mathbf{n},T_0,H = 80\ kOe)$ was determined by subtracting from the experimental resistivity data (e.g, $\rho(\mathbf{n} || [001],T_0,H = 80\ kOe)$ for **H** | | **n** | | [001]) and the dependence $\rho(\mathbf{n} || [111],T_0,H = 80\ kOe)$ for **H** | | **n** | | [111], where the magnetoresistance is minimal. Parameters $T_a{}^{\rho}$ = 13.3–14.8 K found from this approximation in the same range $T \leq 10$ K turn out to be close to $T_a{}^H \approx$ 13.3–13.8 K values and also to T_S~15 K [48], as well as to the activation temperature T_a~14–16 K (see Figure 8a).

3. Discussion

3.1. Multicomponent Analysis of the Contributions to Anomalous Hal Effect (AHE) in the Regime of Ferromagnetic Fluctuations

Previous measurements of the Hall effect in $Ho_{0.5}Lu_{0.5}B_{12}$ in P-phase ($T > T_N \approx 3.5$ K) were carried out in the conventional field-sweep scheme with two opposite field directions ±**H** | | **n** [63]. Ordinary and anomalous components of the Hall effect observed in [63] were described by the general relation

$$\rho_H/B = R_{H0} + R_S\cdot 4\pi M/B, \quad (5)$$

which is usually applied to AHE in ferromagnetic metals [51,64], where R_{H0} and R_S = const(T) are the ordinary and anomalous Hall coefficients, respectively. According to [51], the ferromagnetic AHE regime represented by Equation (5) corresponds to the intrinsic scattering mechanism. Since short-range order effects are observed in the paramagnetic phase of magnetic RB_{12} in the temperature range at least up to $3T_N$ [53,60,65] (see also Figure 4), and that a ferromagnetic component was detected in the magnetically ordered state of HoB_{12} in magnetic fields above 20 kOe, it is of interest to perform the analysis within the framework of Equation (5) of the ordinary and AHE components in the vicinity of T_N. In this case, relying on the above results of angular measurements of $Ho_{0.8}Lu_{0.2}B_{12}$ (Figures 5–7, 9 and 10), one should use isotropic $\rho_{H0}(T, H, n)$ and anisotropic $\rho_H{}^{an}(\varphi, T, H, n)$ components of the Hall signal separated within the framework of Equation (2) (see Figures 2–10). It is worth noting that the analysis performed in [43] within the framework of Equation (5) is applied to the field dependences of Hall resistivity $\rho_H(H)$ obtained in the conventional field-sweep ±**H** | | **n** scheme, leading obviously to mixing of contributions $\rho_{H0}(H)$ and $\rho_H{}^{an}(H)$. As a result, the coefficients R_{H0} and R_S were determined in [43] for the total averaged Hall resistivity, which contains both the isotropic $\rho_{H0}(T, H, n)$ and anisotropic $\rho_H{}^{an}(\varphi, T, H, n)$ components. Actually, each of the angular contributions ρ_{H0} and $\rho_H{}^{an}(\varphi)$ is characterized by two independent coefficients R_0 and R_S, which generally differ in sign. Below, we develop the analysis, considering two ferromagnetic components included in the AHE. To keep generality, we analyze our Hall effect data for all three principal directions of external magnetic field (**H** | | [001], **H** | | [110], and **H** | | [111]), despite the fact that for **H** | | [111], the intrinsic AHE is found to be practically negligible (see Figures 7 and 10).

Thus, for a full description of the Hall effect in $Ho_{0.8}Lu_{0.2}B_{12}$, we used the following relations:

$$\begin{cases} \rho_H^0/B = R_H^0 + R_M^0 * 4\pi M/B \\ \rho_H^{an}/B = R_H^{an} + R_M^{an} * 4\pi M/B \end{cases} \quad (6)$$

where $\rho_H{}^0/B = \rho_{H0}/B(H, T_0, n)$ and $\rho_H{}^{an}/B = \rho_H{}^{an}/B(H, T_0, n)$ are the reduced amplitudes of contributions ρ_{H0} and $\rho_H{}^{an}(\varphi)$ (see Figure 7), which depend on temperature and direction of the normal vector **n** to the sample surface; $R_H{}^0$ and $R_H{}^{an}$ are components of the ordinary Hall effect, connected with magnetic induction B, and $R_M{}^0$, $R_M{}^{an}$ are the coefficients of anomalous Hall effect determined by magnetization M and related to the ferromagnetic component (see Figure 3d–f). Obviously, there are two independent ordinary contributions to Hall resistivity $R_H{}^0 \cdot B$ and $R_H{}^{an} \cdot B$, as well as two independent anomalous (ferromagnetic) components $R_M{}^0 \cdot 4\pi M$ and $R_M{}^{an} \cdot 4\pi M$, which differ for various **H** directions. Note that the last two components $R_H{}^{an} \cdot B$ and $R_M{}^{an} \cdot 4\pi M$ are responsible for the observed Hall effect anisotropy.

Figure 11 shows the linear approximation within the framework of Equation (6) of the reduced amplitudes ρ_{H0}/B and $\rho_H{}^{an}/B$ *vs* $4\pi M/B$ in the range T_0 = 2.1–6.5 K for $Ho_{0.8}Lu_{0.2}B_{12}$ crystals with **n** | | [001], **n** | | [110], and **n** | | [111]. Linear fits are acceptable for directions **n** | | [001] and **n** | | [110] at temperatures of 2.1 K and 4.2 K for both the isotropic ρ_{H0}/B and anisotropic $\rho_H{}^{an}/B$ contributions. This approximation was found to be valid at T = 6.5 K in the interval of small $4\pi M/B$ values (i.e., in high magnetic fields, in more detail see Figure S4 in Supplementary Materials), where the estimated parameters $R_H{}^0$ and $R_H{}^{an}$ are extracted as cutoffs, and $R_M{}^0$ and $R_M{}^{an}$ are the slopes of corresponding straight lines in Figure 11. Since the parameters $R_H{}^0$, $R_H{}^{an}$, $R_M{}^0$, $R_M{}^{an}$ depend weakly on temperature (see Figure S4 in Supplementary Materials), the temperature-averaged values of these coefficients are summarized in Table 1.

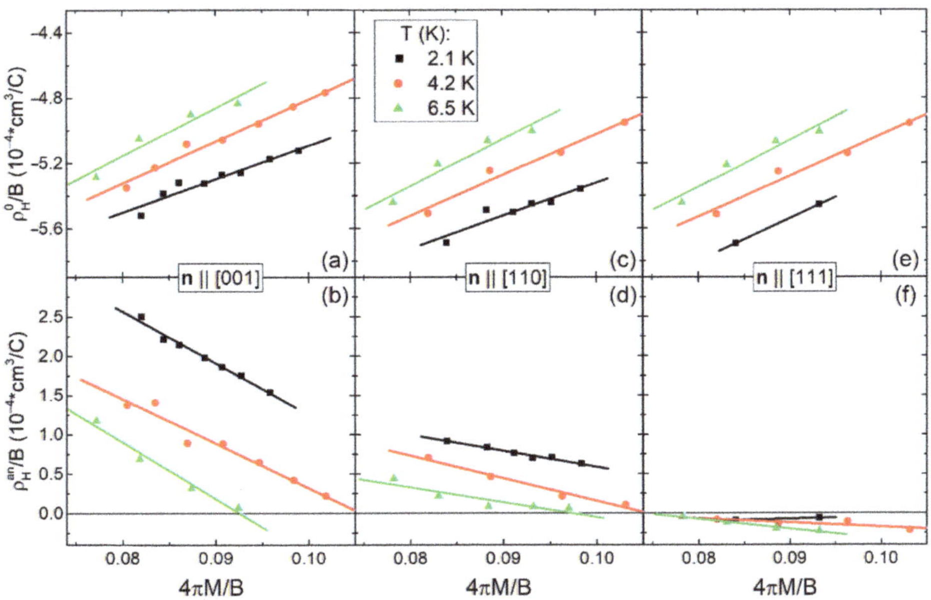

Figure 11. Linear approximation of the isotropic ρ_{H0}/B (**a**,**c**,**e**) and anisotropic $\rho_H{}^{an}/B$ (**b**,**d**,**f**) contributions vs $4\pi M/B$ within the Equation (6) approximation at 2.1–7 K for three principal directions **H** | | [001], **H** | | [110], and **H** | | [111] (see text).

Table 1. Parameters $R_H{}^0$, $R_H{}^{an}$ of the ordinary and anomalous $R_M{}^0$, $R_M{}^{an}$ contributions to Hall effect in $Ho_{0.8}Lu_{0.2}B_{12}$ averaged over temperatures 2.1–7 K.

R_H, $10^{-4} \times cm^3/C$	**H** | | **n** | | [001]	**H** | | **n** | | [110]	**H** | | **n** | | [111]
$R_H{}^0$	−7.3	−7.7	−7.7
$R_H{}^{an}$	6.8	2.3	0
$R_M{}^0$	25.2	26.9	26.5
$R_M{}^{an}$	−64.9	−20.9	0

The analysis based on Equation (6) allows us to conclude that the values of coefficients $R_H{}^{an}$ and $R_M{}^{an}$, which are characteristics of the anisotropic component, turn out to be practically equal to zero for **H** | | [111] (Figure 11f). It can be seen that the Hall coefficient $R_H^0 \sim -7.5 \cdot 10^{-4}$ cm^3/C remains practically invariant for any **H** direction (see Table 1 and Figure S4 in Supplementary Materials), confirming that this ordinary negative component of Hall signal is isotropic. On the contrary, the value of anisotropic positive contribution R_H^{an} changes significantly from $\sim 6.8 \cdot 10^{-4}$ cm^3/C for **H** | | [001] to $2.3 \cdot 10^{-4}$ cm^3/C in **H** | | [110] passing through zero for **H** | | [111] (Table 1). As a result, in $Ho_{0.8}Lu_{0.2}B_{12}$ for strong field **H** | | [001], the anisotropic positive component R_H^{an} B·g(φ,n), which is proportional to magnetic induction, turns out to be comparable in absolute value with the negative isotropic component R_{H0} B of the ordinary Hall effect. Note that the values of coefficients R_M^0, R_M^{an} of the anomalous (ferromagnetic) contributions, which are proportional to magnetization, dramatically exceed the ordinary parameters R_H^0, R_H^{an} that agrees with the result [63] for $Ho_{0.5}Lu_{0.5}B_{12}$.

It is very unusual that a significant isotropic anomalous positive contribution R_M^0 ~25:27·10^{-4} cm^3/C appears in the paramagnetic phase of $Ho_{0.8}Lu_{0.2}B_{12}$, and it may be attributed to the isotropic ferromagnetic component in the Hall signal. We propose that the R_M^0 term depends on the regime of ferromagnetic fluctuations detected in low field magnetic susceptibility above T_N (Figure 4). On the contrary, the anomalous negative contribution R_M^{an} varies strongly in the range (−)(21:65)·10^{-4} cm^3/C depending on **M** direction (see Table 1), and this component appears in strong magnetic field and at low temperatures (Figure 11). To summarize, AHE in the paramagnetic state of $Ho_{0.8}Lu_{0.2}B_{12}$ is proportional to magnetization and is determined both by the positive contribution R_M^0 $\cdot 4\pi M \cdot \cos(\varphi)$ and by the strongly anisotropic negative component $R_M^{an} \cdot 4\pi M \cdot g(\varphi)$. These two contributions compensate each other in the vicinity of **n** | | [110] (dynamic charge stripe direction [35,48]).

When discussing the nature of multicomponent Hall effect in $Ho_{0.8}Lu_{0.2}B_{12}$, it is worth noting the complicated multi-***q*** incommensurate magnetic structure in the Neel state. Magnetic ordering is characterized by propagation vector ***q*** = (1/2±δ, 1/2±δ, 1/2±δ) with δ = 0.035 and detected in [53,66,67] for $Ho^{11}B_{12}$ in the neutron diffraction experiments at low temperatures in low (H < 20 kOe) magnetic field. It was also found for HoB_{12} [53,66,67] that as the strength of external magnetic field increases above 20 kOe, the 4***q***-magnetic structure transforms into a more complex one, in which, apart from the coexistence of two AF 4***q*** and 2***q*** components, there additionally arises some ferromagnetic order parameter. Then, a strong modulation of the diffuse neutron-scattering patterns was observed in HoB_{12} well above T_N [53,67] with broad peaks at positions of former magnetic reflections, e.g., at (3/2, 3/2, 3/2), pointing to strong correlations between the magnetic moments of Ho^{3+} ions. These diffuse scattering patterns in the paramagnetic state were explained in [53,67] by the appearance of correlated 1D spin chains (short chains of Ho^{3+} -ion moments placed on space diagonals <111> of the elementary unit), similar to those detected in low dimensional magnets [68]. It was found that these patterns can be resolved both well above (up to 70 K) and below T_N, where the 1D chains seem to condense into an ordered antiferromagnetic modulated (AFM) structure [53,67,69]. The authors [53] discussed the following scenario for the occurrence of long-range order in HoB_{12}: Far above T_N, strong interactions lead to correlations along [111], they are essentially one-dimensional and would not lead to long-range order at finite temperature. As T_N is approached, the 1D-correlated regions grow in the perpendicular directions, possibly due to other interactions. Cigar-shaped AFM-correlated regions were proposed in [53] that become more spherical when T_N is approached. Within this picture, the ordering temperature is located in the point where spherical symmetry is reached. Only then 3D behavior sets in, and HoB_{12} exhibits long-range AFM order [53]. The refinement of $Ho^{11}B_{12}$ crystal structure was done with high accuracy in the space group $Fm\bar{3}m$, but also small static Jahn–Teller distortions were found in RB_{12} compounds [36,41]. However, the most important factor of symmetry breaking is the dynamic one [36,41], which includes the formation both of vibrationally coupled Ho–Ho dimers and dynamic charge stripes (see [36,48,52] for more details). As a result, twofold symmetry in the (110) plane is conserved as expected for cubic crystal, but the charge stripes and Ho–Ho-coupled vibrations suppress the exchange between nearest neighbored Ho-ions. This result in the emergence of complicated phase diagrams in the AF state with a number of different magnetic phases separated by radial and circular boundaries (Maltese Cross type of angular diagrams in RB_{12} [35,42,44,46]). In this scenario AF magnetic fluctuations develop well above T_N in HoB_{12} along trigonal axis [111], and dynamic charge stripes along <110> suppress dramatically the RKKY indirect exchange between Ho magnetic moments [44] provoking the formation of cigar-shaped AFM-correlated regions proposed in [53]. In our opinion, these effects are responsible both for the emergence of filamentary structures of fluctuating charges in these nonequilibrium metals and for the formation of spin polarization in the conduction band. This also results in the appearance of a complicated multicomponent Hall effect including two (isotropic and anisotropic) anomalous contributions.

3.2. Mechanisms of AHE in $Ho_{0.8}Lu_{0.2}B_{12}$

Returning to commonly used classification [51,70], it is necessary to distinguish between the intrinsic and extrinsic AHE. Intrinsic AHE is related to the transverse velocity addition due to Berry phase contribution in systems with strong spin–orbit interaction (SOI), while the extrinsic AHE associated with scattering of charge carriers by impurity centers. However, AHE also arises in noncollinear ferromagnets, in which a nonzero scalar chirality $S_i(S_j \times S_k) \neq 0$ leads to the appearance of an effective magnetic field even in the absence of SOI [71], and in magnetic metals with a nontrivial topology of spin structures in real space [72–76]. When interpreting experimental data, a problem of identifying the actual mechanisms of AHE arises [51]. Among the extrinsic AHE, skew scattering, for which the scattering angularly depends on the mutual orientation of the charge carrier spin and the magnetic moment of the impurity, predicts a linear relationship between the anomalous component of the resistivity tensor $\rho^{an}{}_H \sim \rho_{xx}$ and usually corresponds to the case of pure metals (ρ_{xx} < 1 μOhm·cm) [77]. In the range of resistances ρ_{xx} = 1–100 μOhm·cm the intrinsic AHE dominates, which is due to the effect of the Berry phase ($\rho^{an}{}_H \sim \rho_{xx}{}^2$) [78]. The contribution to scattering due to another extrinsic AHE, side jumping [79] with a similar scaling ($\rho^{an}{}_H \sim \rho^2{}_{xx0}$, where ρ_{xx0} is the residual resistivity of the metal) is usually neglected [51]. In the "dirty limit" (ρ_{xx} > 100 μOhm*cm), an intermediate behavior is observed with a dependence of the form $\rho^{an}{}_H \sim \rho_{xx}{}^{\beta}$ and with an exponent β = 1.6–1.8, which is associated usually with the transition to hopping conductivity [51].

When identifying the AHE mechanism in $Ho_{0.8}Lu_{0.2}B_{12}$ with a small residual resistivity (ρ_{xx0}~1 μOhm·cm, Figure 2a), it is not possible to follow the traditional classification as no presence of the itinerant AHE with an asymptotic $\rho^{an}{}_{xy} \sim \rho_{xx}{}^2$, or a skew scattering regime ($\rho^{an}{}_H \sim \rho_{xx}$) was found here. However, in order to correctly compare these diagonal and off-diagonal components of the resistivity tensor, it is possible to extract the corresponding anomalous contributions. As can be seen from Figure 2, for **H** | | **n** | | [111], the anisotropic anomalous component of Hall signal is negligible and, as a result, the reduced Hall resistivity measured in angular experiments in direction **H** | | **n** consists of the isotropic contribution only. In addition, according to conclusions made in [43,47], at low temperatures in paramagnetic state the magnetoresistance of $Ho_{0.8}Lu_{0.2}B_{12}$ consists of isotropic negative and anisotropic positive contributions, the latter being close to zero in the direction **H** | | **n** | | [111].

In this situation, for estimating the anisotropic components $\rho^{an}{}_H$ and $\rho^{an}{}_{xx}$, e.g., for the **n** | | [001] sample, it suffices to find the difference $\rho^{an}{}_H$(**n** | | [001]) = ρ_H(**n** | | [001])-ρ_H(**n** | | [111]) (see Figure 2b) and $\rho^{an}{}_{xx}$(**n** | | [001]) = ρ(**n** | | [001])-ρ(**n** | | [111]) (see Figure 2a). Figure 12 demonstrates the scaling relation between these anisotropic components of $\rho^{an}{}_H$ and $\rho^{an}{}_{xx}$ for **H** | | [001] and **H** | | [110] directions, which leads to the following conclusions: (i) For **H** | | [001] a $\rho^{an}{}_H \sim \rho^{an}{}_{xx}{}^{1.7}$ dependence is observed over the entire temperature range $T \leq T^*$~60 K. This regime does not correspond to intrinsic AHE (β < 2), while an onset of hopping conductivity (β = 1.6–1.8) [51] seems to be an unreal scenario in this good metal (ρ_{xx}~1 μOhm·cm, Figure 2a). (ii) On the contrary, for **H** | | [110], two anisotropic components of the resistivity tensor appear in the interval $T < T_S \sim 15$ K and turn out to be related to each other as $\rho^{an}{}_H \sim \rho^{an}{}_{xx}{}^{0.83}$ (Figure 12), which does not favor skew scattering ($\beta \sim 1$) [77]. Note that the exponent β for **H** | | [110] is twice as small as that for **H** | | [100] in $Ho_{0.8}Lu_{0.2}B_{12}$, and these regimes are observed in adjacent $\rho^{an}{}_{xx}$ intervals changing one to another at $\rho^{an}{}_{xx} \sim 0.1$ μOhm·cm (Figure 12). Such a different behavior in charge transport parameters for two different magnetic field directions suggests that the AHE is caused by another scattering mechanism, which, in particular, may result from the influence of external magnetic field on dynamic charge stripes directed along <110> (see Figure 1a).

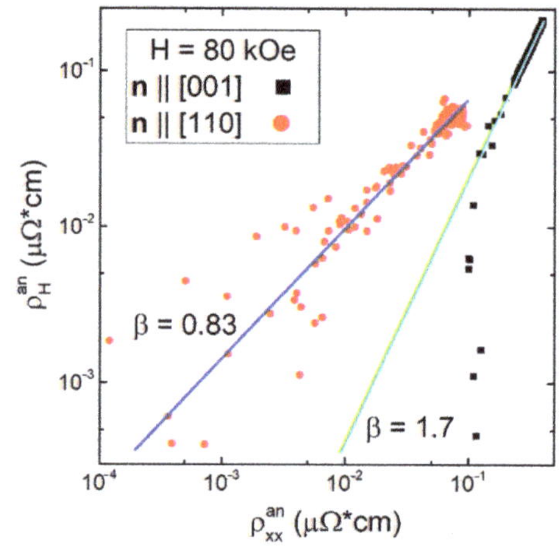

Figure 12. Anisotropic AHE components for directions **H** | | [001] and **H** | | [110] in magnetic field H = 80 kOe scaled in double logarithmic plot. Solid lines display the linear approximations and β denotes the exponent in $\rho^{an}{}_H \sim \rho^{an}{}_{xx}{}^{\beta}$.

In this scenario the appearance of two types of AHE in $Ho_{0.8}Lu_{0.2}B_{12}$ may be interpreted as follows. The first mode of AHE associated with charge scattering in the interval $T < T_S \sim 15$ K is detected when magnetic field is applied along charge stripes (**H** | | [110]), and the regime appears due to the formation of a large size cluster (long chains) in the filamentary structure of fluctuating charges. The second mode of AHE is induced by the order–disorder transition at $T^* \sim 60$ K and corresponds to the magnetic field applied transverse to vibrationally coupled dimers of rare-earth ions (**H** | | [001] ⊥ <110>). In the latter case, when the carrier moves in transverse magnetic field along a complex path, the intrinsic AHE is expected to be influenced by the Berry phase in real space [51,78], but instead, the $\rho^{an}{}_H \sim \rho^{an}{}_{xx}{}^{1.7}$ scaling is observed. This unusual behavior seems to be a challenge to the contemporary AHE theory and has to be clarified in future studies.

3.3. AHE Anisotropy and Dynamic Charge Stripes

The above analysis of Hall effect contributions in $Ho_{0.8}Lu_{0.2}B_{12}$, based (i) on measurements in the conventional field-sweep ±**H** | | **n** scheme (Figures 2, 3a–c and S1a in Supplementary Materials), and (ii) on studies of angular dependences with vector **H** rotating in the (110) plane (Figure 5, Figure 6, Figures S1b, S2 and S3 in Supplementary Materials), allows to obtain a set of AHE coefficients, which characterize the ordinary and anomalous contributions along three principal directions of the magnetic field (**H** | | **n** | | [001], **H** | | **n** | | [110], and **H** | | **n** | | [111]).

In this case, the methodological feature of the performed angular measurements of Hall resistivity shows a cosine modulation of the projection of transverse Hall electric field on the direction that connects two Hall probes and is perpendicular to any of the specific normal vectors: **n** | | [001], **n** | | [110] **n** | | [111], or **n** | | [211]. In this situation, it is more convenient to control the projection of the external field **H** onto the normal vector

$H_n = (\mathbf{H}\cdot\mathbf{n}) = H_0\cdot\cos\varphi$, which is used to determine the amplitude of contributions in the corresponding **n** direction (see inset in Figure 2a). As can be seen from Figure 10, vanishing of the AHE for **H** directed precisely along **n** | | [111] and **n** | | [112] does not mean zero values of $\rho_H^{an}(\varphi)/H$ for these crystals in the entire range of angles. In this case, it is obvious that for **H** in the plane of the sample, near-zero values of $H_n = H_0\cdot\cos$ occur.

For $\varphi = 0$, one should also expect zero values of anomalous contributions to the Hall signal.

In [44,47–49,58], it was found that the angular dependence of magnetoresistance (MR) in RB_{12} is determined by scattering of carriers on dynamic charge stripes. As a result, the maximum positive values of MR are observed for **H** | | [001] perpendicular to the direction of these electron density fluctuations, while the minimum MR is observed for **H** | | [111] (see Figure 13b). To clarify the nature of these anomalies in angular AHE curves, one can restore the angular dependence of the AHE in the entire range of 0–360° and compare the obtained curve with the related MR data. This analysis can be performed by relying on experimental $\rho_H/B(\varphi)$ curves measured at T = 2.1 K in magnetic field H = 80 kOe for four different crystals when **H** is rotated in the same plane (110). Since both the ordinary and anomalous components of Hall signal can be described by cosine dependence $\rho_H(\varphi) = \rho_{H0}\cdot\cos(\varphi - \varphi_1) + \rho_H^{an}\cdot g(\varphi)$, the representation of experimental data shown in Figure 13a in the form of $\rho_H(\varphi)/(H\cdot\cos(\varphi - \varphi_1))$ allows us to separate the isotropic and anisotropic contributions from Hall experiments.

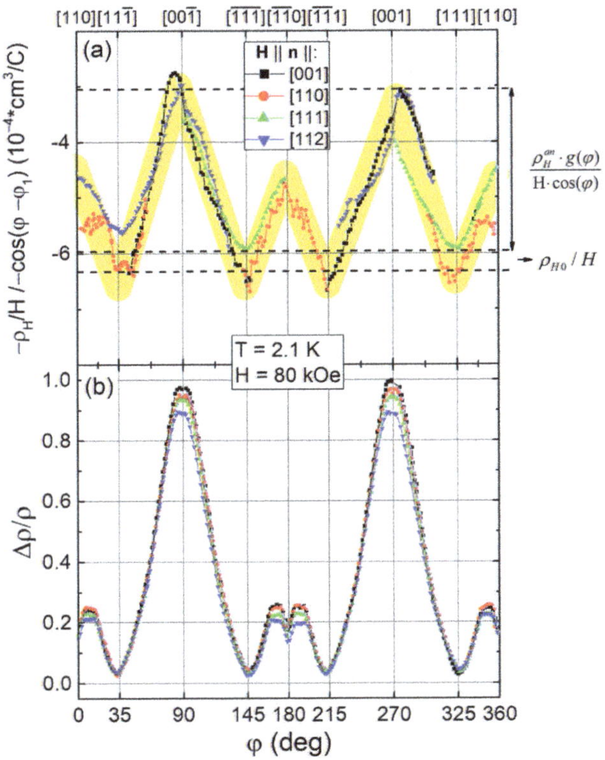

Figure 13. (**a**) Normalized angular Hall resistivity $\rho_H(\varphi)/(H\cdot\cos(\varphi - \varphi_1))$ and (**b**) magnetoresistance curves in field 80 kOe at temperature 2.1 K for four samples with **n** | | [001], **n** | | [110], **n** | | [111], and **n** | | [112]. Yellow shading indicates the common envelope for all four Hall effect measurements (see text).

The averaged envelope (indicated by yellow shading in Figure 13a) was obtained after removing the particular portions of the related angular dependences with singularities associated with division by small values (zeros of cosine), and then averaging the data of these four angular Hall signal dependences. In this case, in accordance with the data in Figure 2b, in **H** | | <111> directions on the resulting envelope curve $\rho_H(\varphi)/(H\cdot\cos(\varphi - \varphi_1))$, the maximum negative values of about -6×10^{-4} cm^3/C correspond to isotropic ordinary component ρ_{H0}/H of Hall effect, which is independent on magnetic field direction. The positive anisotropic component reconstructed from the data of four measurements (the yellow shading in Figure 13a) provide changes of the $\rho_H(\varphi)/(H\cdot\cos(\varphi - \varphi_1))$ in the range (3.2:6)$\cdot10^{-4}$ cm^3/C. Despite the fact that the initial $\rho_H{}^{an}(\varphi)$ dependence is an odd function (see Figures 5, 6, 9 and 10), the result of its division by the odd $\cos(\varphi - \varphi1)$ allows one to obtain the real anisotropic **even** amplitude of Hall effect and compare it with MR. The location of its extrema coincides with the positions of anomalies on the MR curve (Figure 13b). Indeed, the maximum positive contribution to AHE appears synchronously with the MR peak along <001>, while for <110>, a small (if compared with the anomaly along <001>) positive AHE component is recorded (Figure 13b) simultaneously with a small amplitude singularity of MR. We also note that two spatial diagonals <111> on the anisotropic contribution $\rho_H{}^{an}g(\varphi)/(H\cdot\cos(\varphi - \varphi_1))$ seem to be equivalent and show no hysteretic features. The observed behavior of the Hall effect in $Ho_{0.8}Lu_{0.2}B_{12}$ agrees very well with symmetry lowering of the *fcc* structure of $Ho_{0.8}Lu_{0.2}B_{12}$ due to static and dynamic Jahn–Teller distortions [36].

Finally, the comparison of the angular dependences of MR and Hall effect in $Ho_{0.8}Lu_{0.2}B_{12}$ (Figure 13) shows that, along with the normal isotropic contributions to the diagonal (negative MR) and off-diagonal (the ordinary Hall coefficient of negative sign) components of the resistivity tensor, anomalous anisotropic positive components appear both in the MR and in Hall effect at low temperatures. These components reach (i) maximal values in the direction of magnetic field transverse to dynamic charge stripes (**H** | | [001]) and (ii) zero values for **H** | | [111]. This anisotropy arises simultaneously with the transition to the cage glass state at T*~60 K and seems to be related to the formation of vibrationally coupled pairs of rare-earth ions displaced from their centrosymmetric positions in B_{24} cavities of the boron sublattice [48]. A significant increase of this anisotropy is detected at temperatures $T < T_S$~15 K upon the formation of large size clusters (long chains) in the filamentary structure of fluctuating electron density (stripes). Taking into account that, according to the results of room temperature measurements of the dynamic conductivity of LuB_{12}, about 70% of charge carriers participate in the formation of the collective mode (hot electrons) [80], a redistribution of carriers between the nonequilibrium and Drude components should be expected with decreasing temperature.

Apparently, the activation behavior of the Hall concentration of charge carriers observed in the range 60–300 K (T_a~14–17 K, Figure 8a) may be attributed to the involvement of additional conduction electrons in the collective mode. When vibrationally coupled dimers of rare-earth ions are formed below T*~60 K, short and disordered chains of stripes oriented along <110> appear in magnetic field, initiating the emergence of intrinsic AHE (Figure 12). We propose that the AHE in $Ho_{0.8}Lu_{0.2}B_{12}$ is caused by a transverse addition to velocity due to the Berry phase contribution [51], which arises for carriers moving in a complex filamentary structure of the electron density in magnetic field applied transverse to dynamic stripes. During the formation of large size clusters in the structure of stripes (interval $T < T_S$~15 K) in field orientation along the stripes **H** | | [110], no intrinsic AHE is expected. A skew scattering contribution, for which the scattering angle depends on the mutual orientation of the charge carriers spin and the magnetic moment of the impurity, may become noticeable with a linear relationship $\rho^{an}{}_H \sim \rho^{an}{}_{xx}$ between these components of the resistivity tensor (Figure 12). We propose that some geometric factors are responsible for the reduction of β exponent in these two AHE regimes. Approaching the AF transition above T_N, on-site 4*f*-5*d* spin fluctuations in the vicinity of Ho^{3+} ions lead to magnetic polarization of the 5*d* states of the conduction band, which gives rise to ferromagnetic fluctuations in $Ho_{0.8}Lu_{0.2}B_{12}$ (Figure 4). These produce ferromagnetic nanoscale domains (ferrons), and, as a result, initiate the appearance of the ferromagnetic contribution to AHE (Figure 11 and Table 1). We emphasize that such a complex multicomponent AHE in model $Ho_{0.8}Lu_{0.2}B_{12}$ metal with a simple *fcc* lattice turns out to be due to the inhomogeneity and complex filamentary structures of the electron density in the matrix of this SCES with dynamic Jahn–Teller lattice instability and electron phase separation.

4. Experimental Details

$Ho_{0.8}Lu_{0.2}B_{12}$ single-domain crystals were grown by crucible-free induction zone melting in an inert argon gas atmosphere (see, e.g., [34]). Magnetization was measured with the help of a SQUID magnetometer MPMS Quantum Design in fields up to 70 kOe in the temperature range 1.9–10 K. The external magnetic field was applied along the principal crystallographic axes- **H** | | [001], **H** | | [110], and **H** | | [111]. Resistivity, magnetoresistance, and Hall effect were studied on an original setup using the standard DC five-probe technique with excitation current commutation. The angular dependences of transverse magnetoresistance and Hall resistivity were obtained using a sample holder of original design, which enables the rotation of the vector **H** located in the plane perpendicular to fixed current direction **I** | | $[1\bar{1}0]\perp$**H** with a minimum step $\varphi_{step} = 0.4°$ (see the schematic view on the inset of Figure 2a). Measurements were carried out in a wide temperature range 1.9–300 K in magnetic fields up to 80 kOe, the angle φ = **n**^**H** (**n**-normal vector to the lateral sample surface) varied in the range φ = 0–360°. The measuring setup was equipped with a stepper motor, which enabled an automatic control of sample rotation, similar to that one used in [35]. High accuracy of temperature control ($\Delta T \approx 0.002$ K in the range 1.9–7 K) and magnetic field stabilization ($\Delta H \approx 2$ Oe) was ensured, respectively, by LLC Cryotel, (Moscow, Russia) TC 1.5/300 temperature controller and Cryotel SMPS 100 superconducting magnet power supply in combination with a CERNOX 1050 thermometer and n-InSb Hall sensors.

5. Conclusions

In the paramagnetic phase of the cage-cluster high-boron boride $Ho_{0.8}Lu_{0.2}B_{12}$ with a cage-glass state below T*~60 K and electronic phase separation (dynamic charge stripes), magnetotransport was studied at temperatures 1.9–300 K in magnetic fields up to 80 kOe. Field and angular measurements of resistivity and Hall resistivity were performed on single-domain crystals of the $Ho_{0.8}Lu_{0.2}B_{12}$ model metal allowed to separate and analyze several different contributions to the Hall effect. It was shown that, along with the negative ordinary isotropic component of Hall resistivity, an intrinsic AHE of a positive sign arises in the cage-glass state in field direction **H** | | [001], which is perpendicular to the charge stripe chains. This AHE corresponds through the relation $\rho^{an}{}_{H} \sim \rho^{an}{}_{xx}{}^{1.7}$ to the anomalous components of the resistivity tensor. It was also found that at temperatures $T < T_S \sim 15$ K, where long chains prevail in the filamentary structure of fluctuating charges (stripes), a contribution to AHE of the form $\rho^{an}{}_{H} \sim \rho^{an}{}_{xx}{}^{0.83}$ becomes dominant when **H** | | [110]. We propose that these two components are intrinsic (a transverse addition to velocity due to the contribution of the Berry phase) and extrinsic (from the skew scattering mechanism) [51], respectively, and exhibit some decrease of exponents from integers due to the geometric factor. In the paramagnetic phase near Neel temperature, on-site 4*f*-5*d* spin fluctuations in the vicinity of Ho^{3+} ions were found to induce spin-polarized 5*d* states (ferromagnetic nanoscale domains–ferrons) in the conduction band, which result in the appearance of an additional ferromagnetic contribution to AHE, as observed both in the isotropic and anisotropic components of Hall effect. Detailed measurements of the angular dependences of Hall resistivity and MR with vector **H** rotation in the (110) plane, perpendicular to the direction of stripes, made it possible to separate the negative isotropic and positive anisotropic contributions to AHE and MR, and to explain them in terms of charge carriers scattering by dynamic charge stripes.

Supplementary Materials: The following supporting information can be downloaded at: https://www.mdpi.com/article/10.3390/molecules28020676/s1, Figure S1: The direct match between (a) the data obtained in the conventional field-sweep scheme of Hall effect measurements with two opposite directions of ±H| | n and (b) the data extracted in the experiment with step-by-step rotation of the sample around I| | [1–10] with a fixed H direction in the transverse plane; Figures S2 and S3: The approximation of the angular dependences of Hall resistivity ρH(φ) for H = 80 kOe in the temperature range 2.1–30 K for n| | [111] and n| | [112] by Equation (2); Figure S4: Coefficients RH0, RHan of the ordinary and RM0, RMan of the anomalous Hall effect (Equation (6)) depending on temperature for three vectors n| | [001], n| | [110], and n| | [111]; Figure S5: The X-ray powder analysis; Figure S6: Typical X-ray Laue back reflection patterns; Figures S7–S9: Verification of the single-domain crystals by the rocking curve control; Figure S10: The results of the microprobe analysis of the $Ho_{0.8}Lu_{0.2}B_{12}$ single crystals; Table S1: The corrections due to demagnetizing factors and values of the demagnetization factor, depending on the type of experiment; Table S2: The results of the microprobe analysis of the $Ho_{0.8}Lu_{0.2}B_{12}$ single crystals.

Author Contributions: Conceptualization, N.E.S. and V.V.G.; software, A.V.B.; validation, V.V.V.; investigation, A.L.K., K.M.K., A.N.A. and V.N.K.; resources, N.Y.S. and V.B.F.; writing—original draft preparation, A.L.K.; writing—review and editing, N.Y.S.; visualization, A.L.K. and K.M.K.; supervision, S.G. and K.F. All authors have read and agreed to the published version of the manuscript.

Funding: The work in Prokhorov General Physics Institute of RAS was supported by the Russian Science Foundation, Project No. 22-22-00243, and partly performed using the equipment of the Institute of Experimental Physics, Slovak Academy of Sciences. K.F. and S.G. were supported by the Slovak Research and Development Agency under the contract No. APVV-17-0020, and by the projects VEGA 2/0032/20, and VA SR ITMS2014 + 313011W856.

Institutional Review Board Statement: Not applicable.

Informed Consent Statement: Not applicable.

Data Availability Statement: The data presented in this study are available on request from the corresponding author.

Conflicts of Interest: The authors declare no conflict of interest.

References

1. Loudon, J.C.; Mathur, N.D.; Midgley, P.A. Charge-ordered ferromagnetic phase in $La_{0.5}Ca_{0.5}MnO_3$. *Nature* **2002**, *420*, 797. [CrossRef] [PubMed]
2. Moreo, A.; Yunoki, S.; Dagotto, E. Phase separation scenario for manganese oxides and related materials. *Science* **1999**, *283*, 2034. [CrossRef] [PubMed]
3. Salamon, M.B.; Jaime, M. Phase separation scenario for manganese oxides and related materials. *Rev. Mod. Phys.* **2001**, *73*, 583. [CrossRef]
4. Bebenin, N.G.; Loshkareva, N.N.; Makhnev, A.A.; Mostovshchikova, E.V.; Nomerovannaya, L.V.; Gan'Shina, E.A.; Vinogradov, A.N.; Mukovskii, Y.M. Optical and magneto-optical properties of ferromagnetic $La_{1-x}Ba_xMnO_3$ single crystals. *J. Phys. Condens. Matter* **2010**, *22*, 096003. [CrossRef] [PubMed]
5. Rho, H.; Snow, C.S.; Cooper, S.L.; Fisk, Z.; Comment, A.; Ansermet, J.P. Evolution of magnetic polarons and spin-carrier interactions through the metal-insulator transition in $Eu_{1-x}Gd_xO$. *Phys. Rev. Lett.* **2002**, *88*, 4. [CrossRef]
6. Hayden, S.M.; Lander, G.H.; Zarestky, J.; Brown, P.J.; Stassis, C.; Metcalf, P.; Honig, J.M. Incommensurate magnetic correlations in $La_{1.8}Sr_{0.2}NiO_4$. *Phys. Rev. Lett.* **1992**, *68*, 1061. [CrossRef]
7. Cheong, S.W.; Aeppli, G.; Mason, T.E.; Mook, H.; Hayden, S.M.; Canfield, P.C.; Fisk, Z.; Clausen, K.N.; Martinez, J.L. Incommensurate magnetic fluctuations in $La_{2-x}Sr_xCuO_4$. *Phys. Rev. Lett.* **1991**, *67*, 1791. [CrossRef]
8. Tranquada, J.M.; Woo, H.; Perring, T.G.; Goka, H.; Gu, G.D.; Xu, G.; Fujita, M.; Yamada, K. Quantum magnetic excitations from stripes in copper oxide superconductors. *Nature* **2004**, *429*, 534. [CrossRef]
9. Fernandes, R.M.; Chubukov, A.V.; Schmalian, J. What drives nematic order in iron-based superconductors? *Nat. Phys.* **2014**, *10*, 97. [CrossRef]
10. Lee, J.J.; Schmitt, F.T.; Moore, R.G.; Johnston, S.; Cui, Y.T.; Li, W.; Yi, M.; Liu, Z.K.; Hashimoto, M.; Zhang, Y.; et al. Interfacial mode coupling as the origin of the enhancement of Tc in FeSe films on $SrTiO_3$. *Nature* **2014**, *515*, 245. [CrossRef]
11. Ge, J.F.; Liu, Z.L.; Liu, C.; Gao, C.L.; Qian, D.; Xue, Q.K.; Liu, Y.; Jia, J.F. Superconductivity above 100 K in single-layer FeSe films on doped $SrTiO_3$. *Nat. Mater.* **2015**, *14*, 285. [CrossRef] [PubMed]

12. Wang, Q.; Shen, Y.; Pan, B.; Hao, Y.; Ma, M.; Zhou, F.; Steffens, P.; Schmalzl, K.; Forrest, T.R.; Abdel-Hafiez, M.; et al. Strong interplay between stripe spin fluctuations, nematicity and superconductivity in FeSe. *Nat. Mater.* **2016**, *15*, 159. [CrossRef] [PubMed]
13. Si, Q.; Yu, R.; Abrahams, E. High-temperature superconductivity in iron pnictides and chalcogenides. *Nat. Rev. Mater.* **2016**, *1*, 1. [CrossRef]
14. Sachdev, S.; Keimer, B. Quantum criticality. *Phys. Today* **2011**, *64*, 29. [CrossRef]
15. Dagotto, E.; Hotta, T.; Moreo, A. Colossal magnetoresistant materials: The key role of phase separation. *Phys. Rep.* **2001**, *344*, 1. [CrossRef]
16. Ahn, K.H.; Lookman, T.; Bishop, A.R. Strain-induced metal–insulator phase coexistence in perovskite manganites. *Nature* **2004**, *428*, 401. [CrossRef]
17. Uehara, M.; Mori, S.; Chen, C.H.; Cheong, S.W. Percolative phase separation underlies colossal magnetoresistance in mixed-valent manganites. *Nature* **1999**, *399*, 560. [CrossRef]
18. Alvarez, G.; Mayr, M.; Moreo, A.; Dagotto, E. Areas of superconductivity and giant proximity effects in underdoped cuprates. *Phys. Rev. B* **2005**, *71*, 014514. [CrossRef]
19. Aliaga, H.; Magnoux, D.; Moreo, A.; Poilblanc, D.; Yunoki, S.; Dagotto, E. Theoretical study of half-doped models for manganites: Fragility of CE phase with disorder, two types of colossal magnetoresistance, and charge-ordered states for electron-doped materials. *Phys. Rev. B* **2003**, *68*, 104405. [CrossRef]
20. Tokura, Y.; Kuwahara, H.; Moritomo, Y.; Tomioka, Y.; Asamitsu, A. Competing Instabilities and Metastable States in (N d, S m)$_{1/2}$Sr$_{1/2}$MnO$_3$. *Phys. Rev. Lett.* **1996**, *76*, 3184. [CrossRef]
21. Akahoshi, D.; Uchida, M.; Tomioka, Y.; Arima, T.; Matsui, Y.; Tokura, Y. Random potential effect near the bicritical region in perovskite manganites as revealed by comparison with the ordered perovskite analogs. *Phys. Rev. Lett.* **2003**, *90*, 4. [CrossRef]
22. Tomioka, Y.; Tokura, Y. Global phase diagram of perovskite manganites in the plane of quenched disorder versus one-electron bandwidth. *Phys. Rev. B* **2004**, *70*, 014432. [CrossRef]
23. Milward, G.C.; Calderón, M.J.; Littlewood, P.B. Electronically soft phases in manganites. *Nature* **2005**, *433*, 607. [CrossRef] [PubMed]
24. Tranquada, J.M. Spins, Stripes, and Superconductivity in Hole-Doped Cuprates. In *AIP Conference Proceedings*; American Institute of Physics AIP: College Park, ML, USA, 2013; Volume 1550, pp. 114–187.
25. Thampy, V.; Chen, X.M.; Cao, Y.; Mazzoli, C.; Barbour, A.M.; Hu, W.; Miao, H.; Fabbris, G.; Zhong, R.D.; Gu, G.D.; et al. Static charge-density-wave order in the superconducting state of $La_{2-x}Ba_xCuO_4$. *Phys. Rev. B* **2017**, *95*, 241111. [CrossRef]
26. Jacobsen, H.; Holm, S.L.; Lăcătuşu, M.E.; Rømer, A.T.; Bertelsen, M.; Boehm, M.; Toft-Petersen, R.; Grivel, J.C.; Emery, S.B.; Udby, L.; et al. Distinct nature of static and dynamic magnetic stripes in cuprate superconductors. *Phys. Rev. Lett.* **2018**, *120*, 037003. [CrossRef]
27. Anissimova, S.; Parshall, D.; Gu, G.D.; Marty, K.; Lumsden, M.D.; Chi, S.; Fernandez-Baca, J.A.; Abernathy, D.L.; Lamago, D.; Tranquada, J.M.; et al. Direct observation of dynamic charge stripes in $La_{2-x}Sr_xNiO_4$. *Nat. Commun.* **2014**, *5*, 1. [CrossRef]
28. Abeykoon, A.M.M.; Božin, E.S.; Yin, W.G.; Gu, G.; Hill, J.P.; Tranquada, J.M.; Billinge, S.J.L. Evidence for short-range-ordered charge stripes far above the charge-ordering transition in $La_{1.67}Sr_{0.33}NiO_4$. *Phys. Rev. Lett.* **2013**, *111*, 096404. [CrossRef]
29. Kohsaka, Y.; Taylor, C.; Fujita, K.; Schmidt, A.; Lupien, C.; Hanaguri, T.; Azuma, M.; Takano, M.; Eisaki, H.; Takagi, H.; et al. An intrinsic bond-centered electronic glass with unidirectional domains in underdoped cuprates. *Science* **2007**, *315*, 1380. [CrossRef]
30. Gabani, S.; Flachbart, K.; Siemensmeyer, K.; Mori, T. Magnetism and superconductivity of rare earth borides. *J. Alloys Compd.* **2020**, *821*, 153201. [CrossRef]
31. Paderno, Y.; Shitsevalova, N.; Batko, I.; Flahbart, K.; Misiorek, H.; Mucha, J.; Jeżowski, A. Transition and rare earth element dodecaborides. Transition and rare earth element dodecaborides. *J. Alloys Compd.* **1995**, *219*, 215. [CrossRef]
32. Inosov, D.S. (Ed.) *Rare-Earth Borides*, 1st ed.; Jenny Stanford Publishing: Singapore, 2021.
33. Mori, T. Rare Earth Higher Borides. In *Handbook on the Physics and Chemistry of Rare-Earths*; Bunzli, C., Pecharsky, V., Eds.; Elsevier: Amsterdam, The Netherlands, 2008; pp. 39–154.
34. Werheit, H.; Filipov, V.; Shirai, K.; Dekura, H.; Shitsevalova, N.; Schwarz, U.; Armbrüster, M. Raman scattering and isotopic phonon effects in dodecaborides. *J. Phys. Condens. Matter* **2011**, *23*, 065403. [CrossRef] [PubMed]
35. Khoroshilov, A.L.; Krasnorussky, V.N.; Krasikov, K.M.; Bogach, A.V.; Glushkov, V.V.; Demishev, S.V.; Samarin, N.A.; Voronov, V.V.; Shitsevalova, N.Y.; Filipov, V.B.; et al. Maltese cross anisotropy in $Ho_{0.8}Lu_{0.2}B_{12}$ antiferromagnetic metal with dynamic charge stripes. *Phys. Rev. B* **2019**, *99*, 174430. [CrossRef]
36. Bolotina, N.B.; Dudka, A.P.; Khrykina, O.N.; Mironov, V.S. Crystal Structures of Dodecaborides: Complexity in Simplicity. In *Rare-Earth Borides*, 1st ed.; Inosov, D.S., Ed.; Jenny Stanford Publishing: New York, NY, USA, 2021; Chapter 3; pp. 293–330.
37. Sluchanko, N.E.; Azarevich, A.N.; Bogach, A.V.; Vlasov, I.I.; Glushkov, V.V.; Demishev, S.V.; Maksimov, A.A.; Tartakovskii, I.I.; Filatov, E.V.; Flachbart, K.; et al. Effects of disorder and isotopic substitution in the specific heat and Raman scattering in LuB_{12}. *J. Exp. Theor. Phys.* **2011**, *113*, 468. [CrossRef]
38. Sluchanko, N.E.; Azarevich, A.N.; Bogach, A.V.; Bolotina, N.B.; Glushkov, V.V.; Demishev, S.V.; Dudka, A.P.; Khrykina, O.N.; Filipov, V.B.; Shitsevalova, N.Y.; et al. Observation of dynamic charge stripes in $Tm_{0.19}Yb_{0.81}B_{12}$ at the metal–insulator transition. *J. Phys. Condens. Matter* **2019**, *31*, 065604. [CrossRef]

39. Zhukova, E.S.; Gorshunov, B.P.; Komandin, G.A.; Alyabyeva, L.N.; Muratov, A.V.; Aleshchenko, Y.A.; Anisimov, M.A.; Shitsevalova, N.Y.; Polovets, S.E.; Filipov, V.B.; et al. Collective infrared excitation in rare-earth $Gd_xLa_{1-x}B_6$ hexaborides. *Phys. Rev. B* **2019**, *100*, 104302. [CrossRef]
40. Krasikov, K.M.; Bogach, A.V.; Bozhko, A.D.; Glushkov, V.V.; Demishev, S.V.; Khoroshilov, A.L.; Shitsevalova, N.Y.; Filipov, V.; Gabáni, S.; Flachbart, K.; et al. Anisotropy of the charge transport in $Ho_{11}B_{12}$ antiferromagnet with dynamic charge stripes. *Solid State Sci.* **2020**, *104*, 106253. [CrossRef]
41. Khrykina, O.N.; Dudka, A.P.; Bolotina, N.B.; Sluchanko, N.E.; Shitsevalova, N.Y. Structural instability and poorly defined phase transitions in rare-earth dodecaborides RB_{12} (R = Ho-Lu) at intermediate temperatures. *Solid State Sci.* **2020**, *107*, 106273. [CrossRef]
42. Sluchanko, N.; Khoroshilov, A.; Krasnorussky, V.; Krasikov, K. Maltese Cross Anisotropy in Antiferromagnetic State of Metallic $Ho_{0.5}Lu_{0.5}B_{12}$ with Dynamic Charge Stripes. *Acta Phys. Pol. A* **2020**, *137*, 756. [CrossRef]
43. Krasikov, K.M.; Azarevich, A.N.; Glushkov, V.V.; Demishev, S.V.; Khoroshilov, A.L.; Bogach, A.V.; Shitsevalova, N.Y.; Filippov, V.B.; Sluchanko, N.E. Breaking of cubic symmetry in rare-earth dodecaborides with dynamic charge stripes. *JETP Lett.* **2020**, *112*, 413. [CrossRef]
44. Krasikov, K.; Glushkov, V.; Demishev, S.; Khoroshilov, A.; Bogach, A.; Voronov, V.; Shitsevalova, N.; Filipov, V.; Gabáni, S.; Flachbart, K. Suppression of indirect exchange and symmetry breaking in the antiferromagnetic metal HoB_{12} with dynamic charge stripes. *Phys. Rev. B* **2020**, *102*, 214435. [CrossRef]
45. Krasikov, K.M.; Bogach, A.V.; Demishev, S.V.; Voronov, V.V.; Shitsevalova, N.Y.; Filipov, V.B.; Sluchanko, N.E. Magnetoresistance anisotropy in ErB12 antiferromagnetic metal: The evidence for dynamic charge stripes. *J. Magn. Magn. Mater.* **2022**, *545*, 168796. [CrossRef]
46. Azarevich, A.; Glushkov, V.; Demishev, S.; Bogach, A.; Voronov, V.; Gavrilkin, S.; Shitsevalova, N.; Filipov, V.; Gabáni, S.; Kačmarčík, J.; et al. Evidence of symmetry lowering in antiferromagnetic metal TmB_{12} with dynamic charge stripes. *J. Phys. Condens. Matter* **2022**, *34*, 065602. [CrossRef] [PubMed]
47. Sluchanko, N.E.; Khoroshilov, A.L.; Bogach, A.V.; Voronov, V.V.; Glushkov, V.V.; Demishev, S.V.; Krasnorussky, V.N.; Krasikov, K.M.; Shitsevalova, N.Y.; Filippov, V.B. Magnetoresistance scaling and the anisotropy of charge carrier scattering in the paramagnetic phase of $Ho_{0.8}Lu_{0.2}B_{12}$ cage glass. *JETP Lett.* **2018**, *107*, 30. [CrossRef]
48. Sluchanko, N.E. Magnetism, Quantum Criticality, and Metal–Insulator Transitions in RB12. In *Rare-Earth Borides*, 1st ed.; Jenny Stanford Publishing: New York, NY, USA, 2021; Chapter 4; pp. 331–441.
49. Sluchanko, N.; Azarevich, A.; Bogach, A.; Demishev, S.; Krasikov, K.; Voronov, V.; Filipov, V.; Shitsevalova, N.; Glushkov, V. Hall effect and symmetry breaking in the nonmagnetic metal LuB_{12} with dynamic charge stripes. *Phys. Rev. B* **2021**, *103*, 35117. [CrossRef]
50. Khoroshilov, A.L.; Azarevich, A.N.; Bogach, A.V.; Glushkov, V.V.; Demishev, S.V.; Krasnorussky, V.N.; Krasikov, K.M.; Kuznetsov, A.V.; Shitsevalova, N.Y.; Filipov, V.B.; et al. Anisotropy of the Hall Effect in the Paramagnetic Phase of $Ho_{0.8}Lu_{0.2}B_{12}$ Cage Glass. *JETP Lett.* **2021**, *113*, 526. [CrossRef]
51. Nagaosa, N.; Sinova, J.; Onoda, S.; MacDonald, A.H.; Ong, N.P. Anomalous hall effect. *Rev. Mod. Phys.* **2010**, *82*, 1539. [CrossRef]
52. Malkin, B.Z.; Goremychkin, E.A.; Siemensmeyer, K.; Gabáni, S.; Flachbart, K.; Rajňák, M.; Khoroshilov, A.L.; Krasikov, K.M.; Shitsevalova, N.Y.; Filipov, V.B.; et al. Crystal-field potential and short-range order effects in inelastic neutron scattering, magnetization, and heat capacity of the cage-glass compound HoB_{12}. *Phys. Rev. B* **2021**, *104*, 134436. [CrossRef]
53. Siemensmeyer, K.; Habicht, K.; Lonkai, T.; Mat'aš, S.; Gabáni, S.; Shitsevalova, N.; Wulf, E.; Flachbart, K. Magnetic properties of the frustrated fcc–antiferromagnet HoB12 above and below TN. *J. Low Temp. Phys.* **2007**, *146*, 581. [CrossRef]
54. Kalvius, G.M.; Noakes, D.R.; Marcano, N.; Wäppling, R.; Iga, F.; Takabatake, T. Dynamics of the internal field in RB_{12}(R = Er, Yb, Lu). *Phys. B Condens. Matter* **2003**, *326*, 398. [CrossRef]
55. Sluchanko, N.E.; Sluchanko, D.N.; Samarin, N.A.; Glushkov, V.V.; Demishev, S.V.; Kuznetsov, A.V.; Burkhanov, G.S.; Chistyakov, O.D. Hall-effect anomalies near the quantum critical point in Ce $Cu_{6-x}Au_x$. *Low Temp. Phys.* **2009**, *35*, 544. [CrossRef]
56. Baranovskiy, A.E.; Grechnev, G.E.; Shitsevalova, N.Y.; Sluchanko, D.N.; Glushkov, V.V.; Demishev, S.V.; Sluchanko, N.E. Hall effect and magnetic ordering in RB_{12}. *Low Temp. Phys.* **2009**, *35*, 565. [CrossRef]
57. Rybina, A.V.; Nemkovski, K.S.; Alekseev, P.A.; Mignot, J.M.; Clementyev, E.S.; Johnson, M.; Capogna, L.; Dukhnenko, A.V.; Lyashenko, A.B.; Filippov, V.B. Lattice dynamics in ZrB_{12} and LuB_{12}: Ab initio calculations and inelastic neutron scattering measurements. *Phys. Rev. B* **2010**, *82*, 4. [CrossRef]
58. Bolotina, N.B.; Dudka, A.P.; Khrykina, O.N.; Krasnorussky, V.N.; Shitsevalova, N.Y.; Filipov, V.B.; Sluchanko, N.E. The lower symmetry electron-density distribution and the charge transport anisotropy in cubic dodecaboride LuB_{12}. *J. Phys. Condens. Matter* **2018**, *30*, 265402. [CrossRef] [PubMed]
59. Bolotina, N.B.; Dudka, A.P.; Khrykina, O.N.; Glushkov, V.V.; Azarevich, A.N.; Krasnorussky, V.N.; Gabani, S.; Shitsevalova, N.Y.; Dukhnenko, A.V.; Filipov, V.B.; et al. On the role of isotopic composition in crystal structure, thermal and charge-transport characteristics of dodecaborides $LuNB_{12}$ with the Jahn-Teller instability. *J. Phys. Chem. Solids* **2019**, *129*, 434. [CrossRef]
60. Sluchanko, N.E.; Bogach, A.V.; Glushkov, V.V.; Demishev, S.V.; Samarin, N.A.; Sluchanko, D.N.; Dukhnenko, A.V.; Levchenko, A.V. Anomalies of magnetoresistance of compounds with atomic clusters RB12 (R = Ho, Er, Tm, Lu). *J. Exp. Theor. Phys.* **2009**, *108*, 668. [CrossRef]

61. Sluchanko, N.E.; Azarevich, A.N.; Bogach, A.V.; Glushkov, V.V.; Demishev, S.V.; Anisimov, M.A.; Levchenko, A.V.; Filipov, V.B.; Shitsevalova, N.Y. Hall and transverse even effects in the vicinity of a quantum critical point in $Tm_{1-x}Yb_xB_{12}$. *J. Exp. Theor. Phys.* **2012**, *115*, 509. [CrossRef]
62. Koblischka, M.R.; Hewener, B.; Hartmann, U.; Wienss, A.; Christoffer, B.; Persch-Schuy, G. Magnetic force microscopy applied in magnetic data storage technology. *Appl. Phys. A* **2003**, *76*, 879. [CrossRef]
63. Krasnorussky, V.; Bogach, A.; Sluchanko, N.; Demishev, S.; Shitsevalova, N.; Filipov, V.; Pristas, G.; Gabani, S.; Flachbart, K.; Glushkov, V.V. Anomalous Magnetic Contributions to Hall Effect in $Ho_{0.5}Lu_{0.5}B_{12}$. *Acta Phys. Pol. A* **2020**, *137*, 767. [CrossRef]
64. Karplus, R.; Luttinger, J.M. Hall effect in ferromagnetics. *Phys. Rev.* **1954**, *95*, 1154. [CrossRef]
65. Sluchanko, N.E.; Sluchanko, D.N.; Glushkov, V.V.; Demishev, S.V.; Samarin, N.A.; Shitsevalova, N.Y. Anomalous hall effect in HoB_{12}. *JETP Lett.* **2008**, *86*, 604. [CrossRef]
66. Kohout, A.; Batko, I.; Czopnik, A.; Flachbart, K.; Matas, S.; Meissner, M.; Paderno, Y.; Shitsevalova, N.; Siemensmeyer, K. Phase diagram and magnetic structure investigation of the fcc antiferromagnet HoB_{12}. *Phys. Rev. B* **2004**, *70*, 224416. [CrossRef]
67. Siemensmeyer, K.; Flachbart, K.; Gabáni, S.; Mat'aš, S.; Paderno, Y.; Shitsevalova, N. Magnetic structure of rare-earth dodecaborides. *J. Solid State Chem.* **2006**, *179*, 2748. [CrossRef]
68. Tennant, D.A.; Nagler, S.E.; Garrett, A.W.; Barnes, T.; Torardi, C.C. Excitation Spectrum and Superexchange Pathways in the Spin Dimer $VODPO_4{\cdot}{}_{1/2}D_2O$. *Phys. Rev. Lett.* **1997**, *78*, 4998. [CrossRef]
69. Flachbart, K.; Bauer, E.; Gabáni, S.; Kaldarar, H.; Lonkai, T.; Mat'aš, S.; Pavlík, V.; Priputen, P.; Shitsevalova, N.; Siemensmeyer, K.; et al. Magnetic ordering in HoB_{12} below and above TN. *J. Magn. Magn. Mater.* **2007**, *310*, 1727. [CrossRef]
70. Dyakonov, M. Basics of semiconductor and spin physics. In *Spin Physics in Semiconductors*; Springer: Berlin/Heidelberg, Germany, 2008.
71. Batista, C.D.; Lin, S.-Z.; Hayami, S.; Kamiya, Y. Frustration and chiral orderings in correlated electron systems. *Rep. Prog. Phys.* **2016**, *79*, 084504. [CrossRef]
72. Taguchi, Y.; Oohara, Y.; Yoshizawa, H.; Nagaosa, N.; Tokura, Y. Spin chirality, Berry phase, and anomalous Hall effect in a frustrated ferromagnet. *Science* **2001**, *291*, 2573. [CrossRef]
73. Neubauer, A.; Pfleiderer, C.; Binz, B.; Rosch, A.; Ritz, R.; Niklowitz, P.G.; Böni, P. Topological Hall effect in the A phase of MnSi. *Phys. Rev. Lett.* **2009**, *102*, 186602. [CrossRef]
74. Pureur, P.; Fabris, F.W.; Schaf, J.; Campbell, I.A. Chiral susceptibility in canonical spin glass and re-entrant alloys from Hall effect measurements. *Europhys. Lett.* **2004**, *67*, 123. [CrossRef]
75. Machida, Y.; Nakatsuji, S.; Onoda, S.; Tayama, T.; Sakakibara, T. Time-reversal symmetry breaking and spontaneous Hall effect without magnetic dipole order. *Nature* **2010**, *463*, 210. [CrossRef]
76. Liu, Z.H.; Zhang, Y.J.; Liu, G.D.; Ding, B.; Liu, E.K.; Jafri, H.M.; Hou, Z.P.; Wang, W.H.; Ma, X.Q.; Wu, G.H. Transition from anomalous Hall effect to topological Hall effect in hexagonal non-collinear magnet Mn_3Ga. *Sci. Rep.* **2017**, *7*, 515. [CrossRef]
77. Smit, J. The spontaneous Hall effect in ferromagnetics II. *Physica* **1958**, *24*, 39. [CrossRef]
78. Jungwirth, T.; Niu, Q.; MacDonald, A.H. Anomalous Hall effect in ferromagnetic semiconductors. *Phys. Rev. Lett.* **2002**, *88*, 207208. [CrossRef] [PubMed]
79. Berger, L. Side-jump mechanism for the Hall effect of ferromagnets. *Phys. Rev. B* **1970**, *2*, 4559. [CrossRef]
80. Gorshunov, B.P.; Zhukova, E.S.; Komandin, G.A.; Torgashev, V.I.; Muratov, A.V.; Aleshchenko, Y.A.; Demishev, S.V.; Shitsevalova, N.Y.; Filipov, V.B.; Sluchanko, N.E. Collective infrared excitation in LuB_{12} cage-glass. *JETP Lett.* **2018**, *107*, 100. [CrossRef]

MDPI
St. Alban-Anlage 66
4052 Basel
Switzerland
Tel. +41 61 683 77 34
Fax +41 61 302 89 18
www.mdpi.com

Molecules Editorial Office
E-mail: molecules@mdpi.com
www.mdpi.com/journal/molecules

www.ingramcontent.com/pod-product-compliance
Lightning Source LLC
Chambersburg PA
CBHW061522210926
51850CB00008B/2004
* 9 7 8 3 0 3 6 5 8 3 2 7 3 *